The Computer Revolution in Canada

History of Computing

I. Bernard Cohen and William Aspray, editors

William Aspray, *John von Neumann and the Origins of Modern Computing*

Charles J. Bashe, Lyle R. Johnson, John H. Palmer, and Emerson W. Pugh, *IBM's Early Computers*

Paul E. Ceruzzi, *A History of Modern Computing*

I. Bernard Cohen, *Howard Aiken: Portrait of a Computer Pioneer*

I. Bernard Cohen and Gregory W. Welch, editors, *Makin' Numbers: Howard Aiken and the Computer*

John Hendry, *Innovating for Failure: Government Policy and the Early British Computer Industry*

Michael Lindgren, *Glory and Failure: The Difference Engines of Johann Müller, Charles Babbage, and Georg and Edvard Scheutz*

David E. Lundstrom, *A Few Good Men from Univac*

R. Moreau, *The Computer Comes of Age: The People, the Hardware, and the Software*

Emerson W. Pugh, *Building IBM: Shaping an Industry and Its Technology*

Emerson W. Pugh, *Memories That Shaped an Industry*

Emerson W. Pugh, Lyle R. Johnson, and John H. Palmer, *IBM's 360 and Early 370 Systems*

Kent C. Redmond and Thomas M. Smith, *From Whirlwind to MITRE: The R&D Story of the SAGE Air Defense Computer*

Raúl Rojas and Ulf Hashagen, editors, *The First Computers—History and Architectures*

Dorothy Stein, *Ada: A Life and a Legacy*

John N. Vardalas, *The Computer Revolution in Canada: Building National Technological Competence*

Maurice V. Wilkes, *Memoirs of a Computer Pioneer*

The Computer Revolution in Canada
Building National Technological Competence

John N. Vardalas

The MIT Press
Cambridge, Massachusetts
London, England

© 2001 Massachusetts Institute of Technology

All rights reserved. No part of this book may be reproduced in any form by any electronic or mechanical means (including photocopying, recording, or information storage and retrieval) without permission in writing from the publisher.

Set in New Baskerville by The MIT Press.
Printed and bound in the United States of America.

Library of Congress Cataloging-in-Publication Data

Vardalas, John N.
The computer revolution in Canada : building national technological competence / John N. Vardalas.
p. cm. — (History of computing)
Includes index.
ISBN 0-262-22064-4 (hc. : alk. paper)
1. Electronic data processing—Canada—History. 2. Computers—Canada—History. I. Title. 2. Series.
QA76.17 .V37   2001
004'.0971'09045—dc21   2001030234

Contents

Acknowledgements *vii*

Introduction *1*

1
Canadian Military Enterprise and the University 15

2
The Navy's Pursuit of Self-Reliance in Digital Electronics 45

3
Complexity and the Military Imperative to Miniaturize 79

4
Civilian Public Enterprise Encourages Domestic R&D in Digital Electronics 105

5
The Effort to Create a Canadian Computer Industry 143

6
The Sperry Gyroscope Company of Canada and Computer Numerical Control 181

7
The Dilemma of "Buying" Mandated Subsidiaries: The Case of the Control Data Corporation 223

Conclusion 275

Notes 303
Index 405

Acknowledgements

Many people and institutions played important roles in helping me research and write this book. First I would like to thank the National Archives of Canada, the Hagley Museum and Library, the Charles Babbage Institute, the University of Toronto, and the Federal Reserve Bank of New York. The archival records stored at these institutions were the raw materials for this book's historical narrative. I am particularly grateful for the excellent assistance provided by the staffs of these institutions.

An important portion of this research rests on extensive interviews. I want to thank all the engineers, businessmen, and government officials who so graciously agreed to be interviewed and who generously shared their memories with me. The names of these individuals appear throughout the notes.

I would like to give special thanks to Donald Davis and William Aspray. A considerable portion of this book comes from my doctoral dissertation. As my supervisor, Donald Davis played an invaluable role in helping me sharpen my skills as an historian. I also want to thank William Aspray for his constructive critiques of several drafts. William Aspray was also instrumental in bringing my work to the attention of The MIT Press.

Without the generous financial support of the Social Sciences and Humanities Research Council of Canada this book could not have been written. I also want to thank the Interuniversity Centre for Research on Science at Technology, in Montreal, for welcoming me as a Research Associate.

Finally, on an emotional level, I owe a great deal to my wife, Karen. Her patience, optimism, and love give me strength.

The Computer Revolution in Canada

Introduction

In December 1945, ENIAC, the first fully functional digital electronic computer, was put into operation. Known only to a small circle of people, and understood by even fewer, ENIAC made scarcely a fold in the socio-economic and industrial fabric of the day. Yet today the paradigm it pioneered has touched almost every aspect of existence. Whether or not one wants to go so far as to argue that "being digital" has become a measure of modernity itself, one fact remains incontestable: a nation's capacity to innovate products on the frontiers of digital electronic technology has become the crucial measure of its techno-economic development. Not only has it driven post-World War II growth, but this "transformative" technology has profoundly redefined the economic production function across the entire economy.

The foundation of each nation's present capacity to exploit the commercial opportunities of the digital paradigm was laid in the postwar years. Between 1945 and 1975, a new generation of entrepreneurs created new industries to exploit digital electronic techniques. It was also during this period that each nation created the core of academic, governmental, and industrial expertise in which present-day technical competence in the digital universe is rooted. It was during those years that the major industrialized nations of the world, challenged by the technological and industrial hegemony of the United States, first scrambled to assert domestic design and manufacturing competence within the paradigm of digital electronics.

Canada was one of the scramblers, and this book will tell the story of Canada's quest from 1945 through the 1970s to create a self-reliant capacity to use, adapt, invent, and commercialize digital electronic technology. In so doing, it will illuminate the role that the Canadian military played in shaping radical technological change during the postwar economic boom. Canada's experience during World War II and the geopolitics of the ensuing Cold War fostered the emergence of an assertive Canadian military. In defense of its sovereignty and in quest of a more commanding voice in postwar international affairs than it had before the war, Canada flirted with military self-reliance.[1] A cornerstone of this self-reliance was the determination to be an active participant within the "North Atlantic Triangle" (Great Britain, Canada, and the United States) in the research, development, and testing of advanced weapons systems.

My narrative will also extend into the civilian sector, adding needed balance to the current Canadian nationalist historiography of technological change, in which foreign subsidiaries are depicted broadly as passive agents, stripped of any desire or capacity to pursue leading-edge innovation. It will provide a richer and more nuanced account of how national economic, political, and corporate forces shaped the specific contents, extent, direction, and rhythm with which digital innovation propagated through Canada.

In the chapters that follow, the reader will also learn, through a study of Ferranti-Packard Ltd., the Sperry Gyroscope Company of Canada Ltd., and Control Data of Canada, the contributions that foreign-owned subsidiaries made to Canada's capacity to exploit domestic and international opportunities in the new and rapidly expanding market for products based on digital electronics.[2] These stories reveal a far more complex portrait of foreign-owned subsidiaries than the one painted by Canada's economic nationalists during the 1970s and the 1980s. Far from being passive puppets, the two case studies portray remarkably innovative firms with aggressive programs to develop an all-Canadian, in-house research and development and manufacturing capacity of international stature in computer design, digital communications, real-time distributed data processing equipment, and real-time digital electronic automation of batch manufacturing. The case of Control Data Canada demonstrates a parent firm willing to transfer an important portion its core technological competency to Canada in the pursuit of international competitive advantages. Very quickly, the Canadian subsidiary became a crucial center of excellence in Control Data's world operations. But, more important, this latter case study reveals that, even though Control Data Canada's capacity for product development was initiated and managed by the parent firm, in time the subsidiary proved quite able to win its own new product mandates.

The reader is cautioned not to extrapolate from these three corporate case studies a pattern of behavior typical of all foreign multinationals and their Canadian branch plants. But these studies do underscore the distortions of truth promulgated by the nationalist historians who argue that "branch plants" were the agents of Canada's de-industrialization.

An important subtext of my book is the importance for a small open economy of keeping technologically busy during the onset of an innovation wave. The legacy of Canada's experience with the wave of innovation in digital electronics does not lie in any specific commercial success story. Rather, it resides in the early creation of a national pool of technical know-how that served to keep the nation always within reach of the leading edge of technical change and always looking for market niches to exploit.

Schumpeter, Neo-Schumpeterians, and Innovation Waves

An important metaphor underlying my historical narrative is Joseph Schumpeter's notion of global innovation waves. In the depths of the Great Depression, this Harvard economist argued that the economic development of the major industrialized nations since the end of the eighteenth century had occurred as a series of bursts, or "waves," marked by periods of high growth rates and the concentrated appearance of radically new innovations and products.[3] The agents of these bursts were entrepreneurs who saw opportunities for monopoly profits. But innovations and profits encourage imitation, and a bandwagon effect occurs as new competitors swarm in to share in the profits. It is this "swarming" around a cluster of radical innovations that leads to the emergence of what Schumpeter called the new "leading sector" of economic growth. For Schumpeter, the rapid expansion of new industries was not simply a consequence of economic growth but the cause of it.[4] With time, however, the advantage of technological exclusivity disappears, competition intensifies, profits and the rate of new fixed capital investment decline, and the leading sector ceases to be the engine of economic expansion. Graphed, this evolution has the S shape that is typical of the life cycles of new products: slow acceptance, accelerated market demand, and then relative decline in market interest. As the once-radical innovations saturate the market, refinements to them in subsequent years bring diminishing returns. The more basic the innovation underlying the new products, the longer the life cycle.

Considerable debate swirls around the Schumpeterian argument about the role of innovation in long-term economic growth. Some argue that there is insufficient evidence to demonstrate a temporal synchronization over a wide range of economic variables.[5] Or they dispute whether radical innovations do indeed come in waves.[6] Against such arguments, Christopher Freeman, John Clark, and Luc Soete, economists from the University of Sussex's Science Research Policy Unit, have put together a convincing empirical study to support the argument that radical innovations have indeed clustered since the Industrial Revolution.[7] Based on a disaggregated analysis of an extensive list of innovations, Freeman, Clark, and Soete divide innovations into those of "product" and "process." "Product" innovations introduce fundamentally new products or services. Driven by supply, they engender a fundamental shift in economic development, and they create a new leading sector (or sectors) that forms the basis of the expansionary upswing. "Process" innovations are aimed at producing an existing set of products and services in a more efficient way. Process innovations, argue Freeman, Clark, and Soete, tend to dominate during the intense competition that follows "swarming." Driven by demand, they arise from competitive pressure to rationalize production, lower costs, improve products, or create further product differentiation.[8] With this classification of innovations, Freeman, Clark, and Soete have provided a bridge between Schumpeter's supply-side view of autonomous innovation and Jacob Schmookler's demand-side perspective that innovation "is as much an economic activity as is the production of bread."[9]

One can identify four innovation waves since the beginning of the industrial revolution: steam mechanization of production (starting at the turn of the nineteenth century); the constellation of innovations (beginning in the middle of the nineteenth century) that revolved around railways; the electrical, chemical, and automobile industries at the close of the nineteenth century; and the cluster of innovations in computer, solid-state electronics, telecommunications, and synthetic materials that occurred during the postwar economic boom from 1945 to about 1970.[10] The onset of each of these innovation waves was not a spatially homogeneous process. Each first appeared in one or two countries and then over time spread to others. Because substantial monopoly-like economic advantages accrued to the first nations to implement the new cluster of innovations, tensions arose as other nations tried to break the monopoly.[11] Such tension manifests itself internationally as the process of innovation diffusion. The rate of the diffusion depends on the complexity of the technologies, on the ability of the recipient nations' economic, corporate, social, cultural, political, and educational institutions to appropriate these technologies, and on the ability of the nation within which the original innovations took place to block the diffusion.[12]

Innovation in the Canadian Political Economy: A Discourse of Passivity

Until the 1960s, Canadian economic historians and political economists portrayed Canada's march to modernity in terms of the rise and fall of one staple export after another. From cod fisheries, fur, timber, gold, wheat, base metals, pulp, and paper to oil and gas, Canada's economic development involved an incessant search for new global opportunities in natural resource exports. Underlying this "staples" model of development was an assumption, first promulgated in the late 1920s by W. A. Mackintosh, that Canada would be able to use its natural resource exports as the springboard to a more technologically sophisticated and diversified manufacturing-based economy, as the United States had done.

At the start of the 1970s, as the postwar economic expansion ended, an important intellectual shift occurred in the interpretation of Canada's historical development. A new generation of nationalist historians and political economists started to argue that Canada had failed to break away from its dependence on raw material exports as the engine of growth. The new historical discourse on Canada's economic development looked to Third World models of dependency to explain Canada's "truncated" or "arrested" pattern of economic development.[13] Armed with macroeconomic indicators drawn from data on foreign investment, export and import trade, and manufacturing, the new political economists created a profile of Canada resembling that of semi-industrialized nations. Inspired by the contemporary discourse of underdevelopment, the new political economy described Canada's postwar history as a process of "de-industrialization." Daniel Drache concluded in the early 1970s that Canada had moved from a semi-center economy to a semi-peripheral one. In the 1980s, Melissa Clark-Jones similarly framed the de-industrialization thesis within a model of "continental resource capitalism."[14]

The nationalist discourse pointed to the dominance of foreign-owned branch plants as the chief culprit for de-industrialization.[15] Borrowing from the literature on Latin America, political economists argued that dependency arising from the branch plant had blocked the creation of a domestic entrepreneurial class.[16] Instead, the branch plant had created a class that merely acted as agents for foreign-based entrepreneurs.[17] Thus, Mel Watkins asserted, the dominance of foreign-owned branch plants had "arrested" the emergence of a Canadian industrialist class.[18] R. T. Naylor asserted that the old mercantile elite, beneficiaries of Canada's early staple economy, subsequently had blocked the emergence of an entrepreneurial class through their control of Canada's financial and political institutions.[19] Kari Levitt asserted in *Silent Surrender* that the dominance of the subsidiaries of foreign multinationals had blocked the emergence of a class of innovators and risk takers who could create new leading sectors. Wallace Clement's sociological study of the Canadian corporate elite similarly sought to prove that an unequal alliance had created a class of Canadian capitalists that controlled the distribution and finance aspects of the economy while abdicating control over industrial development to the US-based multinationals.[20]

The nationalist political economists and historians never undertook a full examination of the relationship of industrialization to the process of invention, research, and development.[21] In particular, there has never been a study of how Canada created and sustained a capacity to innovate in the most important cluster of technologies driving the postwar economic boom: digital electronics and computers.

When the nationalist political economic discourse first emerged in the late 1960s, J. J. Brown had just published his grim assessment of the history of Canadian technology.[22] "The paradox that enlivens the history of Canadian invention," wrote Brown in 1967, "is that Canada is a great producer of ideas, yet it has virtually no native technical industries."[23] Brown's resolution of the paradox was a scathing condemnation of Canadian industrial culture, if not of Canadian society: "We tend to be terrified little men, clutching frantically at what we have and afraid to take any risk whatsoever even for large rewards."[24] Brown suggested that fear of losing the comfortable life had led Canadians to forfeit their ability to risk and create, and to accept integration into the US economic empire.

Nearly 15 years after Brown's condemnation of Canada's technological record, Christian DeBresson rebutted Brown's idea that innovation had died in Canada.[25] DeBresson accused Brown of having falsely shaped public opinion and of fostering the myth that Canadians lacked the nerve to be innovators and entrepreneurs. The thrust of DeBresson's argument was that Canada's use of foreign technology has not been a process of imitation. Rather, the import of technology has entailed a creative adaptation to the existing economic and social context.[26] In a detailed inter-industry study of how innovation flowed between suppliers and users, DeBresson concluded that, except for a few industries, innovative activity was alive and well and driven by Canada's staple industries.[27] Reinforcing Schmookler's model of demand-driven, incremental innovation, DeBresson's data underscored the vitality of "process" innovations within Canada's staple economy. Lacking access to detailed primary case studies of the firms that created the fourth innovation wave, however, DeBresson was unable to explore the vitality of "product innovations" in Canada, particularly during the critical onset of the wave. The cases examined in this book underscore both the simplistic nature of Brown's interpretation of the history of Canadian technology and the incompleteness of DeBresson's rebuttal.

At the start of the 1970s, the newly created Science Council of Canada refurbished Brown's assertions within a more encompassing political economic model of Canada that saw foreign ownership as the principal obstacle to innovation. Throughout the 1970s and the 1980s, the Science Council expanded the debate on Canada's "truncated" development to include the nation's capacity to carry out scientific research and advanced experimental engineering.[28] "Technological underdevelopment," argued the Council, was the cause of Canada's de-industrialization.[29] For the Council, the absence of "technological sovereignty"[30] undermined Canada's ability to ensure its future prosperity.[31]

Recognizing the strategic importance of digital electronics and computers, the Science Council voiced concern over the absence of a national plan to create a Canadian capacity for design and manufacture within this cluster of technologies.[32] Since it did not refer to Canada's experience in trying to appropriate the digital electronics revolution, the Council's studies presented a snapshot of Canada's predicament that was stripped of any historical context. In its critique of corporate R&D, the Council did not offer any in-depth studies of how the innovation process had actually unfolded within Canadian companies identified with the fourth global innovation wave.

Long before the Science Council of Canada advocated its nationalist R&D agenda, however, Canada's Departments of National Defence and Defence Production had been fostering domestic R&D and industrial capacity in all things digital. Had the Science Council been aware of this military-funded research by foreign subsidiaries in Canada, its advice might have been more pertinent to decision makers.

Canadian Military Enterprise and the Fourth Innovation Wave

The historian Merritt Roe Smith argues that the military establishments of the industrialized nations have played an important historical role in shaping technological and economic change by linking national defense with national welfare. In developing this argument, Smith underscores the importance of the "complementarity of military and economic forces" in the development of the American industrial system.[33] His own social history of armory practice at Harper's Ferry, West Virginia, during the first half of the nineteenth century shows how military enterprise affected the design, manufacturing technology, and labor process of an entire industrial sector.[34] As Smith and David Hounshell have demonstrated, interchangeable parts and specialized machine tools, which later became the technical basis of mass production, had deep roots in the US Army's preoccupation with standardized performance and construction down to the smallest subassembly.[35] Furthermore, Smith's narrative reveals that the introduction of interchangeable parts was not simply a neutral add-on to existing shop floor production techniques. Rather, interchangeable parts entailed a redefinition of the nature of work and of who controlled the labor process.[36] The international diffusion of the "American System of Manufactures" was relatively slow until it became an urgent matter for the military of Europe's major powers. It took World War I, argues William McNeill,[37] to provide the crisis whereby military imperatives pushed Europe away from its older socio-economic patterns and closer to the mass-production society.

The sociologist Gordon Laxer suggested in 1986 (surprisingly, in view of his intellectual debts to the 1970s radical political economists) that the absence of a dynamic military was a significant contributing factor to Canada's arrested political economy and truncated technological capabilities. "Instead of developing a domestically owned and innovative engineering industry under the protective care of an independent military policy," wrote Laxer, "it relied almost exclusively on foreign technology, which was a major factor in the emergence of the branch-plant structure."[38] Other scholars described the Canadian military as ineffective in promoting economic growth or industrialization. But they, like Laxer, did not move beyond generalities to study the details of any industry.

Although historians have written about how World War II led to an expansion of Canada's technological and industrial capacity, few have examined this subject for the postwar era.[39] David Zimmerman, in a fascinating study of the Royal Canadian Navy, describes how, at the outbreak of World War II, this institution was totally unprepared to manage the scientific and technological demands of modern warfare. By the end of the war, however, the RCN had learned to integrate the scientific-technical mind with the military mind. What happened to this technological capacity after demobilization? In an epilogue to his narrative, Zimmerman suggested that "much of [RCN's] accumulated wartime scientific and technical expertise was lost as a result" of Prime Minister William Lyon Mackenzie King's "ill-conceived" policy of slashing military budgets.

Though Zimmerman's wartime examination of the RCN draws on considerable primary source evidence, his postwar speculations rest on no more than the first few months after the war.[40] His assessment echoes that of the political scientist James Eayrs,[41] and it certainly captures Mackenzie King's intentions in late 1945. But then came the Cold War. The consequent shift in geopolitics undermined Mackenzie King's anti-military agenda and allowed the RCN, along with the other services, to pursue a technologically mediated vision of self-reliance.

Of course, for those who saw only dependence and de-industrialization in Canada's postwar defense planning it was natural to assume that military enterprise did little to promote domestic research and development on the frontiers of technology. They argued that the US military-industrial machine's need to secure access to natural resources reinforced the historical dominance of staple exports in the Canadian economy. This may have been true, but it was not the totality of the military's impact on Canada. The search for petroleum, nickel, and uranium did not preclude a quest for semiconductors.

Lawrence Aronsen is the only historian to have examined in some detail the extent to which a Canadian military-industrial complex emerged during the immediate postwar years. In some respects, his work offers a counterbalance to the nationalist model. While recognizing that the continental military alliance accelerated Canada's economic integration with the United States, Aronsen has also underscored the benefits of this connection to Canadian manufacturing.[42] Through the vehicle of the branch plant, he argues, Canada was able "to maintain state-of-the-art production through technology transfer." He goes on to explain: "American integration of Canada's resources and industrial defence production . . . was particularly beneficial to the host country. . . . The American-controlled high-technology industries, such as automobiles, electronics and oil refining, recorded the highest rates of productivity in North America as measured by output per man-hour."[43] Although Aronsen thus understands that the framework of continental defense allowed Canada to become good at copying sophisticated US products and techniques, his work does not show the extent to which Canadian military enterprise fostered domestic design capacity.

On the question most pertinent to this book, whether Canada's military was able after World War II to generate and sustain its own advanced weapons R&D program, Aronsen slips into a nationalist discourse: "Canadian defence industries had a closer liaison with American military officials on questions concerning equipment design and production than they did with the Canadian military."[44] He does not, however, present any evidence to substantiate his contention that Canadian military enterprise played a minor role in shaping a national industrial R&D agenda. Furthermore, his analysis does not reveal if the creation of a domestic design capacity accompanied the growth of Canada's defense-oriented high-tech industries. Were these Canadian manufacturers of high-tech military equipment to which Aronsen refers merely passive copiers of US designs, or did they also promote a domestic capacity for R&D in leading-edge technologies?

Jack Granatstein's 1993 argument that a widespread "envy" of US military equipment drove Canadian postwar defense planning reinforced Aronsen's speculations.[45] Granatstein brought up the demise of the AVRO Arrow interceptor (unviable because it was unsellable in the United States) to illustrate the insurmountable problems that Canadian technology faced within the context of a continental defense driven by the large and powerful US military-industrial complex.[46] The Arrow is often the illustrative example when historians discuss Canada's postwar relationship with the US government and military.[47] Yet there has been little or no attempt to determine whether the technological ambitions that drove the Arrow were an isolated exception to the general rule of buying defense materials off the American shelf or whether they reflected a broader and more systematic defense initiative to nurture a domestic R&D capacity. If the latter, then it is important to look past the Arrow to Canada's military-industrial complex, whose nature has been obscured by the Arrow's political notoriety.[48]

The end of World War II did not bring a complete dismantling of Canada's military, as had been the case after World War I.[49] New geopolitical realities infused Canada's military with an unprecedented peacetime purpose. Rising political tensions between the Soviet Union and the United States raised the frightening possibility of another world war. With the misfortune of being on the flight path of a possible nuclear attack by Soviet bombers and missiles, peacetime defense took on a new meaning for Canadians. In addition to defending Canada, the military also participated in the defense of Europe. For the Canadian government, participation in a multilateral military alliance, such as the North Atlantic Treaty Organization (NATO), was a way to counterbalance the hegemony of the United States in the postwar era. Canadian historiography has yet to fully explore the impact of the unprecedented peacetime defense role on Canada's postwar technological and industrial development.

The Canadian defense establishment's efforts to shape technological development in Canada after World War II constituted part of what the historian Keith Krause has called Canada's first and only "flirtation with an independent [arms] production policy."[50] The new postwar complementarities between military enterprise and industrial development encouraged Canada to attempt to leapfrog up the international hierarchy of arms production. In 1939, Canada had been stuck on the "fourth tier" of arms-producing states. This terminology, indeed the idea of tiers in general, comes from Krause, who describes the historical structure of arms production and trade in the world in terms of five levels. Tier 1 suppliers innovate at the technological frontier. Tier 2 suppliers produce weapons at the technological frontier and adapt them to specific market needs. Tier 3 suppliers copy and reproduce existing technologies, but do not capture the underlying process of innovation or adaptation.[51] Tier 4 consists of "strong customers" who can buy and use sophisticated weapons. In tier 5 are "weak customers," who either do not know how to use the weapons they purchase or do not purchase advanced weaponry. An important feature of this hierarchy, according to Krause, has been its relative stability. Mobility up and down the hierarchy is difficult.

Before World War I, Canada was essentially in tier 5—a weak customer. During World War I, it moved up to tier 4 to become a "strong customer."[52] In *The Great Naval Battle of Ottawa*, Zimmerman recounts the difficulties Canada had moving from strong customer to a tier three supplier during World War II. At the end of World War II, Canada sought to jump to tier 2, and on occasion to tier 1. Arctic research, DATAR, the Mid-Canada Line, the Velvet Glove missile, the AVRO Arrow, George Bull's super-cannon research, transistor circuit design research in the mid 1950s at the DRTE, and research into communications in the high Arctic are examples of upward jumps. These efforts, however, contained fundamental contradictions. Krause points out that tier 1 and tier 2 suppliers find their strength in a domestic market. The fundamental dilemma faced by the Canadian military was how to get local industry to invest in R&D and production technology even though the volume of Canadian procurement was not sufficient to support this investment. "The real problem," noted one Department of National Defence memo in 1958, "is not establishing a development group in industry but keeping it in business."[53]

Overview: The Seven Case Studies

Whereas US historiography on the subject is rich, the role of peacetime military enterprise in shaping postwar Canadian technological and industrial development remains inadequately explored. Most work in this area has revolved around Canada's efforts to design and manufacture one of the most advanced all-weather fighter interceptors of the 1950s, the CF-105 or AVRO Arrow. Joseph Jockel's political and institutional analysis of the Mid-Canada Line and the DEW Line, *No Boundaries Upstairs*, alludes to Canada's efforts to establish a high-tech electronics industry but does not explore its articulation.[54] Through three case studies, the first half of this book examines how, within the emerging digital electronics revolution of the late 1940s and the 1950s, Canadian military enterprise shaped the direction and rhythm of domestic research and development.

Chapter 2 argues that the quest for military self-reliance required a domestic capacity for large-scale, automated, high-speed computations. In 1949, the need for a national computational facility resulted in the military's support of a university effort to design and build Canada's first electronic digital computer. Chapter 3 examines how the Royal Canadian Navy, in its attempts to define a unique postwar role within the North Atlantic Triangle alliance, became from 1948 to 1956 the single most important institutional force within Canada pushing to create a domestic industrial capacity to design and manufacture digital electronic technology.

Chapter 4's theme is the centrality of the transistor in resolving the complexity-vs.-reliability impasse that faced efforts to design large-scale military digital electronic systems. The focus of that chapter is the role of a government defense laboratory in advancing and diffusing transistor circuit design know-how in Canada.

An important theme that emerges from the narrative in chapters 2–4 is how Canada's position within the North Atlantic Triangle served as both an impetus and a deterrent to the Canadian military's program of technological self-reliance.[55] John Bartlet Brebner's triangle metaphor aptly places Canadian history at the crossroads of two powerful geopolitical and economic currents: the east-west transatlantic interaction with Great Britain and a north-south continental interaction with the United States. The Canadian military enterprise's entire policy of technological and industrial self-reliance had to be defined within the shifting tensions that bound the Canadian vertex to the two vertices of the North Atlantic Triangle during World War II and the immediate postwar years. Linked to two industrial and military powers, the Canadian military became a leading advocate of trans-national standardization as a way to ensure that Canada's small and open economy could play an active role within the alliance. Through chapters 2, 3, and 4, the reader will learn how Canadian military enterprise, in its pursuit of technologically mediated self-reliance, created and sustained Canada's first pool of expertise in digital electronics and computers.

Chapter 5 discusses how two public institutions, faced with mounting public and political pressure to improve service, turned to radical innovation and the paradigm of digital electronics. The Post Office Department set out to be the first in the world to sort mail automatically by means of an "electronic brain." Trans-Canada Air Lines (now Air Canada) decided to merge computation and communications and to pioneer the development of a national automated seat reservation system. An important theme of this chapter is how two public enterprises, with very different technological cultures, managed innovation. Chapter 5 underscores the crucial role civilian public enterprise played, when Canada's flirtation with military self-reliance wavered, in sustaining the core of know-how in digital electronics that had been built up.

Chapter 6 examines the first effort in Canada to develop a commercially viable computer system. Ferranti-Canada, a subsidiary of a British company, tried to use the R&D that was being done for the Post Office and Trans-Canada Air Lines projects to advance its ambitions of becoming Canada's first domestic computer designer and manufacturer. Ferranti-Canada had been one of the Royal Canadian Navy's key industrial partners in the exploitation of digital technology for naval warfare. The first section of chapter 6 shows how this Canadian subsidiary tried to create a niche for itself in a market dominated by IBM but was stymied by international and internal corporate forces. The collapse of this effort underscored the absence of national industrial policy in computer technology. The second section of the chapter explores the Canadian government's first efforts to develop a civilian-driven policy for computer technology to fill the void left when the military retreated from the pursuit of technological self-reliance. This part of the narrative casts the Canadian government's policy in the context of the late 1960s and the early 1970s, when the governments of all the major industrialized nations scrambled to strengthen their domestic computer industries to counter the first-entry advantages of US companies. These governments viewed a strong domestic computer industry as an essential condition for national well-being. Faced with a weak domestic capacity to design and manufacture computers, the Canadian government also got caught up in the international rush to nurture a domestic competency in these areas.

Whereas chapters 5 and 6 deal with a British-owned Canadian subsidiary, chapter 7 focuses on the role of a US-owned subsidiary in creating a domestic capacity to design and commercialize the "product" innovations of digital electronics. Here the story revolves around the efforts of Sperry Gyroscope of Canada Ltd., a subsidiary of Sperry Rand, to break its dependence on the uncertain defense market that had appeared after the Korean War. Its strategy was to diversify into civilian-oriented products. In a classic Schumpeterian initiative, Sperry Canada tried to move itself through radical innovation into an exclusive market niche based on digital electronic technology: the numerically based control of machine tools. As it saw this market being eroded by the swarm of imitators, the company tried to move into an even more exclusive digital niche, becoming the first company in the world to market computer numerical control (CNC) by introducing computers onto the shop floor to reprogram machine tools. Chapter 7 examines the role of Canadian public enterprise and parent-firm relationships in Sperry Gyroscope's effort to define a world product mandate. An important theme here, as in all the other chapters, is how the process of innovation was filtered through Canada's political economy.

Believing that all strategic initiatives in subsidiaries emanated from their parent firms, the Department of Industry, Trade and Commerce had difficulty in responding in a timely manner to the needs of Ferranti-Packard and those of Sperry Gyroscope of Canada. Though its response to these two subsidiaries turned out to be too little too late, the department was quick to grant a great deal of money to another multinational, Control Data Corporation, which was headquartered in Minneapolis. In a move that was very atypical of the US computer industry, CDC moved the development of its next generation of computer to Canada. CDC's entry into Canada played a pivotal role in the future of the company. Chapter 7 describes the journey that led CDC to establish an important design and manufacturing capacity in Canada, and the subsequent fate of this enterprise. As chapter 7 shows, no matter how well intentioned a transfer of product mandates can be, a subsidiary's long-term success can still be fraught with considerable uncertainty.

1
Canadian Military Enterprise and the University

In 1936, Canada did not possess a single anti-aircraft gun. The armaments guarding Esquimalt, British Columbia, and Halifax, Nova Scotia, were defective. Furthermore, there were no bombs for the Canadian Air Force's 25 outdated aircraft.[1] And Canada's military unpreparedness extended beyond equipment inadequacies and into the ranks of its meager leadership. "By disbanding the hard-won expertise of the Canadian Corps so thoroughly after 1919, by starving the tiny Permanent Force and the larger but weaker Non Permanent Active Militia during the 1920s and 1930s," writes the historian Jack Granatstein, "Canada's governments ensured that when war came in September 1939 there was nothing in place to meet its demands but good wishes."[2] The Chiefs of Staff concluded in 1945 that "Canada had not had to pay the awful price of defeat, but still had to pay dearly for lack of preparedness."[3]

As Canada planned for reconstruction after World War II, the memory of its unpreparedness on the eve of that war haunted the minds of its military leaders. In 1945, in the midst of a massive demobilization program, the leaders of the Army, the Navy, and the Air Force wanted to retain a sizable peacetime defense force to serve as a nucleus around which the nation could mobilize in a future war.[4] Peacetime military preparedness, however, demanded far more than retaining high levels of recruitment and conventional equipment procurement. The pivotal role that scientific research and technological development had played in the latter years of World War II had made a profound impression on the Canadian Chiefs of Staff. Rockets, automatic fire control, guidance systems, radar, atomic energy, synthetic materials, operations research, and the electronic digital computer were the more dramatic developments arising from wartime imperatives. To ensure Canada's capacity to design, adapt where necessary, test, manufacture, and use the most advanced weapons, military leaders set out to define a national scientific, technological, and industrial policy. Director General of Defence Research Omond M. Solandt summed up the new attitude: "The defence of Canada requires that she be continually prepared to make and use the most up-to-date weapons of war."[5]

The military's determination to expand Canada's R&D capacity reflected the nation's efforts to gain a voice within the inner circle of postwar Western military powers. Canada, to be sure, realized that it could not carry out the same scale and scope of defense development as the United States or the United Kingdom. Hence, Canada pursued a policy of specialization and cooperation. Canada set its contribution to the alliance after 1945 as a small set of research, development, and testing programs of interest to itself and the entire alliance. Though the Canadian military understood the financial necessity to limit diversity in its R&D program, it still had very high ambitions: the military wanted Canada to situate itself at the leading edge of some facets of military technology in order to have something to "horse trade" with its allies.

Whether the military chose guided missiles, automated fire control, digital radio communications transmission, or supersonic fighter aircraft, the fact remained that access to fast, large-scale, computational facilities was essential to any weapons development program. Until the early 1950s, electronic digital computers were not standardized products that could be ordered from commercial concerns. Instead, they were one-of-a-kind prototypes developed by defense agencies or in military-funded university projects. Thus, when the University of Toronto proposed to establish a national computational center, the Canadian military saw one possible way to get access to computational power.

Though Canada had lost many men in battle, it was one of the few nations to emerge from World War II with an intact industrial infrastructure and economy. In fact, the war had dramatically expanded Canada's technological and industrial capacity. Filled with an unprecedented sense of possibilities, the Canadian military purposively led the nation into its first peacetime weapons development program. This chapter investigates how these ambitions fostered Canada's first experiments in digital electronic computers. Computers were central to plans for Canadian autonomy: without these "engines of the mind," any program to exploit the intensive use of science to develop modern weapons system would quickly grind to a halt.

Before proceeding to the computer research program at the University of Toronto, I must discuss how the military's pursuit of self-reliance created conditions that would allow the University of Toronto to finance its computer project.[6]

Peacetime Military Enterprise Offers a New Outlet For Canadian Academic Research

In a 1945 brief to the Cabinet Committee on Research for Defence, the Canadian General Staff underscored "the modern need for progressive research and development in the interests of national defence."[7] "The new weapons of tomorrow," warned the brief, "may decide the outcome of a war in a matter of days by destroying the industrial capacity of a country, its system of communication, and the will of the people to resist."[8] The accelerated rate at which science and technology had been applied during World War II demonstrated, in the view of the Canadian Chiefs of Staff, that a well-orchestrated national military R&D program was necessary in peacetime. An essential element in this program was the creation of a new partnership among academic, industrial, and military research.[9] For the Chiefs of Staff, it was imperative that science and technology become an integral part of the military mind.[10]

In their 31 October 1945 memorandum to the Cabinet Committee on Research for Defense, the Chiefs of Staff underscored the fact that Canada had neither the financial resources, the industrial infrastructure, nor sufficiently large armed forces to support a research, development, and production program across the entire spectrum of modern defense needs. To ensure access to the latest weapons technology, Canada would have to nurture the interchange of scientific and technical know-how and equipment within the North Atlantic Triangle.[11] But any tripartite technology-sharing arrangement would collapse if Canada could not pull its own weight. To have access to the military technology of its allies, Canada would also have to make its own valuable scientific and technological contributions.[12] This insight left open this question: In which facets of defense research and production should Canada seek to develop a comparative advantage?

In 1946, in a policy brief to the Cabinet Committee on Research for Defense, Omond Solandt, the first Director General of Defence Research, reiterated the strategic importance of mounting an independent defense research initiative. He argued that only through judicious specialization of arms research, development, and production could Canada be an equal partner in the alliance.[13] Although Solandt emphasized the importance of exploiting new Canadian ideas, he was keenly aware of the pitfalls of pursuing uniquely Canadian designs. Ultimately, the product of Canadian weapons research had to be in harmony with the military practices of Canada's allies and their equipment standards. But Canada was faced with the dilemma of having to reconcile two different standards, one of which arose from Canada's historic transatlantic military ties with Britain and one of which was a consequence of the growing US pressure for a unified continental defense system.

The pursuit of interchangeability and standardization within the North Atlantic Triangle became the cardinal principle of Canada's postwar defense research, development, and production policy. During World War II, most of Canada's production of military equipment was destined for British and Commonwealth use. Tuned to US design and manufacturing standards and practices, Canadian industry ran into serious problems when it tried to gear its production to British-designed military equipment. "That these difficulties were overcome," observed the Chiefs of Staff, "was not due to planning but to the achievement of our industry and the fortuity of time for the task."[14] In any future war, reasoned the Chiefs of Staff, Canada would once again be called upon to become Britain's arsenal, but without the "fortuity of time" to make the conversion to British production standards.

By 1946, however, the US military had concluded that the next war would be decided in North America, and not Europe. Because advances in military technology were making the United States increasingly vulnerable to heavy attack from its northern flank, the US military wanted to establish an integrated defense system that included Canadian territory. Fearful for its sovereignty, Canada could not agree to this unless it had an active part in the defense of North America. Collaboration with the United States in defending North America further intensified the harmonization of Canadian and American technological and industrial standards. In view of these conflicting objectives, it was in Canada's interest to promote greater standardization across future British and American weapons development. To be a credible broker between the two systems, however, Canada would have to "share more actively in research, development and design with both the United Kingdom and the United States."[15]

The military's efforts to be a voice in peacetime technological, industrial, and economic planning was not well received in some circles of the government. Prime Minister William Lyon Mackenzie King, along with most of his senior civilian policy planners, wanted to return the military to its prewar obscurity. According to the political scientist James Eayrs, Mackenzie King had a deep-seated and lifelong mistrust of the military.[16] Very early on in World War II, he had become convinced that the "unless [the military] was carefully watched the ambition of [its] officers was likely to lead Cabinet into difficulties."[17]

Mackenzie King was not alone in his suspicion of the military. Canada's civil service elites, argues Eayrs, saw the military as intellectually unprepared to tackle the complexities and subtleties of postwar policy issues. Whereas in the United States the military was actively brought into the federal government's planning after World War II, in Canada the political and civil service elites were anxious to remove the military from any role in postwar policy formulation. According to Eayrs, the lack of formal university education within the ranks of Canada's most senior officers was seen by the elite of Canada's Civil Service as proof of the military's inability to appreciate the complex nuances of postwar national policy. Yet the exigencies of war had catapulted the military into prominent policy roles. With the end of the war, Canada's civilian policy community sought to end this intrusion on its power and to return the military to the periphery.[18]

On 12 December 1946, Mackenzie King gave Brooke Claxton the task of pulling the reins in hard on the postwar ambitions of the armed services. Mackenzie King and his cabinet had already forced two large reductions in the manpower and armament requests of the Chiefs of Staff for Canada's postwar defense forces when Claxton took over as Minister of Defence.[19] In offering Claxton the defense portfolio, Mackenzie King emphasized the need to bring the independence and extravagance of the armed services back under control. The services "were continuing to go too far," Mackenzie King warned.[20] Referring to Claxton's former post as Minister of National Health, Mackenzie King presented the need to cut back defense as a guns-or-butter issue. The government could not target social needs, Mackenzie King argued, if its armed services continued to spend at current levels.

In choosing Claxton to be the "hatchet man," Mackenzie had not reckoned with Claxton's conviction that the shifting postwar geopolitical realities called for strong defense policies. Although Claxton did proceed to make considerable savings in the 1946–47 defense budget, and twice force down the Chiefs of Staff's 1947–48 budget estimate of $339.8 million, Claxton balked at cabinet's request for an additional 10 percent across-the-board cut.[21] As Eayrs points out, Claxton's experience with Soviet negotiating techniques at the 1946 Paris Peace Conference "led him to believe that the time had come for Canada, in the company of friends and neighbours, to turn for protection to the military."[22]

Canada's Peacetime Defense Research Program and National Computational Capacity

The Allies' pursuit of victory in World War II and the United States' pursuit of power in the immediate postwar years drove the early development of high-speed, large-scale computational technologies. ENIAC—the first fully functional electronic digital computer—arose out of the necessity to automate complex ballistic computations in order to accelerate the production of bombing tables. The construction of the first atomic bomb entailed even more complex mathematical computations. ENIAC was developed too late to be used in the Manhattan Project. Only through considerable manpower and ingenuity did the scientists at Los Alamos succeed in using the existing manual desktop tabulators and semi-automatic, electro-mechanical punch-card machines to carry out the complex scientific calculations needed. Such Herculean efforts would not have produced timely design answers for the hydrogen bomb, but ENIAC proved very useful for that purpose.

The computational demands arising from intensive and systematic application of science to weapons development went beyond atomic and thermonuclear bombs to include development of guided missiles and supersonic aircraft. The equations of hydrodynamics and aerodynamics that these technologies required had long been impenetrable to traditional abstract mathematical analysis. Indeed, John von Neumann, an eminent mathematician and scientist who had the ear of the US military, proclaimed that only the electronic computer could open up the laws of hydrodynamics to the requisite analysis. From the moment ENIAC was completed, von Neumann declared the machine inadequate. He called for the construction of a new generation of performance-driven computers. In the 1940s there were no commercial incentives to justify industrial R&D in large-scale electronic computers. It took the escalating political tensions between the United States and the Soviet Union to provide the funding for this costly development.

By 1948, increasing tensions between the Soviet Union and the North Atlantic Triangle alliance and accelerated exploitation of science in the development of military systems had created the circumstances and politics to support the Canadian military's push for technological self-reliance and the accelerated entry of digital technology into Canada. In April 1947, the Canadian government established the Defence Research Board,[23] a fourth independent service. DRB's mandate was to run a research program that would respond to the common needs of the other three services. The chairman of DRB sat as an equal member on the Chiefs of Staff, an arrangement that distinguished Canada from most other nations. According to Goodspeed: "Instead of leaving defence research to be conducted by a separate government department or by a number of government departments, as is the case in Britain, France and Russia, or instead of having of having it conducted by the Armed Forces themselves, as is broadly the case in the United States, Canada decided that defense research should be the responsibility of a separate organization, serving the Navy, the Army, and the Air Force and closely integrated with them but concerned with the scientific aspects of war."[24] By 1948–49, DRB's budget had grown from 1.6 percent to 7.2 percent of total military expenditures.[25]

Created to rationalize and consolidate the research for the three services, DRB set about defining Canada's participation in the creation and diffusion of defense technology within the North Atlantic Triangle alliance. In his 1946 brief to the Cabinet Committee on Research for Defense, Solandt, in consultation with the Chiefs of Staff, identified atomic warfare, bacteriological warfare, chemical warfare, guided missiles, rockets, supersonic aerodynamics, gas turbines, jet propulsion, electronics (radar and communications), and the Arctic as high priorities of Canada's defense R&D program. Though computers were omitted from DRB's original list, it soon became apparent that the success of the military's weapons development programs depended on a capacity to solve large-scale computational problems at high speeds. For example, much of the 40 person-years required to design a new aircraft resulted from having to do extensive calculations by manual means.[26]

Without computational self-reliance there would be no military self-reliance. The power to reduce the design time of new military technology now made the nascent electronic computer crucial to the arms race. "The electronic computer," noted B. G. Ballard, the director of NRC's Electrical and Radio Engineering Division, "provides a tool which increases tremendously the capacity of a given staff . . . and enables the design to be completed in a sufficiently short period of time so that the data employed does not become obsolete before the design is completed."[27] On 8 July 1947, DRB's Electronic Advisory Committee discussed "the need for an electronic computer in Canada."[28] Though the committee had no direct experience with large-scale digital computers, it nevertheless felt that these machines were of strategic importance to Canada's military research programs—particularly the Canadian Armaments Research and Development Establishment (CARDE).[29] Before bringing up the matter with the Chiefs of Staff, the committee wanted to have a more accurate picture of the military's future requirements for such a machine. An informal survey of the three services on this matter revealed the belief that "Canada should have an electronic computer to enable her to pursue her own research program independent of foreign countries. . . . Until Canada is equipped with a suitable computer centre, she will be obliged to rely upon foreign aid for many of the designs which require large-scale calculations. If it is agreed that Canada should be independent in this respect, then the time has arrived to initiate a computer centre in Canada."[30]

In the context of the Canadian military's concern over the absence of digital electronic computer technology, a group of University of Toronto mathematicians, scientists, and engineers found the opportunity to advance interests and ambitions they had harbored since 1946. Word of US research on large-scale automatic digital computation began to filter into Canada in 1946. Curious to learn more about this remarkable but as yet unseen technology, a group of University of Toronto professors set up an interdepartmental Committee on Computing Machines.[31] In the summer of 1946, the committee organized a fact-finding mission to the United States with the goal of learning firsthand what had been achieved in "the construction of large-scale high-speed computing machines" and "enquir[ing] into the nature of the computations undertaken by existing machines."[32] In its subsequent report to the university, the committee concluded that purchasing or even replicating any of the existing American computers would be premature. "Both those using [these] machines and those designing them, without exception," observed the committee, "agreed that all existing machines are obsolescent in the sense that no copies of the large machines will be made in the future."[33] The rapid flux and unpredictable metamorphosis in computer technology called for a wait-and-see approach, the committee reasoned; a clearer paradigm of computer design was needed before the university should build or acquire its own machine.[34] The committee shared the view that "any actual construction [of a computer] to be undertaken in Canada should be suspended until the results of at least the pilot version of Professor von Neumann's machine [at Princeton University's Institute for Advanced Study] are at hand." Though the construction of a computer in the near future was ruled out, the committee did call for the creation of a small computational laboratory at the University of Toronto as an initial training ground in scientific computational techniques.

Nearly a year after its initial report to the university, the committee's ambitions expanded dramatically as it called for the creation of a Canadian computation center to handle all of the nation's scientific computational needs.[35] In an effort to enlist financial backing, the committee draped its ambitions in the language of national imperatives. The absence of a large-scale high-speed computation facility, the committee asserted, seriously impeded scientific and technological progress in Canada. Extracting further knowledge about the physical world from abstract mathematical models was becoming increasingly elusive.[36] Important mathematical problems essential to scientific and technological advance, the committee argued, were "treated as insoluble, solved imperfectly, prepared for solution at a computing centre in the United States, or solved at a cost of very tedious calculations carried out with existing facilities and involving months or even years of effort."[37] Taking a clue from von Neumann's argumentation in the United States, the committee saw a purely electronic machine as the only effective way for Canada to break through the computation impasse. The committee then turned to the only two agencies with money available to help create a national computation center at the university itself: the National Research Council (NRC) and the Department of National Defence (DND). Both institutions had an urgent need for computational power, the military for its weapons development program and the NRC for its atomic energy research program at Chalk River, Ontario.

E. G. Cullwick, DRB's Director of Electrical Research, informed the University of Toronto's Committee on Computing Machines that its proposal would undoubtedly be of value to the Defence Research Board.[38] At a separate meeting with university representatives on 8 January 1948, in Ottawa, NRC and DRB agreed that Canada needed computational capabilities such as those proposed by the university and expressed a willingness to fund a five-year program to establish a "first-rate" computing center.[39] The general objective, agreed upon by all at this meeting, was to have a full-scale electronic digital computer in operation by 1953 and a highly skilled professional staff capable of exploiting the potential of such a machine.[40]

The strategy adopted was an incremental one. From 1948 to 1953, the Computation Centre's capabilities were to be expanded in three distinct phases. The first and most immediate step was to lease IBM punched-card machines and start training a professional staff. Not only was this staff to be competent in one or more areas of applied mathematics, it also had to understand how to translate science and engineering problems to machine-based numerical methods. The next phase called for the construction of an interim large-scale electro-mechanical computer based on relays, the switching technology used in telephone exchanges. Though much slower than a purely electronic computer, the relay computer offered fast, risk-free entry into large-scale computational technology. The university recommended copying Bell Laboratories' latest design at a cost of $45,000. This estimate was based on the committee's expectation, which appears to have arisen from assurances from George Stibitz at Bell Laboratories, that there would be no licensing fees. While the first two phases were being carried out, the plan called for the gradual development of an electronics research team to construct an electronic computer after a period of experimentation.[41]

The decision of DRB and NRC to support the creation of a computational center at the University of Toronto was no doubt influenced by the absence of any other site contenders. The university had the intellectual resources to develop an effective, multi-disciplinary team in scientific computation. But was the university the best place to design and build a full-scale electronic digital computer? If any single Canadian group had the theoretical and production engineering experience to build a computer in 1948, it was the NRC. Indeed, the NRC had concluded the previous summer that it was "the organization best qualified to build an instrument of this nature."[42] The Electronics Advisory Committee, agreeing, recommended that, if the Chiefs of Staff were in agreement, "a proposal should be forwarded to D. Mackenzie [president of NRC], with an offer of financial support from the Services and DRB."[43] However, nearly a year later, the head of NRC's Radio and Electrical Engineering Division, B. Ballard, wrote to Douglas Hartree (a noted British physicist and computer pioneer) that, although "it would be profitable for us to undertake the development of an electronic computer at the National Research Council," "we feel that we could not justify a computing machine in the laboratory for work in this district."[44] The existing historical record is silent on whether DRB ever made an offer to NRC. Neither is there any information on how Mackenzie felt about the idea of NRC's embarking on such an undertaking. For whatever reasons, NRC did not venture into computer technology.

Just days before Christmas of 1948, the university computer group received the unexpected news that they had been misinformed by Stibitz and that the licensing fees for the Bell relay computer would be $25,000.[45] The cost estimates for copying Bell's "X75320 Network Computer" thus jumped from $45,000 to $70,000. At a meeting held in Ottawa on 16 March 1949, attended by Davies (from DRB), C. J. Mackenzie, W. B. Lewis, E. C. Bullard, and C. C. Gotlieb, the decision was made to drop the relay computer and to accelerate the program to build an electronic computer.[46] DRB agreed to increase its annual funding of electronics R&D at the university from $20,000 to $30,000. NRC's contribution remained at the original stated level of $20,000 per annum. In addition, DRB promised to pay "an undetermined larger amount when the time [came] to commence construction of the computer."[47]

The rationale for dropping the relay computer remains unclear. In retrospect the decision to bypass this electro-mechanical technology and to go directly to a purely electronic computer may seem obvious, but in the context of 1948–49 the relay computer had distinct advantages. It was easier to design, build, and maintain. Unlike ENIAC and other electronic computers of the day, the relay machine neither consumed vast quantities of electrical power nor required elaborate climate control systems to keep it from overheating. Relay computers were no more cumbersome in size than their purely electronic counterparts. Furthermore, they were far more reliable and required less servicing. The relay computer's principal limitation arose from the slower switching times of the telephone relays used because the mechanical motion of the armature in the relay limited the overall speed of the computer. Purely electronic switching offered the prospect of enormously greater speeds.

If quick access to a useful large-scale computational facility had been the military's prime reason for funding the University of Toronto project, then, in the context of 1949, the relay computer offered a prudent approach. Despite speed limitations, relay computers still offered a cheaper and quicker technological solution for a wide range of large-scale scientific computations than purely electronic approaches. In the period 1945–1951, a range of relay computers, from small to very large, found extensive use in military and academic settings. The US Army's Ordnance Department had three relay computers running at the Aberdeen Proving Ground in Maryland and another at Army Ground Forces in Texas. In 1949, the (US) National Advisory Committee on Aeronautics had just completed construction of a relay computer to be used in laboratories at Langley Air Force Base. The US Navy had three relay computers in operation at its Bureau of Ordnance and another at its Office of Naval Research.[48] Installed in 1948 at the US Naval Proving Ground, the Mark II relay computer was running 24 hours a day, 6 days a week, on ballistic research problems.[49] In 1951, the Naval Proving Grounds built the Mark III, another relay computer.[50] At the Aberdeen Proving Ground, where both the electronic ENIAC and a Bell Telephone Laboratories relay computer were used for scientific computations, one L. Alt observed that the "Bell Laboratories' machine and ENIAC represent two extremes." Alt continued: "The ENIAC is adapted to very long problems, provided they are not too complicated and do not need too much storage capacity. The Bell Laboratories' will handle the most complicated problems, requiring considerable number storage, provided they are not too long. The ENIAC prefers continuous runs, the Bell Laboratories machine does not mind 'on-and-off' problems."[51]

Though its slow switching would limit the long-term utility of the relay computer, it nevertheless offered Canada a valuable 2–3-year interim solution to its computational research needs, thereby affording the university's Computation Centre an earlier opportunity to develop its computational expertise. In addition, a relay computer would have also provided the Computation Centre's hardware engineering team with more time to grow in size and experience before tackling the construction of an electronic computer. However, the decision was not made to go with the surer but lower-performance relay computer.

Since DRB was to carry the entire costs of building both the Bell Relay computer and the electronic computer, one can reasonably assume that DRB was the prime mover behind the change of plans. The added licensing fees may have caused DRB to stop and ask if the relay computer, an interim solution, was worth the expense. Yet it is doubtful that the relay computer would have been dropped had the university insisted on using it. Though it meant postponing the introduction of large-scale computational techniques, the promise of accelerating the development of an electronic computer seduced the university. The Computation Centre Committee's thinking on this matter was quite different when it met on 6 January 1949 to reassess the role of the relay computer in light of the new costs. Anxious to get onto the learning curve of large-scale computation technology quickly, the committee reaffirmed the short-term to mid-term importance of the relay computer. Because designing and testing its own relay computer in a short time was impracticable, the committee felt it was important to convince DRB of the merits of following the original plan and to buy the Bell relay computer.

DRB's willingness to reallocate funds to an accelerated hardware development program underscores an additional facet that electronic computer R&D presented to Canada's military. Digital electronics was just starting to appear at the very frontier of weapons-system research. Designing and building computers was thus a useful vehicle for fostering a national capacity in digital circuit design.

The cancellation of the relay computer at the 16 March meeting raised the stakes in the Computation Centre project dramatically. Those who needed computational power clung to the expectation that an electronic computer would be constructed quickly as the university committee had so confidently promised.⁵² Yet Canada's entire computational program now rested on the shoulders of a small, inexperienced team of engineers. N. L. Kusters, a senior engineer in NRC's Electrical and Radio Engineering Division, expressed serious reservations about dropping the relay computer and accelerating the electronic computer program.⁵³ In a memo to his director, Kusters wrote:

> While the University of Toronto might be an ideal place for the location and operational responsibilities of Canada's computing center, there is considerable doubt in my mind whether it is the appropriate place for the development of a large scale electronic digital computing machine. . . . Most of the people [engaged in the development program] have a limited amount of practical experience which seems to be out of proportion with the magnitude of the project. A project of this size requires a large amount of experimental background, a relatively large staff which is permanently assigned to the project, and sufficient [sic] close contact with production methods to make sure that none of the advantages achieved by tedious design are lost between the bread-board model and the final machine.⁵⁴

The cancellation of the relay computer heightened the need to develop an electronic computer as quickly as possible. Yet this sense of urgency did not translate into a practical design approach. No real effort was made to reproduce a design proven elsewhere. Neither was the team content to use less advanced but nevertheless workable technological solutions. Copying a design that had proved itself elsewhere was one way to make the task easier. By 1949 the Institute for Advanced Studies computer was serving as the genetic code for a whole generation of computers in the United States. Some were identical clones; others were close adaptations. Though members of the University of Toronto's Committee on Computing Machines had contact with the IAS computer group, no attempt was made to copy their machine.⁵⁵ Instead, the lure of ever-higher performance and the desire for originality drove the designers to a project called the UTEC (University of Toronto Electronic Computer) Mark I.

From the moment Josef Kates was brought in on the University of Toronto's electronic computer project in 1948, his ideas, ambition, and assertive drive were the dominant elements shaping it.⁵⁶ After the Nazis rose to power, Kates, a secondary-level student, fled Vienna and made his way to Canada. While working at the Imperial Optical Company in Toronto on wartime ordnance work, Kates managed to earn a high school degree and to complete a correspondence course in radio electronics. In 1948, after 4 years of study, Kates completed a Bachelor of Science degree in physics at the University of Toronto. Before he enrolled in the university, Kates's efforts to pursue an invention had brought him to the attention of the Rogers Majestic Corporation, and during his entire undergraduate studies, Kates worked for Rogers Majestic, first on high-power tubes and then, after the war, on receiving radio tubes.

Kates was a graduate student in electrical engineering in 1948, when the Committee on Computing Machines asked him to join the project. At about the same time, another graduate student in electrical engineering, Alfred Ratz, was also enlisted. By 1949, the development of Canada's computer hardware was resting completely on the shoulders of these two graduate students.[57] Before DRB would release any money for the construction of a full-scale electronic computer, a small successful prototype had to be built. In the summer of 1949, Kates and Ratz, along with one technician and five summer students, set out to design and build the UTEC Mark I. The plan was to have that machine "handle numbers of approximately 4 decimal digit accuracy," "be able to add, subtract and choose among different order sequences," "contain about 600 vacuum tubes," and "occupy a space of approximately 4 feet × 4 feet × 6 feet."[58]

Calvin C. Gotlieb, who had recently completed his doctorate in physics, was given the job of overseeing the development of a program in large-scale, high-speed computation. Gotlieb's interest and experience lay in the broader questions of devising computational techniques and applying them to large scientific calculations. Until UTEC was ready, he spent most of his time leading a team of mathematicians and scientists in the development of numerical methods for IBM punched-card machines and desk tabulators. Lacking an electrical engineering background, he left the details of hardware design to Kates and Ratz. Despite their junior status, Kates and Ratz became the de facto hardware experts of the computer project. The details of how Kates and Ratz worked out the design between themselves are not known, but evidently it was Kates's forcefulness and persuasiveness that continually convinced the Committee on Computing Machines, DRB, and NRC that he and Ratz had the right approach.

The military's ambiguity over the primary objective of the UTEC project fueled Kates's and Ratz's pursuit of novelty. The inexperienced team's pursuit of new and faster circuits, a radically new arithmetic tube, and parallel electrostatic memory continually pushed back the completion date of UTEC Mark I. After 2½ years, UTEC Mark I still did not completely satisfy the design goals specified in 1949. This delay set the stage for conflict between the engineers who were absorbed by the challenge, excitement, and glory of building a computer and the scientists and mathematicians who were impatient for a computational tool. This conflict of interests was what caused the project's demise.

Efforts to design and manufacture a new class of vacuum tubes for digital electronic circuits illustrate how Kates's ambitions and pursuit of performance delayed the construction of UTEC Mark I. Though computation was being revolutionized by digital electronics, the vacuum tube itself had quickly become a "reverse salient" in the development of large-scale digital systems.[59] The pursuit of greater computational power demanded a continual increase in the overall number and complexity of computer circuits. In the late 1940s and the early 1950s, the drive toward higher circuit complexity ran up against the vacuum tube's economic and technological limitations. Circuits composed of thousands of vacuum tubes consumed great amounts of power and occupied considerable physical space. The IBM 702, which had 5562 tubes, consumed 87 kilowatts of power and occupied 250 square feet. The UNIVAC, with 5600 tubes, consumed 120 kilowatts and occupied 1250 square feet.[60]

Kates wanted to design and build the university computer around a new special vacuum tube of his own design. He argued that his invention not only would dramatically reduce the number of tubes needed to achieve a given computational capacity but also could produce substantial increases in speed.[61] Kates's invention implemented the entire input-output functional table of binary addition in one tube through the internal manipulation of electric potential gradients. Kates argued that this manipulation was attainable through the proper placement and shaping of electrodes within the tube. His ongoing doctoral work in the physics of vacuum tubes had convinced him of the theoretical correctness of his idea. His experience with vacuum tube technology at Rogers Majestic convinced him that such a tube could be mass produced.

In February 1949, V. G. Smith, a member of the university's Committee on Computing Machines and a professor of electrical engineering, wrote to the director of NRC's Radio and Electrical Engineering Division asking if NRC would undertake the construction of a prototype of Kates's binary adder tube concept.[62] But in April 1949, after a tour of US computer centers, Smith had second thoughts about Kates's idea. "We sense," Smith wrote to Ballard, "a general theorem that the same results may be obtained by clever circuit design and ordinary tubes and that the computing centers have chosen this path. While we have not abandoned the idea it does not have a high priority with us just now."[63]

Smith's sense of the word "we" may have referred to other members of the committee, but it did not include Josef Kates. In May 1949, Kates wrote a report in which he argued that, in comparison to conventional electronic, digital, design practices, his binary adder tube promised "greater reliability, greater simplicity, greater speed, a smaller fraction of components required, and a more compact and sturdier tube than conventional grid-controlled tubes."[64] While the senior members of the Committee on Computing Machines vacillated over the merits of the binary adder tube concept, Kates promoted the idea to the military and industry. Despite numerous difficulties, P. A. Redhead of the Microwave Section of NRC's Radio and Electrical Division succeeded in building the first working model of the binary adder tube in early June 1949.[65] Within days of receiving the tube, Kates had constructed a demonstration of a single-digit binary adder. He immediately declared the crude prototype a success. With better prototypes coming from Redhead's laboratory, Kates set out to convince the industry of the commercial potential of his invention.

Kates did not confine the scope of his idea to the arithmetical operation of addition. He also designed a tube that could do subtraction.[66] In a letter to Redhead, Kates confided that the principles of the prototype adder tube could also be used to reduce the speed of multiplication considerably. Because the operation of multiplication in computers was implemented as a succession of many additions, Kates saw this operation as a serious bottleneck in computational speed. The fastest computer of the day multiplied two numbers in about 250 microseconds. Kates suggested that he could design a tube that could perform multiplication in one step and in only a few microseconds.[67]

Was Kates wasting his time? Hindsight reveals that transistor technology overcame the vacuum tube's reverse salient. However, in the context of the late 1940s and the early 1950s, the ascendancy of the transistor was not at all evident. The technical and institutional "inertia" of the vacuum tube was considerable, and the superiority of transistors was very much in doubt.[68] Any vacuum tube innovation that could alleviate the size and complexity bottleneck in electronic digital circuit design still had the glitter of great technical achievement and commercial success. In July 1949, discussions started between Kates and Canadian Patents and Development Limited over the issues of patent rights and commercialization. Canadian Patents and Development Limited, a crown corporation and a wholly owned subsidiary of the NRC, was created in 1947 to patent and promote the commercialization of promising inventions from NRC and other government financed research. CPDL agreed to pursue patents rights for Kates's invention; it also agreed to facilitate the necessary industrial development of the tube and to handle the licensing of the patent.

CPDL asked six Canadian companies if they were willing to develop the binary adder tube beyond the prototype stage.[69] After examining the prototype, Rogers Majestic, Canadian Westinghouse, and Canadian Marconi expressed interest. To get a better idea of what the development of the tube would entail, each of these three manufacturers produced some samples based on Kates's specifications. Kates found the Canadian Marconi samples "wholly unsuitable."[70] By the end of the summer of 1949, Rogers Majestic and Canadian Westinghouse each agreed to undertake the development of the tube at its own expense on the condition of being given exclusive worldwide rights. However, with a patent application still not filed, CPDL was in no position to offer any party an exclusive license.

Efforts to move Kates's patent application through the American, French, British, Dutch, German, and Canadian systems proved extremely frustrating. The difficulty lay in the characterization of Kates's tube. What were the unique elements of the tube that could withstand prior patent claims? Answering this question turned out to be far harder than anyone expected and delayed the actual filing of the patent. Kates was certainly not the first to propose vacuum tube switching devices. High-speed switching in telegraphy,[71] in telephony,[72] and even in radio[73] had been the focus of several patents. RCA's wartime work on automated digitally based anti-aircraft fire control had produced several important patents in vacuum-tube electronic switching.[74] The art of ascribing claims to a patent is to cast as wide a net as possible without pulling in any prior patent claims. Unable to circumscribe all the subtleties of this new technology, the Ottawa law firm handling CPDL patent claims, Smart & Biggar, snagged its net on one objection after another.

In a letter to J. W. MacDonald, the General Secretary of CPDL, E. G. Bullard, chairman of the University of Toronto's Committee on Computing Machines expressed concern about "the long time [CPDL] people were taking in getting the protection of a provisional patent for Kates's adder tube": "I hear rumours of a similar development in Holland and whether this is true or not, I am quite certain that someone else will come upon the idea in the very near future."[75] MacDonald replied that CPDL was pursuing the matter vigorously, "but that, in all matters electronics, it is necessary to make very detailed searches in order that the patent application, when drawn may meet in advance any objections which could be raised in respect of prior patents in the same or similar fields."[76] MacDonald failed to keep his promise that a patent would be filed by the end of 1949. Not until 17 January 1951, 20 months after CPDL first agreed to promote Kates's invention, was a patent filed in the United States.[77]

Delays in filing a patent application threatened to stop further industrial development of the binary adder tube, until the Royal Canadian Navy's (RCN) DATAR project entered the scene.[78] DATAR, which called for the marriage of computer, radar, sonar, and communications into one digital medium, called for circuit complexity, scale, and reliability that could not be easily supported by the existing vacuum tube technology. Kates's claim that his invention was applicable to a wide range of digital switching problems in computers and computers caught the Royal Canadian Navy's attention, because of the considerable miniaturization that would follow if Kates was right.

In September 1949, the Navy agreed to fund further research and development of the Binary Adder Tube at Rogers Majestic, with Kates retained by the company as a consultant. Three promising types of switching tubes emerged from the Navy's R&D contract.[79] As RCN funding to Rogers Majestic started to run low, the former proposed that DRB also contribute to the tube's further development. W. B. Lewis's favorable evaluation of Kates's invention added considerable weight to the RCN's suggestion. Lewis corroborated Kates's contention that the applications of the switching tube went beyond computers. Kates's invention, suggested Lewis, could "prove of great importance to the major communication concerns that policy should be carefully considered." He went on to warn: "If the Canadian effort is too small the large firms will probably discover certain vital features for which they would secure patents. This would be least harmful to Canadian interests if either the basic features are patented and sold to such company before any large expenditure is made in Canada, or detailed publication is made so as to make future patent action difficult."[80]

To accelerate testing and practical application of the tubes, DRB and Rogers Majestic decided to focus on the binary adder tube and to freeze its design. According to M. L. Card, the Royal Canadian Air Force's representative on the military's Electronics Standards Sub-Committee (ESSC), the prior RCN contract had "produced several improvements in the original design, involving an analysis of the tube theory, multiple-beam construction, rod elements and double-ended and subminiature packaging, to the extent that an early production model of a satisfactory adder tube could be made available during May 1950."[81] The goal of the DRB X-17 contract was to develop reliable, rugged, and standardized subminiature models of Kates's tube that would be suitable for military applications.

When it comes to the development and commercialization of inventions, timing is critical. If there ever was a door of opportunity for Kates's switching tubes, the inability to file a patent quickly and to accelerate its industrial development closed the door. After Kates's paper was presented to the International Radio Engineering Convention, Jan Rajchman of RCA's computer research group raised "the possibility of obtaining a non-exclusive license with the right to grant sub-licenses under [Kates's] US applications and patents."[82] In exchange for 20 switch tubes, RCA offered Kates a prototype of the Selectron memory tube that it was pioneering. No working laboratory prototypes of Kates's other kinds of switching tubes were ever built.

Delays and skirmishes characterized the pursuit of patent rights as Kates and CPDL lawyers responded to a litany of objections from patent examiners. Lewis had warned earlier that "if patent action is taken and an attempt is made to obtain a monetary gain[,] the forces called into play might be too great for any peace of mind of the inventor."[83] As the quest for patent rights dragged on, Lewis's words were prophetic. Finally, in 1955, approvals for Kates's Binary Adder Tube were won in the United Kingdom, France, Holland, and Germany. But in the United States, the center of postwar technological and commercial development of the computer, Kates's patent was rejected. In 1956, Canadian Patents and Development Limited decided to let these patents lapse because "this vacuum tube [was] now obsolete."[84]

Despite all the time and energy spent pursuing this new switching device, it was never incorporated into any of UTEC Mark I's circuits. The only application of Kates's switching tube occurred at Toronto's 1950 Canadian National Exhibition (CNE). After working on the binary adder tube for nearly a year, Rogers Majestic wanted to publicize some concrete use of this tube. Kates came up with a tick-tack-toe game called "Bertie the Brain," possibly the world's first computer game.[85] Paid for by Rogers Majestic, "Bertie the Brain" was finished just in time for the CNE. Quite compact because of the use of the binary adder tube, Bertie was quite an attraction. However, Bertie was the only consequence of Kates's ambition to design a faster and better electronic digital computer using his invention. Development of the switching tubes advanced too slowly for them to be seriously integrated into UTEC's design and testing. Without an early resolution of the patent process, there was little incentive for small Canadian firms such as Rogers Majestic to pursue extensive development of all the switching tubes. The window of opportunity for the concept of the switching tube was quite narrow; by the early 1950s it had closed.

The switching tube is not the only example of the non-conservative design ethos that dominated the UTEC project. Another example (probably more fatal) was the advocacy of parallel electrostatic memory based on new principles. Fascination with ultimate performance led the inexperienced University of Toronto design team to pursue a parallel rather than serial architecture. In serial machines, bits of data move in and out of memory one at a time along the same channel. In parallel architectures, all the bits of a word move simultaneously, each along its own channel. The parallel machine promised much faster processing, but it was far more complex and costly to design, build, and test. For a computer using 40-bit words, as was the case with IAS computer, a parallel architecture required a 40-fold increase in the number of logic circuits and fast memory units.[86] The design of IBM's first commercial venture into electronic computers was deeply influenced by von Neumann's idea of parallel architecture.[87] Called the IBM 701, this computer was originally conceived of as a "Defence Calculator" for the US military.[88] Except for one installation at IBM's World Headquarters in New York, all of the nineteen 701s were delivered to customers within the swelling US military-industrial complex.[89] The explosion of defense spending that accompanied the Korean War provided the financial climate to justify the $8000-per-month rental fee for the IBM 701. Ultimately, the willingness of the US military to push the limits of performance at any price sustained the early development of parallel architectures.

Despite the performance advantages of parallel machines, serial computers, because of their simplicity, were still being actively pursued. All the contemporary British projects, including Pilot ACE at the National Physical Laboratory, EDSAC at Cambridge University, and Mark I at the University of Manchester, were serial machines. The first computer to be commercialized in Britain in 1950 by Ferranti Limited was a serial machine. In the United States, J. Presper Eckert and John Mauchly, who had pioneered the design and development of ENIAC, the first full-scale electronic computer, used a serial architecture for their first commercial venture, the UNIVAC. Engineering Research Associates' ERA 1101 was another commercial serial machine that appeared in 1950. The machine that first secured IBM's dominance of the commercial general-purpose computer market was the IBM 650. Whereas its predecessor, the IBM 701, was a large parallel machine, the 650 was a serial machine designed as a low-cost alternative. The 650 proved to be a gold mine for IBM throughout much of the 1950s, and it established IBM as the leader in the computer business.[90] In 1949, a serial architecture offered a relatively inexpensive and rapid technological solution to Canada's search for computational capacity.

The high risk of the parallel design choice was compounded by Josef Kates's new approach to electrostatic memory. During the period 1946–1950, considerable effort was devoted in Britain and the United States to finding a purely electronic method for fast memory (i.e., what today would be called RAM). Mercury delay lines, which stored numbers as recirculating acoustic waves, undermined the performance gains proposed by the advocates of parallel machines. From the initial conception of the IAS computer, John von Neumann insisted that fast access to memory would be critical to the project's success. Based on Rajchman's and Snyder's wartime work, RCA joined the IAS project proposing to design the first purely electronic memory. As delays in RCA's development of the Selectron kept increasing, von Neumann's group looked around desperately for an alternative and found the solution in Britain.

Unlike RCA's expensive custom-made Selectron, F. C. Williams and T. Kilburn's approach to memory storage used conventional, inexpensive, low-voltage cathode ray tubes, such as those found in oscilloscopes.[91] The "Williams tube" stored binary numbers as electrostatic charges on the surface of the cathode ray tube's screen. With some difficulties, the IAS group adapted the Williams tube to its parallel machine. Williams and Kilburn, on the other hand, used their electrostatic memory approach as the basis for a *serial* computer they were designing at the University of Manchester. This computer later became the basis of Britain's first commercial computer.[92] After some teething problems, Williams's and Kilburn's approach to fast memory became the mainstay of the world's computer industry and remained so well into the late 1950s.

Instead of simply copying the Williams electrostatic memory, Kates proposed a different and novel approach to the implementation of electrostatic memory on conventional cathode ray tubes based on space charge.[93] Kates argued that the existing theoretical model used by his contemporaries to explain how electrostatic memory worked was wrong. Rather than looking to the properties of secondary electron emission to understand the limits of storage, he suggested that the physics of space charge was the critical factor. Kates contended that such a reinterpretation opened the way for faster, more reliable, higher-density storage systems.

The University of Toronto's Committee on Computing Machines had initially expressed concerns about the wisdom of trying to implement parallel electrostatic memory. But Kates, as Calvin Gotlieb later recalled, "convinced us and himself that the parallel cathode ray tube memory was possible."[94] After all, Kates was writing his physics doctoral dissertation in this area. No one on the committee was in a position to seriously challenge the soundness of Kates's arguments, for he had made himself the de facto expert on memory tubes. Moreover, the experimental verification of the space charge concept at Los Alamos reinforced Kates's position.[95] J. Richardson's experimental collaboration with Kates contributed to Los Alamos's decision to abandon the IAS approach to Williams memory and "pursue an independent course for memory, using a 2 inch tube rather than the 5 inch one considered by Princeton."[96] Los Alamos's success in developing a 40-digit memory, which made use of Kates's theoretical interpretation of electrostatic storage, gave Kates's approach considerable credibility.[97]

But could the University of Toronto team engineer a reliable full-scale parallel computer based on Kates's theory soon enough to meet the growing pressure for a computational device? The Toronto group did not have the depth of engineering experience of Los Alamos. Furthermore, by its own volition Toronto did not benefit from the very close technology transfer that characterized the relationship between the Princeton group and Los Alamos.[98] It took the small Toronto team a year and half to go from a one-digit stage of Kates's memory to a 12-digit memory.[99] All the component parts of UTEC Mark I were finally assembled in the fall of 1951. And yet, even after 2½ years of work on this small experimental prototype, a progress report stated that before UTEC Mark I could be put to full use further tests would be needed to allow a careful assessment of its reliability.[100] Despite the slow progress, the computer group expressed confidence in its ability to build a full-scale computer. "The period of profitable development work on Model I is now at an end," wrote Alfred Ratz. "The work," he continued, "has gained for the project the experience necessary to produce a full-scale machine, and the time to start on this computer has arrived. A reliable and useful instrument can be produced in a reasonable length of time, if the project is suitably staffed and organized."[101] The university computer group estimated that it would take an additional 2½ years to get a full-scale computer designed, built, and tested.[102] As long as the UTEC project was the only game in town, those people and institutions anxious for a computational tool could do little but wait.

But new options for obtaining computational power finally overtook the UTEC project. Even as the University of Toronto computer group was confidently asserting its abilities to build Canada's first computer, W. B. Lewis, Director of Research at Chalk River, had set his eyes on the new Ferranti Ltd. machine. In 1948, the British government issued Ferranti a five-year contract, worth about $510,000, to produce a commercial version of the computer that Williams and Kilburn were developing at the University of Manchester.¹⁰³ Ferranti's sales strategy consisted of (Lord) Vivian Bowden going on the road in search of orders while the Ferranti board used its considerable contacts in the British government and military to promote sales at home.¹⁰⁴ During the summer of 1951, Lewis learned from his old friend Bowden that Ferranti Ltd. was selling computers.¹⁰⁵ Bowden told him that Ferranti would install a complete computer system anywhere in Canada for $220,000, plus freight charges from some British port.¹⁰⁶ Resorting to the usual salesman's ruse, Bowden wrote that, because Ferranti's production lines were being quickly tied up with future orders, Lewis would have to act immediately if Canada wanted delivery within a year.¹⁰⁷ Lewis was delighted with Bowden's offer. "This is better than I hoped for in the matter of price," wrote Lewis to K. F. Tupper, Dean of Engineering at the University of Toronto. As chairman of the Committee on Computing Machines, Tupper no doubt sensed the disastrous implications for UTEC behind Lewis's excitement.

The lack of high-speed computation facilities had become a serious problem for Canada's atomic energy program. Efforts to develop a commercial reactor at Chalk River were generating an ever-escalating demand for high-speed, large-scale computational tools. Lewis had warned C. J. Mackenzie, president of NRC: "Unless we get a machine in the near future, Chalk River, at considerable inconvenience and expense, will have to arrange for needed calculations to be made in England and the United States."¹⁰⁸ He added that the university's punch-card machines were inadequate.¹⁰⁹ Lewis's patience with the university computer project had run out.

Mackenzie was very sensitive to Lewis's complaint. The success of Canada's quest to commercialize atomic energy was paramount to Mackenzie. Canada had emerged from World War II as one of the best-placed nations in the world to develop nuclear reactor technology. The United States and the United Kingdom were the other two. For Mackenzie, this capacity was an opportunity for Canada "to get on the ground floor of a great technological process for the first time."¹¹⁰ From the beginning, Chalk River's computational needs had been a major motivating factor in NRC's support of the University of Toronto's ambitions to become Canada's center for large-scale scientific computation. Until Bowden's letter, hardware development had been an organic part of the development of a computation center. With the availability of the Ferranti machine at a price that was less than the cost of building a machine, Mackenzie wanted to remove hardware R&D from the Computation Centre's mandate.

In a telephone call to Dean Tupper, Mackenzie suggested buying the Ferranti computer and locating it at the university. The university's Computation Centre Committee voted unanimously to reject Mackenzie's suggestion.[111] For the committee, the Ferranti machine's performance paled before what UTEC could achieve if built. The committee expressed its fears that the Ferranti machine would soon be inadequate, if not obsolete, and thus would condemn the university to a role of secondary importance in the computing field.

By December 1951, Mackenzie was exasperated with what he felt were the Computation Centre Committee's illogical and contradictory objections to purchasing the Ferranti computer. He once again emphasized the need for early access to computational power. As to the university's ability to build the computer, Mackenzie was unequivocal: "To be brutally frank, I don't think the group at Toronto have sufficient staff or experience to be able to guarantee a first class rugged, dependable, workable commercial machine in any where near the time contemplated, and I think it is certainly not in the interests of Toronto to court that sort of disaster."[112] Since Mackenzie knew little of digital electronics and computer technology, his harsh criticism probably reflected the views of W. B. Lewis. Mackenzie, however, was also sensitive to Gotlieb's fears that the purchase of the Ferranti machine and the cancellation of the UTEC project would demoralize the electronics team and lead to its dissolution. He suggested to Tupper that "Gotlieb's [electronics] group would be better employed doing more fundamental research than trying to manufacture production equipment."[113]

In arguing for the Ferranti computer, Mackenzie, as president of NRC, had to tread the fine line separating Chalk River's interests and academic freedom. He knew it would be very difficult to impose a solution on the university, particularly since funding had already been approved.[114] Mackenzie raised the Toronto dilemma with DRB's chairman, Omond Solandt. Solandt's initial response reflected DRB's view of the strategic importance of a digital design capacity. DRB, wrote Solandt, was "more interested in having the electronics group carry on in the development of the machine as a computer as they would undoubtedly gain much electronics experience that might be useful in an emergency."[115] But at the same time, Solandt was sympathetic to Chalk River's needs. At a meeting of DRB's Committee on Extra-mural Research, on 8 December 1951, Solandt outlined the issues dividing Mackenzie and the Computation Centre. The committee backed a compromise presented by Solandt. DRB had already promised $300,000 toward the construction of the university computer. If

NRC agreed to pay for half of the Ferranti machine ($150,000), DRB would pay the other half. Furthermore, DRB agreed to allocate the remaining $150,000 to support basic research into computer components and digital electronics. In effect, NRC had to pay an additional $120,000 to get its computer and placate the university. To drive home its position to the university, DRB's Extra-mural Committee recommended that "in the event that it was decided to continue with the construction of the Toronto computer, steps should be taken to ensure that the work was contracted out to industry, and not carried out by the university."[116]

As pressure mounted from without, those within Computation Centre who were interested in the art of machine-based scientific computation realized where their future interests lay. On 11 January 1952, the Computation Centre Committee concluded that "it was important to work with Chalk River since the theoretical physics group there would certainly become one of the largest customers for computation."[117] The committee voted unanimously to accept the offer NRC and DRB made to buy the Ferranti Computer.[118] The moment this decision was taken, support for the electronics group within the Computation Centre collapsed completely. Tupper informed Mackenzie that, other than putting some finishing touches on its small experimental prototype computer, UTEC Mark I, the university would probably "not wish to continue electronic work under Computation Centre auspices."[119] Ratz's attempts to salvage the research experience in digital electronics that had been built up over the preceding 4 years fell on deaf ears.[120] A new ethos had taken over the Computation Centre: "From here on it will be computing rather designing, mathematics rather than machine."[121]

In his capacity as a member of DRB's Electronic Advisory Committee, Lewis asked Tupper to reconsider killing the digital electronics R&D program. "I feel it would be a mistake," wrote Lewis, "to disband the Toronto group when there is such a big future for the techniques they have learnt. Also there is a real defence interest in many applications of this technique."[122] Lewis went on to suggest that the electronics group broaden its R&D interests to include the very strategic area of digital communication. Tupper was "distinctly cool towards the suggestion that [the Computation Centre's] electronics work should be tied to similar secret projects elsewhere." "The Computation Centre," he said, "should not fall under the dead hand of 'Security.'"[123] Tupper's aversion to soiling the Computation Centre's hands with secret projects may have been somewhat misplaced. After all, much of the Centre's future computational work would have its raison d'être in secret weapons development.

Installation of the British computer started at the end of April 1952. By October, FERUT (FERranti University of Toronto), as the computer became known, was fully tested and operational. In his history of Canada's pursuit of atomic energy, Robert Bothwell writes that "in the United Kingdom, Lewis was regarded as a great Englishman, and his Chalk River laboratory was considered, at least in an intellectual sense and certainly in a sentimental one, to be an extension of British science."[124] The birth of Canada's first large-scale, high-speed computational facility was also an extension of British scientific establishment.

Life after UTEC

Despite its failure, UTEC provided a training ground for Canada's first pool of expertise in computer technology. Though the cancellation of UTEC disheartened the young electronics team, it was surely the Computation Centre's refusal to entertain any further military-funded basic research program in digital electronic techniques that ultimately led to the group's abrupt departure from the university. In 1953, three of the team members—Len Casciato, Josef Kates, and Bob Johnston—set out to form a new company called Digitronics and to design their own commercial computer.[125] Their UTEC experience had convinced them that they could design and build a very compact general-purpose computer for business, industry, and government.[126] They estimated that it would take 12 months and cost $100,000 to develop a production-ready prototype, and that the subsequent commercial units would cost $50,000 to manufacture.

Armed with a business plan and with letters of endorsements from K. F. Tupper (the man who had killed any further research in digital electronics at the university), from the director of the Computation Centre, and from others, Casciato, Kates, and Johnston went looking for $246,000 of startup venture capital to see their new company through the first 18 months. Their supply-driven market previsions cast an ambitious net over all the sectors of the economy. They argued that the potential use of computers in such areas as inventory control, automated reservations systems, accounting, automatic process control in manufacturing, and office filing systems represented in 1953 a potential for $50 million of sales in Canada.

There is no record of all the potential investors Kates approached,[127] but it must have been quite frustrating trying to find investors in such a radical venture. At some point they approached Sir Robert Watson-Watt, one of the pioneers of radar, who had a high-tech consulting company in Montreal called Adalia Ltd.. He did not bite, but he did hold out a carrot.

If they joined forces with him, he would find a way to carry out their ambitions to build and sell data processing technology. Tired of knocking on doors and out of money, they went to work for Watson-Watt and became Adalia's expert computer consultants. After some important work on a feasibility study to introduce the world's first computerized systems in Trans-Canada Air Lines' operations, the relationship between Watson-Watt and the trio fell apart because they grew to feel that he had no intention of getting into computer hardware.

In 1954 or early 1955, Kates and Casciato, along with a new partner, Joe Shapiro, formed KCS Data Control Ltd., which became the most successful computer consulting firm in Canada in the late 1950s and the early 1960s. KCS's know-how played an important role in the computer technology decisions taken by companies and government agencies. This company also developed real-time process control for Canadian petrochemical companies.[128] But KCS's biggest prize was the design and deployment of the world's first computerized urban traffic control system (for Toronto, in 1962).[129] In 1964, to celebrate the KCS achievement, the American Society of Traffic Engineers held its annual conference in Toronto—the society's first meeting outside the United States. Not only was the construction of the computerized traffic control system a superb technical achievement; it was also a remarkable piece of urban political work for Kates, who later went onto become chairman of the Science Council of Canada. KCS Data Control was subsequently bought by the large Toronto firm of Peat Marwick. In the 1980s, Kates formed Teleride/Sage, a company that created digital systems for public transportation networks.[130] Also working on the design of computer and communications systems for transportation networks, Casciato contributed to the design of the prototype TRUMP (Transit Universal Microprocessor) system, which monitored the operation of the Toronto Transit Commission's buses in real time.[131]

Equally important, the loss of the University of Toronto's ability to design computers, which the purchase of FERUT embodied, was more than compensated by the accelerated development of the capacity to design software and make innovative use of the computer. Although FERUT contributed to the demise of UTEC, it also accelerated the diffusion of computer technology in Canada. FERUT allowed the University of Toronto to become an important training center and innovation pole in the application of computers. Though in charge of the University of Toronto's research in large-scale automated computational techniques, C. C. Gotlieb played a nominal part in the design of UTEC. But once

FERUT was installed Gotlieb came into his own. Under his leadership, the Computation Centre grew into Canada's leading center of research on the use of computers. It was from this center that data processing techniques and technology moved to Toronto's largest insurance companies in the late 1950s. With FERUT, Gotlieb was able to launch Canada's first department of computer science. Because of FERUT, a generation of Canadian scientists and engineers got to use computers at least 2 years before they otherwise might have (assuming, of course, that UTEC would have been ready in 2 years). [132]

The most striking example of the skillful use to which the University of Toronto's Computation Centre put FERUT came within a year of the machine's installation. The Centre's development of sophisticated computer-based hydrological models served to advance Canada's territorial interests in negotiations with the United States. In 1952, both the Canada and the United States agreed that an extensive public works program of dams, locks, and channels was needed to improve the navigability and efficiency of power stations along the St. Lawrence River from Lachine, Quebec westward. Because any modification in the hydraulic structure of the river was bound to flood some areas, it was important to understand the relationship between a particular modification in the natural flow of the river and its consequences on the landscape. Each country, of course, wanted to minimize the negative effects of changes to its territories while reaping the most benefits from changes to the river's flow. Unable to agree on terms, the two nations submitted their case to an International Joint Commission. Effective presentations to the Joint Commission hinged on prior knowledge of all the possible scenarios. Ontario Hydro went to Gotlieb's group for help. Gotlieb later recalled that "considerable political consideration was given as to whether the seaway would be an all-Canadian one or [one shared with the United States]," and that "the Hydro, with the aid of [FERUT] did massive calculations, more voluminous than anything done in New York."[133] With FERUT, what would have taken 20 person-years of manual computation took only 21 days. Thus, Ontario Hydro always had a calculation for a credible all-Canadian Seaway up its sleeve before the United States presented its various Seaway scenarios.

The Canadian military's support of university computation centers stemmed from its continued interest in promoting computational self-reliance. There was a need to ensure that an adequate number of Canadian mathematicians, scientists, and engineers were trained in the use of general-purpose computers. Without such a national pool of expertise, the military knew that it could never launch or sustain a semblance of R&D self-reliance. The military also wanted to ensure that it had direct access to a sufficient amount of machine time to meets its day-to-day computational needs. Related to these two factors was the need for an adequate regional decentralization of computational facilities to match the coast-to-coast geographical dispersal of military R&D centers.[134] Before 1952, the military's need for computational power required the promotion of a domestic capacity to design and build computers. After 1952, access to computational power became a question of being able to buy or lease commercial computers, to ensure an adequate number of people trained in the use of computers for scientific research and experimental engineering, and to support university computation centers. The disappearance of the military's interest in promulgating a domestic capacity to design and build machines to do large-scale, high-speed scientific computation did not mean that it abandoned interest in digital electronics. As the next two chapters will show, the purchase of FERUT did not remove the need to develop a domestic capacity to design advanced weapons systems using digital electronics. The next chapter examines how and why, during the late 1940s and the early 1950s, the Royal Canadian Navy became one of the world's leading promoters of automated digital warfare.

2
The Navy's Pursuit of Self-Reliance in Digital Electronics

In 1952, with military support, the University of Toronto became Canada's first high-speed, large-scale computational facility. The commercial availability of the Ferranti Mark I computer had removed the military's imperative to design and build high-speed computational devices. The intensive mathematical simulations and calculations needed to accelerate the prototyping of new weapons systems could now be done on an off-the-shelf commercial product. Yet the purchase of a Ferranti Mark I did not diminish the military's desire to create a national competence in the design of digital electronic hardware, for the impact of the digital paradigm on military technology went far beyond the need for off-line, batch-oriented, scientific computation. During the 1950s, real-time digital electronic computation became embedded in the actual operation of navigation, communications, early warning, target tracking, and automatic fire control systems. The Royal Canadian Navy's efforts in the immediate postwar years to automate naval warfare offer the most striking example of how "being digital" had become a vehicle to military self-reliance.

In 1948 the Royal Canadian Navy turned to the still-unexplored world of digital electronics in an effort to assure itself of a sovereign and important role in the postwar era. David Zimmerman's analysis of the wartime development of ASDIC suggests that the role of the self-reliant technological pioneer had not hitherto been part of Royal Canadian Navy culture.[1] Yet that is exactly the role that the RCN chose in the period 1945–1955, during which it became the world's leading exponent of electronic digital information-processing systems in modern naval warfare. The reason for the RCN's sudden embrace of leading-edge research and development stemmed from its anti-submarine and convoy escort role during World War II, when serendipity had brought Canada a historically unprecedented degree of autonomy in its military relations within the North Atlantic

Triangle. After the war, the RCN looked to technological self-reliance as a way to make its wartime gains permanent. This chapter explores how the RCN, in pursuit of this R&D agenda, played an important role in advancing Canada's industrial capacity up the fourth innovation wave. More specifically, the chapter focuses on the RCN's part in the birth and subsequent nurturing of Ferranti Electric Ltd.'s electronics group. For more than 10 years, even after the relationship between the RCN and this company ended, Ferranti Electric (later to become Ferranti-Packard) remained Canada's best hope of creating a domestic computer industry. Finally, this chapter explores the dilemma that Canada's small, open economy faced when it tried to pursue military self-reliance through costly, complex, large-scale technological systems. The story begins in the naval battles of World War II.

Anti-Submarine Warfare as a Cornerstone of Canada's Naval Self-Reliance

In World War II, Germany reasoned that if it could choke off all the transatlantic resupply lines to Great Britain from Canada and the United States then Great Britain's demise would be only a matter of time. The failure of Germany's surface fleet to sever Great Britain's lifeline to North America led to the promotion of the submarine as Germany's principal form of naval warfare.[2] "By the end of 1940," writes W. G. D. Lund, "Germany's strategy of starving Britain into submission . . . became evident and the long struggle by Commonwealth forces to maintain the vital lines of communication began."[3] Unless it was a very fast luxury liner, such as the *Queen Mary*, sending a solitary supply ship across the Atlantic was sheer folly. The low speeds of lesser ships made them easy prey for German submarines. To assign a naval vessel to escort each supply ship was impracticable.

To counteract the threat of the submarine, the Allies instituted convoys. There was safety in numbers. Large convoys lost proportionately fewer ships. Thus, a convoy of supply ships would be assembled and moved across the Atlantic by a small protective escort made up mostly of corvettes. But protecting a slow-moving convoy that extended over many square miles proved extremely difficult. By the end of 1942, the German submarine "wolf packs" were taking a devastating toll on Allied shipping. In November 1942 alone, 720,000 tons of supplies were sunk by German submarines. With the rate of shipping losses exceeding the rate of production, the Allied leaders gave the submarine problem top priority at their January 1943 meeting in Casablanca. "If the menace [from submarines] could not be conquered," explains Gerhard Weinberg, "the steady diminution of Allied tonnage would immobilize the Western Allies."[4] By 1943, the shipping lanes had become World War II's pivotal "battlefield." For Hitler, the submarine campaign had taken on an importance second only to the Eastern front.[5]

Canada devoted all its naval resources to protecting the transatlantic convoys.[6] By the end of 1942, Canada was providing 48 percent of all the convoy escorts.[7] Yet Canada had no say in the strategic use of its considerable anti-submarine resources. The dismissive attitude of British and American naval authorities toward Canada's views on the disposition of its own naval resources frustrated and angered senior officers of the Royal Canadian Navy. "The British Admiralty," concluded Canadian Admiral Victor Brodeur, "still looked upon the RCN as the naval child to be seen and not heard when no outsider [the US Navy] looked on or listened in."[8] American naval authority was no different, added Brodeur. Japan's entry into the war provided the RCN with unexpected bargaining leverage to pry the command of Atlantic convoy escort operations away from the United States because it forced the Americans to pour all their naval resources into the Pacific. The time was ripe for the RCN to win important concessions from its patronizing allies.

The RCN decided to use the high-level Atlantic Convoy Conference to push the British Royal Navy and the US Navy to give it a primary command role in the North Atlantic. The conference started on 1 March 1943. At its close on 6 March, the RCN emerged with a historically unprecedented military role in the North Atlantic Triangle. Canada had won control of all surface and anti-submarine escorts in the western half of the North Atlantic. In addition, Canada now shared control, with the United Kingdom, of all convoys running between the British Isles and North America, including those originating in New York. The senior RCN officer corps had achieved what the MacKenzie King government could not do: the assertion of Canadian autonomy in the military sphere. Through a commitment to anti-submarine warfare, Canada had gained a key command role in the battle for the North Atlantic. "No other small power," Lund argues, "enjoyed such a position."[9]

Determined to preserve its hard-won special status in the postwar era, the RCN wanted to cast itself as the Western alliance's anti-submarine and escort navy. By 1947, however, the RCN realized that advances in submarine technology would soon render its World War II anti-submarine (A/S) equipment obsolete.[10] If the RCN wanted to be taken seriously in its efforts to be recognized as an anti-submarine force, then it would have to embark on its own R&D program. The RCN understood that the intensive exploitation of science and technology had become the sine qua non of military competency in the postwar era. In 1946 the RCN had yet to articulate its R&D ambitions. By the end of 1948, however, its R&D agenda started to take shape. The most daring item on this agenda was the proposal to move naval warfare into the electronic digital universe. The impetus for this came from the technical mid-level officer corps within the Electrical Engineer-in-Chief's Directorate (EECD), but it was a receptive senior officer corps that provided the essential political and financial support. The senior officers who had fought so hard to win military sovereignty in the battle for the North Atlantic were ready to finance any bold technological undertaking that would bolster the RCN's claim to be the backbone of the Western Alliance's postwar anti-submarine efforts.

The inability to capture, extract, display, communicate, and share accurate tactical information in a timely manner had been a central limitation of anti-submarine operations. The movement of large convoys across the Atlantic during World War II had presented monumental logistical and tactical challenges. Keeping track of the positions of all the ships in a convoy that stretched over many square miles proved problematic. In a battle situation, the difficulty was compounded by the need to monitor the movements of all the enemy submarines. The long human chain needed to convert ASDIC data, radar data, and other tactical data into useful information for command and control was slow and often unreliable. With a new postwar generation of submarines and aircraft, the RCN realized that the slow, human-intensive conversion of input data to a coordinated tactical response had become a serious weakness in anti-submarine operations.[11]

Aside from the fact that the flow of tactical information on a single anti-submarine ship was slow, precious little information flowed *between* escort ships. The inability of ships to exchange tactical data in real time severely limited the ability of the escort fleet to respond as a unit. The RCN came to believe that there was an "urgent operational requirement" for new systems that would allow "for closely co-ordinated tactics by convoy escort and hunter-killer anti-submarine groups."[12] The Development Section of the Electrical Engineer-in-Chief's Directorate advocated an integrated and automated information system that could simultaneously provide the Command of each anti-submarine escort with a complete, concise, up-to-date picture of the tactical situation, provide the necessary information to weapons-control systems to assist in target designation, and incorporate tactical information on all aircraft into one battle picture. Unless the RCN could automate the production, exchange, and use of tactical data, EECD's technical people reasoned, Canada's anti-submarine escorts would prove ineffective against the coordinated attacks of a new generation of submarines and aircraft. But how could this technological breakthrough be achieved? To the technically minded mid-level officers in the EECD's Development Section, the ENIAC computer offered the answer.

A top-secret project during World War II, ENIAC had been built to accelerate the calculation of ballistic and bombing tables. Still the only fully electronic digital computer in the world in 1948, ENIAC was an obscure technology known and understood by a very small circle of people. But the members of the Development Section had followed the military reports on ENIAC with great interest. The speed and precision of electronic digital computation had made a dramatic impact on their technical imaginations. They saw more than just a radical advance in calculating technology. In these officers' minds, the new electronic digital paradigm offered a revolutionary way to unify the collection, interpretation, communication, and representation of tactical information into one automatic, interactive, and decentralized network linking all the escort ships of a convoy. They called this first comprehensive digital perspective of naval warfare Digital Automated Tracking And Resolving (DATAR). Expecting pack attacks by fast submarines armed with sophisticated homing torpedoes, the RCN proponents of DATAR argued that there was an "urgent operational requirement for an electronic system that will allow closely coordinated tactics by convoy escort and hunter-killer anti-submarine groups."[13]

The idea of using digital techniques to automate warfare was not new. For example, during World War II the Radio Corporation of America had worked on an electronic system intended to convert optical and radar target data into precise information that would allow a naval gun's servo-control unit to track a target. Because of the high precision required, traditional analog computational techniques were inadequate. Digital techniques, as one 1943 RCA patent application explained, allowed "any required accuracy to be obtained with the numerical method of computation merely by using the proper number of digits. Therefore, the accuracy of one part in 10,000 needed in the computer with the present day optical and radar tracking instruments for the present day medium and long range guns can be achieved and easily exceeded. . . . Furthermore, the great speed inherent in electronic devices makes it possible to perform complicated numerical computations without setting any limit on the firing speed of the gun."[14] Within the DATAR framework, automatic fire control became one element in a more sophisticated, comprehensive, and global use of tactical information.

By the fall of 1948, DATAR was still only an idea. Eager to test its technical feasibility, the Development Section of EECD grappled with the question of how to do the R&D. The RCN had neither the electronic laboratories nor the technical manpower to tackle DATAR by itself. In fact, in 1948 few people in the world had any knowledge or experience with the digital electronics, communications, and computer issues that DATAR raised, none of them in Canada. The UTEC project had not yet started. The search for an industrial partner for DATAR led the RCN to Ferranti Ltd. and its subsidiary in Toronto, Ferranti Electric Ltd. The meeting between the RCN and Ferranti Ltd. was a serendipitous byproduct of the Defence Research Board's search for an industrial base on which to rest a national defense R&D capacity.

Any self-reliant program to develop new military equipment was contingent on the existence of a sophisticated industrial electronics sector equipped with the facilities and the staff to carry out leading-edge R&D. Immediately after World War II, the Canadian military started to campaign for the inseparability of peacetime defense planning and national economic and industrial issues. Canada's Chiefs of Staff argued that peacetime military preparedness demanded a national capacity to create, build, and use science-based weapons systems. In 1947, the head of the Chiefs of Staff, Lieutenant General Charles Foulkes, placed the technological issue within a broader industrial and economic framework. Foulkes reminded Minister of National Defence Brooke Claxton that it had taken 4 years for Canada's industrial mobilization to be felt in the active military theaters.[15] This experience convinced Foulkes of the need to establish a peacetime economic and industrial infrastructure around which Canada could mobilize rapidly in any future conflict.[16] "It is highly unlikely," explained Foulkes to Claxton, "that we shall ever again have as long a period as we had before the last war to mobilize the resources of the nation."[17] If need be, Foulkes added, Canada had to be ready to "subsidize industry [in peacetime] when in the interests of National Security."[18] Though it is not clear whether Claxton agreed with the organizational implications of this plan, he was receptive to the overall thrust of Foulkes's ideas on postwar industrial planning, despite King's strong anti-military sentiments.[19]

In a 1948 speech to the Canadian Manufacturing Association, the chairman of the Defence Research Board, Omond Solandt, reiterated the military's concern for a postwar industrial strategy: "Industrial preparedness in peacetime is just as important to victory as is military preparedness. ... If the full potential of Canadian victory is to be reached quickly in the event of war, we must start now to form plans for the use of industry to strengthen the research and development side of industry and foster the partnership of science, industry, and the Armed Forces that is so essential to victory."[20]

Victory in war was not the only issue underlying the call for a stronger industrial base. With much of Europe in ruins, Canada had emerged from World War II as the third strongest nation in the industrialized world, behind the United States and the United Kingdom. Canada could not pursue anywhere near as much military R&D as either of those countries. Even so, if Canada did not at least "pull its own weight" in the triangular alliance, the Americans and the British would surely shunt it aside in postwar planning. Marginality within the alliance would have seriously jeopardized Canada's access to the latest military equipment.[21] In the minds of the Chiefs of Staff, if Canada wanted to dip into the weapons reservoir of its allies then it would itself have to pour money into designing, developing, and testing weapons. That strategy, however, had little chance of success if Canada could not cultivate a domestic industrial capacity for peacetime defense research and development. The Canadian military hoped to entice industry to invest in a more self-reliant R&D capacity by hinting at broader commercial spinoffs for them. Peacetime industrial preparedness, Solandt told Canada's manufacturers, was not only "desirable from the military point of view" but also "for the normal expansion of industry."[22] However, he added, "from the Defence point of view there is a need to accelerate the process, especially where the civilian interest is not great."[23] Though DRB was aware of the need to increase Canada's industrial R&D capacity, it did not comprehend the extent to which Canadian industry was unprepared to develop and apply the state-of-the-art electronics needed for national defense.[24]

The seriousness of the problem became apparent in the summer of 1947, when, after comparing their wish list of defense-oriented electronics research, the Canadian armed forces concluded that the paucity of corporate R&D compromised the nation's postwar military preparedness.[25] The Electronics Advisory Committee recommended immediate action:

It is essential to the Canadian defence programme to enlist the services of Canadian industrial laboratories. . . . Since few Canadian industries are now equipped with the necessary laboratory facilities, it is believed that the Defence Research Board should foster and support the establishment and maintenance of such laboratories by appropriate means with the industries . . . [DRB must] emphasize the imperative demand for such facilities and recommend a mutually satisfactory basis for a solution of this problem. Such laboratories would form the nuclei of highly qualified groups so essential in the event of future conflicts, and without which Canada was so handicapped in World War II.[26]

Soon after the appearance of the electronics R&D wish list of the Canadian Armed Forces, the three industrial representatives on DRB's Electronics Advisory Committee—Northern Electric, Canadian Marconi, and RCA Victor—advised the committee that their industry could develop the capacity to tackle these problems if assured adequate defense contracts.[27] They tabled a proposal showing how the military's electronic R&D requirements could be neatly partitioned out to what they called Canada's "primary" and "secondary" electronics industries.[28] Their self-serving proposal conveniently distributed all R&D contracts to their own three companies. Their industrial representatives on the Electronic Advisory Committee then convinced their colleagues to recommend that "DRB give immediate consideration to placing the [high-priority] projects upon these firms in order that these important requirements for research and development may be undertaken without further delay."[29]

DRB, however, was unwilling to give Northern Electric, RCA Victor, and Canadian Marconi any contracts without open tendering. In August 1948, DRB asked for proposals from 53 Canadian companies.[30] In a letter to each of these companies, E. G. Cullwick, DRB's Director of Electrical Research, announced: "The Defence Research Board intends to initiate a programme of applied research in Canadian industry in fields related to the use of electronic equipment in the armed services. Such a programme would balance the Board's programme of basic research in Canadian Universities, and would, it is felt, prove a valuable foundation for the Canadian production of Service equipment."[31]

Accompanying Cullwick's letter was a description of the electronics problems that DRB wanted industry to tackle. Four broad classes of electronics R&D were listed: precision servo-systems for computers (including automatic following devices for radar systems), frequency control (including crystals), unattended power supplies, and radar systems. For each of these classes, DRB specified a number of more specific problems. This list, however, comprised the 34 highest-priority items from the armed services' original electronics requirements. Of the 53 companies solicited, only nine responded, submitting proposals for 16 R&D problems. It was this call for industrial assistance that prompted the delegation from Ferranti Ltd. to see if it could set up an electronics group in its Canadian subsidiary to respond to DRB's R&D needs.

Enter Ferranti-Canada

The Canadian subsidiary Ferranti Electric Ltd. had started in 1908 as a simple distribution agency, run by the prominent Royce family of West Toronto Junction, to sell the watt-hour meters of the British electrical manufacturer Ferranti Limited (Ferranti UK).[32] The rapid and widespread adoption of hydroelectric power that accompanied Canada's rapid economic expansion during the boom period 1896–1912 created an enormous potential for electrical capital goods. In 1912, at the peak of this buoyant electrical market, a full-fledged trading company, Ferranti Electrical Co. of Canada Limited (Ferranti-Canada), replaced the Royce family agency. Over the next 80 years, three factors shaped the growth of the Ferranti-Canada: the need to survive in a North American electrical capital goods market dominated by the large US multinationals General Electric and Westinghouse, the Canadian subsidiary's continual pursuit of greater manufacturing and design autonomy, and the transfer of the parent firm's technology-driven corporate culture to the subsidiary. As of 1948, other than an X-ray department that produced machines for industrial inspection and small portable units for the field hospitals, Ferranti-Canada had no experience in electronics. But its parent firm had made considerable inroads in this area and was interested in getting a foothold in the burgeoning North American defense market.

Considerable discussion ensued within the parent firm over how to respond to DRB's call for R&D proposals. What electronics technology should Ferranti go after? What would be the best corporate response? Senior managers and engineers convinced Vincent Ziani de Ferranti (Ferranti UK's CEO, chairman of the board, and owner) to propose an R&D program in precision servo-systems for computers, including automatic following devices for radar systems. The second question the Ferranti group had to resolve in response to DRB was one of corporate strategy. This question was a bit more complex because DRB had stipulated that "the value of such work to the problems of defence will of course depend on the degree to which the programme can be carried out with Canadian facilities and personnel."[33] The Canadian subsidiary had neither the financial, the managerial, nor the technical resources to undertake defense electronics R&D. Though the parent firm was willing to offer the necessary technology transfer, it was unwilling to underwrite the capital and operating costs of such an endeavor. These would have to be written into the defense contract. Confident that Canada had good defense contracts to offer, Vincent Ziani de Ferranti and his people sailed for Canada to meet with DRB.

The 18 October 1948 meeting between DRB and the Ferranti delegation, which consisted of senior executives from the parent firm and its Canadian subsidiary, was a dismal failure. While DRB's Electronics Advisory Committee supported the British company's proposal to transfer a vital defense R&D capacity to its Canadian subsidiary, DRB's management rejected the Ferranti bid. Omond Solandt, the chairman of DRB, explained to Vincent Ziani de Ferranti that the board had not "yet reached the stage where we could undertake to support a research venture such as the one which [Ferranti] propose[d] which ha[d] no direct commercial value but is entirely dependent on government support."[34]

Ferranti-Canada's response to DRB's non-offer was scornful but prudent: "Solandt leaves a small loophole for us to be associated with the Board, provided we will work for nothing," wrote A. B. Cooper, president of the Canadian subsidiary, to Vincent Ziani de Ferranti.[35] Behind this unwillingness to fund R&D at Ferranti-Canada lay the reality that in 1948 DRB did not have adequate financial resources to foster industrial R&D effectively.[36] The general manager of Ferranti-Canada, John Thomson, was inclined to "discredit" the value of DRB contracts to Ferranti-Canada. With funding limited and already allocated to Northern Electric, to Canadian Marconi, and to RCA Victor in Montreal, DRB had nothing left for Ferranti-Canada. However, Cooper shrewdly acknowledged that it might be in the company's "general interest not to ignore Solandt's suggestion of unpaid co-operation."[37] Increasing international political tensions, Cooper intimated, would lead to expanded defense spending in the future. Richard Davies of Ferranti's New York office, who also attended the 18 October meeting, best zeroed in on the contradictions of DRB's approach: if the venture had direct commercial value, why would any company approach the military in the first place?[38]

Sustaining a strong industrial defense capacity posed serious problems because Canada lacked the peacetime military economies of scale of its allies. In the United States and the United Kingdom, Solandt observed, "the productive capacity of industry is adequate both quantitatively and qualitatively to meet the needs of the Armed Forces" and "it is therefore possible to have a very close relationship between industry and the Armed Forces since the Armed Forces are virtually the only customer of war industry."[39] Unable to completely sustain a Canadian defense industry, DRB rested its strategy on the hope that civilian spinoffs could underwrite a healthy portion of military R&D. DRB never elaborated on how this strategy would work in practice.[40]

E. G. Cullwick, DRB's Director of Electrical Research and a former RCN officer, was aware of the EECD's search for an industrial R&D base from which to build DATAR. In October 1948, Cullwick called Lieutenant Jim Belyea, one of the leading advocates of the DATAR concept within the EECD, and informed him of the Ferranti Ltd. delegation's visit. Ferranti UK's experience in digital R&D dovetailed well with the technical challenges raised by DATAR. Ferranti UK had supplied 30 percent of the British Army's requirements in servo-control equipment, which was an integral part of automated fire-control technology.[41] After the war, Ferranti UK took up the "design of naval fire control equipment involving electronic computers, regenerative tracking & automatic following" for the Royal Navy.[42] Ferranti UK's work on the "Admiralty Flyplane" embodied electronic computing.[43] Dietrich Prinz, an eminent research scientist on the staff at Ferranti UK, was studying methods of high-speed data transmission.[44] Finally, Ferranti UK had just received a contract from the British government to do the necessary design and production engineering to turn the Manchester University Mark I computer into a commercial venture.[45] After hearing Cullwick's profile of Ferranti UK, Belyea seized the moment and visited the delegation at the Chateau Laurier.

It is not known whether Belyea went to the see the Ferranti group with the clear plan to enlist its help. But during the meeting, Belyea became convinced that the Ferranti organization was the company to develop DATAR. At the same time, the RCN's project appealed instantly to the technology-driven Ferranti group. Not only did the work fit well into the company's current R&D activities, but nothing in England came close to the scope of DATAR. "It seemed to our group," reflected Kenyon Taylor, who was a member of that delegation, "that what [Belyea] had in mind was very much the proper thing to be doing. . . . It was a first step in push-button warfare. Lt. Belyea was thinking 15 years ahead of his time and Sir Vincent de Ferranti and the rest of our party were well in tune with him."[46] With RCN support, Ferranti was ready to assemble Canada's first leading-edge industrial R&D group in digital electronic computers and communications.

The Development Section understood that, unless it could prove that tactical information could indeed be exchanged reliably between two distant points through digitally encoded radio communication, there was little hope that an understandably skeptical RCN hierarchy would embrace this radical concept. Belyea offered Ferranti-Canada the challenge to develop completely novel digital electronic circuitry based on a concept first proposed by H. A. Reeves in 1939 but never really explored.

The idea behind Reeves's pulse-coded modulation (PCM) was to convert analog signals into a series of binary values for the purpose of communication. Today PCM is an ubiquitous element in all communications technology, but in 1948 it was unexplored territory.

Too late to obtain a formal budget, Belyea funded the first year's work by siphoning money from other projects.[47] What should have been one relatively large contract was instead broken down into smaller contracts. By keeping each contract under $5000 and thus within the realm of his discretionary spending authority, Belyea bypassed the need for formal spending approval from outside the EECD. In a span of 3 weeks, the Development Section issued four contracts. The first dealt with the study of digital transmission methods and devices, the second with the design and construction of binary digital components for transmission systems, the third with the design and construction of PCM components and devices, and the fourth with the design of an experimental PCM transmission system.[48]

Though Belyea's contracts were quite modest, they did demonstrate the RCN's good faith to Vincent Ziani de Ferranti. In January 1949, Ferranti sent Kenyon Taylor, his most imaginative inventor, to set up an electronics R&D team in Canada.[49] Taylor had started working for Ferranti as a "lab boy" in 1931, when textile machinery still occupied an important place in the pantheon of Ferranti's technological interests. The company's founder, Sebastian Ziani de Ferranti, had devoted considerable effort to designing high-speed textile machinery.[50] Kenyon Taylor's success in applying electronic techniques to improving the operation of Ferranti textile machines not only led to his first patent[51] but also brought his inventive talents to Sir Vincent's attention. Over the years, they became close friends. By the time Taylor set sail for Canada, he had acquired 55 patents, covering both consumer and military electronics. As the leader of the new Ferranti-Canada electronics group charged with the responsibility of turning the RCN's DATAR concept into concrete engineering, Taylor offered a combination of great inventive talent and a wealth of practical experience.

Though the RCN had nothing to do with the Ferranti delegation's visit to Canada, it was quick to seize the opportunity. While the inability to fund Ferranti's proposal underscored the financial limitations on DRB in executing its mandate, Belyea's ability to issue contracts in rapid succession revealed that the services had a stronger hand to play and could nurture industrial R&D capacity. The RCN's independent support of the DATAR project also reveals that the line demarcating DRB's and the armed services' responsibilities in weapon's research was not as clear-cut as Solandt may have wished. Motivated by the Mackenzie King government's fear of wasteful duplication, DRB's mandate was to coordinate all the basic research common to the three other services. However, each service could initiate research that was unique to its own particular military requirements. Despite this rationalization, the services still guarded their R&D territory quite jealously wherever that was possible. After all, the desire of each service to protect its autonomy, which had been a factor in the initial reluctance of the Army, the Navy, and the Air Force to create DRB as a fourth service, did not just evaporate with the policy of R&D consolidation. Where the difference between what R&D was unique to the service and what was common to all services became blurred, there were bound to be jurisdictional conflicts. The DATAR project was framed in the context of naval needs, but digital information processing had obvious and far-reaching implications for all the armed services.

Since Ferranti-Canada's initial contract arrangements with Belyea were carried out without DRB's knowledge, the company's management worried that its secret dealings with the RCN would prove to be a source of conflict with DRB. Their fears did not materialize; after learning about DATAR, Solandt rationalized DRB's non-participation on the grounds that the project entailed "engineering development" rather than "research," the latter being DRB's actual mandate.[52] With regards to DATAR, Solandt's distinction between "engineering development" and "research" was nothing more than face saving. The research for DATAR was at least as basic as any of the electronics research DRB was currently funding at universities and in industry.

The money from Belyea's four small initial contracts with Ferranti-Canada quickly ran out. The Development Section squeezed out an additional $15,000 in May 1949, but considerably more money was needed to demonstrate the technical feasibility of PCM.[53] To release more funds, the Electrical Engineer-in-Chief's Directorate now had to seek more formal approval of the project from higher levels within the RCN. The series of small contracts to Ferranti-Canada had already given DATAR a momentum of its own that made it difficult for senior RCN officials to refuse the additional funds needed to complete the demonstration. On 7 October 1949, the Chiefs of Naval Staff approved this initial phase of DATAR. Three weeks later, an additional $50,000 was awarded to complete the final design and construction of the experimental PCM transmission equipment.[54] However, any further financial support for DATAR hinged on the success of the proposed communications demonstration.

The idea of the demonstration was to transmit simulated tracking data reliably from Ferranti-Canada's laboratories, via radio PCM, and to display the targets' movements in the RCN's Ottawa laboratory. Computing Devices of Canada was to develop cathode-ray-tube display equipment for the experiment.[55] Computing Devices of Canada, founded in 1949 by George Glinski (a professor of electrical engineering at the University of Ottawa) and P. F. Mahoney, became the other strategic industrial element in the RCN's efforts to enter the digital age.[56]

The demonstration held in February 1950 made a vivid impression on the RCN's senior staff. A small Canadian team had shown the technical feasibility of digitally sharing target information in real time among a fleet of anti-submarine escorts. Euphoric over the success of the small PCM demonstration, the DATAR group decided to take a giant leap and call for the development of a large-scale naval weapons system. On 5 April 1950, Captain W. H. G. Roger, the Electrical Engineer-in-Chief, went to the RCN's Research Control Committee with a request for a $1.5 million, 30-month program to design, develop, build, and test a DATAR system prototype.[57]

From Concept to Prototype: A Journey That Took Too Long

Captain Roger's belief that it was possible to jump from a small PCM transmission experiment to a large-scale production version for only $1.5 million was naive. Digital displays, servo-control elements, miniature digital circuitry, data processing units, digital trackers, special input devices, digital storage devices, digital computer units, and navigational devices all had to be designed de novo. Then there was the question of systems integration. With digital electronics still in an embryonic state, it is understandable that, with the RCN's relatively modest financial resources, the committee was cautious about jumping into a full-scale development program when no one had a clear picture of the technical hurdles that lay ahead. Roger's proposal also failed to show how the RCN's needs should inform the specifics of the DATAR design program. Short on details, Roger's proposal left the design of DATAR wide open. There were no functional specifications describing the operation of DATAR's subsystems. An even more serious flaw in Roger's request for $1.5 million was the absence of any specific analysis of how the design was to dovetail with the RCN's operational requirements. The issue for the committee was not the generic concept of DATAR, but rather the specific form of DATAR technology appropriate for the RCN. Until the RCN's

DATAR requirements were clearly understood and delineated, the committee saw no point in talking about prototypes ready for production. Thus, although the Research Control Committee accepted the view that some form of DATAR technology had become an operational necessity for the RCN's anti-submarine operations, it considered Roger's proposal poorly conceived.[58]

Six weeks after his first submission, Roger returned with a humbler and more methodologically incremental proposal, which the Research Control Committee quickly approved. Instead of proceeding immediately to a build a full-scale prototype, Roger's directorate received $279,000 for a detailed 13-month engineering study. Most of the money went to underwrite Ferranti-Canada's further experimental exploration of the questions about digital circuit design that DATAR had raised.[59] The new proposal also responded to the committee's call for a coordinated, comprehensive analysis of the RCN's future DATAR requirements. As an automated information-processing system, the DATAR concept required close harmonization of needs and practices across several of the RCN's command jurisdictions. Yet, surprisingly, Roger's original submission made no reference to any intra-service coordination. At the Research Control Committee's insistence, elaboration of the RCN's specific DATAR requirements was to be a collaborative effort of the Electrical Engineering, Telecommunications, Weapons, and Tactics Directorates.

A cornerstone of Canada's postwar military policy of self-reliance was avoiding head-on competition with either the United States or the United Kingdom in the research, development, testing, and production of new defense systems. Within this policy framework, Canada sought to pursue R&D that would meet Canada's needs and would complement the work of Canada's allies. Despite the relatively modest nature of Roger's proposal, by its very nature DATAR was likely to be a large, complex, costly undertaking that would challenge British and American R&D programs.

The RCN assured the Minister of Defence that DATAR would fit very well within the constraints of the government's policy on military R&D. Not only was it a pivotal technological component of a specifically Canadian responsibility (anti-submarine warfare and convoy escort); it was also a uniquely Canadian concept. As the Deputy Minister of Defence explained to Claxton, no one in the United States or in Great Britain had yet to embark on anything resembling DATAR.[60] Canada stood apart, the Deputy Minister continued, in its vision and achievements in the area of digital naval tactical information systems. But the Deputy

Minister reassured the Minister that the RCN's program to develop an automated tactical information system would nevertheless be carried out in close cooperation with Canada's allies to ensure that there would be no wasteful "reinvention of the wheel." Underneath this commitment to cooperate was the RCN's firm conviction that, because of DATAR's superior technical vision, any transnational harmonization of R&D in this area should be on Canadian terms.

By the time of the successful PCM transmission (February 1950), Ferranti-Canada had already recruited a few bright young Canadian electrical engineers. Its reliance on a succession of small contracts made it very difficult for Ferranti-Canada to nurture a stable design team. Accordingly, the company advised the RCN that further efforts to build the kind of development engineering staff that DATAR required called for a more substantial and longer-term commitment from the military. Ferranti-Canada demanded a minimum of $250,000 worth of contracts annually over 3 years,[61] but Roger's proposal assured them of only $234,000 for one year.[62] However, by October 1951 Ferranti-Canada was able to extract an additional $200,000 from the RCN to continue the engineering study and to experiment with circuits.[63] One factor that contributed to the contract increases was the unexpected extension of Ferranti-Canada's technical responsibilities.

The Electrical Engineer-in-Chief's Directorate had originally wanted Ferranti-Canada to explore all aspects of DATAR other than the computer unit. Since Computing Devices of Canada (a small company based in Ottawa) had already been contracted to design and build a digital computerized battle simulator, it seemed only natural that the same company design the DATAR computer; therefore, the RCN allocated $45,000 to Computing Devices of Canada for a 9-month preliminary engineering study of DATAR's computer needs.

The simulator turned out to be a very complex undertaking for Computing Devices of Canada.[64] In early 1951, fearful that that firm would not be able to handle both projects, the RCN asked Ferranti-Canada to build a small 12-bit experimental computer as part of its DATAR work. This new assignment had a dramatic effect on Ferranti-Canada's subsequent technological and corporate objectives. Suddenly the design and manufacture of electronic computers became a powerful and pervasive ambition among Ferranti-Canada's electronics engineers. Parallels were drawn between Ferranti-Canada and its parent firm in the United Kingdom. In 1951, Sir Vivian Bowden of Ferranti UK, while visiting Canada's atomic energy project, asked his friend W. B. Lewis "if there was any possibility that Ferranti, Toronto, might be given a contract to build a computer on the same sort of terms as Ferranti UK had."[65] With a contract from the British government, Ferranti UK had transformed the prototype computer developed at Manchester University into a commercial product. Could Ferranti-Canada do the same with the research going on at the University of Toronto? The subsequent collapse of the UTEC project ended this prospect abruptly.[66]

In 1952, buoyed by the rapid progress being made on DATAR, Arthur Porter[67] boasted to E. L. Davies, vice-chairman of DRB, that the Canadian subsidiary "could produce a computing machine as efficient if not better than the present Ferranti equipment [Ferranti Mark I], in approximately 12 months for roughly $150,000."[68] Porter, a respected British physicist, had done pioneering work in analog computers (i.e., differential analyzers) and was a noted authority on control theory, servo-control mechanisms, and operations research. During World War II, Porter had been associated with "Blackett's Circus," the British group that had first demonstrated how mathematical and scientific techniques could be applied to military operations. Operations Research became a major element in the postwar quest for military strategic advantage. In a move calculated to give the Canadian group greater scientific credibility and prestige, Vincent Ziani de Ferranti asked Porter to head up Ferranti-Canada's Research Department. Davies offered Porter an opportunity to make good on his boast. "I am suggesting a method by which Ferranti-Canada could effectively and cleanly cut the throat of Ferranti England," he said. "This," he elaborated, "depends on whether or not your statement at the very pleasant lunch last Thursday was affected by the liquid refreshments! . . . We do not have in Canada, at present, any need for a further computer but we have heard that Dr. Ellis Johnson, Director of Operations Research [in Washington] . . . is considering purchasing one of the Ferranti machines for roughly $300,000. This is your chance to go to it and earn some US dollars for Canada, our contribution being the know-how we paid for in your development of DATAR."[69] Ferranti-Canada never built this computer. Porter's belief that Ferranti-Canada could make a full-scale general-purpose computer for only $150,000 constituted wishful thinking. The giant engineering leap from a small experimental 12-bit computer to a general-purpose computer would have required far more than $150,000. Nevertheless, by replacing Computer Devices of Canada, Ferranti-Canada's engineers had tasted the excitement of participating in a new and revolutionary technology. Nothing less than building computers would satisfy their future ambitions.

Although the Electrical Engineer-in-Chief's Directorate had succeeded in articulating a consensus view of what Canada's anti-submarine DATAR requirements should be, and Ferranti-Canada had made considerable progress in its development work, the RCN seemed content to proceed in a very slow and conservative manner. Then, in the fall of 1951, a new sense of urgency invaded the DATAR project. The RCN's Electrical Engineer-in-Chief's Directorate learned that the British Royal Navy's Automatic Surface Plot (ASP) system, a rival to DATAR, would be ready for testing as early as 1953.[70] The RCN's schedule for DATAR called for field tests in 1954–55. Captain Roger asked the Research Control Committee for substantial funding increases to speed up the program in order to field test DATAR at the same time as the British tests.[71] The committee balked at Roger's request and increased spending by only $100,000. Roger reminded the committee what was at stake. Britain was pushing hard to get its allies to accept the ASP system. If Canada did not field its own demonstration prototype to coincide with the British field trials, the DATAR program might well collapse. Unless Canada sold the DATAR technology to its allies, the RCN had little hope of underwriting the costs of DATAR through its own procurement. If the British field trials were to go unchallenged, Roger argued, the ASP system would become the de facto standard. What galled the RCN's DATAR proponents was the prospect of having to buy a technology they considered intrinsically inferior to the Canadian system in versatility, capacity, accuracy, and operational performance.[72] Roger's arguments eventually found resonance within the RCN's nationalist senior officer corps, and the race between the Canadian and British systems commenced.

With the new pressure of a June 1953 deadline, the pace accelerated and the costs climbed. By January 1953, the cost of designing and building the demonstration DATAR equipment had risen to $1.57 million.[73] Despite the RCN's technological nationalism, this kind of expenditure would not have been possible had it not been for escalating political tensions and the outbreak of the Korean War. After the outbreak of the Korean War, Canada committed itself to an accelerated rearmament program. From $387 million in fiscal year 1949–50, the Department of National Defence's budget increased to $784 million in 1950–51.[74] In 1951–52, military expenditures were $1.6 billion.[75] In the 1949–50, Navy appropriations stood at $73 million. By 1953–54, they had swelled to $332 million.[76] This huge swell of defense spending gave the RCN an opportunity to get funding for daring weapons-system work.

Even with the deeper military pockets of the early 1950s, the continual procession of cost increases to the DATAR contracts raised eyebrows within the RCN. Stan Knights, the project's technical officer, questioned Ferranti-Canada's ability to manage the large project. Ferranti-Canada's costs were excessive, Knights told the Research Control Committee, because the company had "been lax in the preparation of its estimates" and had "not exercised due economy in its methods."[77] Yet the RCN, committed to getting a demonstration ready by the summer of 1953, felt it had little choice but to pay for management it considered inefficient.

The demonstrations of DATAR took place on Lake Ontario from September into November of 1953. Originally, three Canadian Bangor-class minesweepers were to be used in the demonstration, but pulling three ships out of service at the height of the Korean War proved impossible. Instead, two ships were used and a shore station was outfitted to simulate the third. All ASDIC and radar data, which came in as analog signals, were immediately filtered and converted to digital. Sophisticated cathode-ray-tube displays on each ship depicted aircraft and submarines as distinctly different graphical icons. Information about any of these targets could be called up instantly by placing a cursor over its image on the screen.[78] All, or any portion, of the information stored in the computer was shared via pulse-coded modulation with selected escort ships or with all of them. By taking into account the relative motions of all the escort ships, DATAR presented the picture of the tactical situation that was appropriate to each ship's own reference frame, even though the ASDIC or radar data displayed on one ship may have been collected by another. If the target was masked by such things as sea turbulence, rain, clouds, or radar windows, DATAR's computer supplied clues to the probable location of the target.[79]

In the rush to get a DATAR demonstration ready, compromises were made to the design of the electronic digital computer. The Ferranti-Canada design team chose a magnetic drum over more elaborate memory technologies as the computer's main memory. The reliability problems of electrostatic memory systems would have unnecessarily risked the overall success of the demonstration. In view of the overall complexity of the project, drum memory offered the most reliable option for the demonstration. Even so, both the RCN and Ferranti-Canada knew that a better solution to memory had to be found for the final DATAR prototype. Drum memory was too slow and electrostatic memory too bulky, and both were quite vulnerable to vibration and motion. Ferranti-Canada had wanted to pursue the idea of ferrite cores that Jay Forrester of MIT had come up with for the SAGE air defense project, but this idea still required more development work than the race to get a demonstration ready permitted.

The Battle of the Systems: Canada's Effort to Set Standards

As one would expect, the Canadians and the British had different interpretations of the results of the DATAR demonstration. The Royal Canadian Navy was elated that, although the demonstration equipment was small in scale and experimental, it proved that DATAR technology had the ability to coordinate, with remarkable speed and accuracy, the collective actions of escort ships and aircraft in an anti-submarine operation.[80] Commodore H. N. Lay, who had played an important part in Canada's quest for military autonomy in the battle for the North Atlantic, announced to his fellow members of the Permanent Joint Board on Defence (a board composed of American and Canadian senior military officials) that DATAR had shown the superiority of the Canadian Navy's digital tactical information system.[81] On the other hand, representatives of the British Admiralty, though impressed with the technical advances in data handling made by DATAR project, concluded that the United Kingdom's ASP system was still cheaper, simpler, and more effective than DATAR.

The British criticism of DATAR reflected a fundamental disagreement with Canada over air defense. The RCN insisted that any naval tactical information system would have to integrate air defense and anti-submarine data seamlessly. The modern convoy escort, the RCN felt, had to defend against both aircraft and submarines. If the escort ships' weapons could not directly attack enemy aircraft, DATAR could at least coordinate the use of friendly aircraft to fend off an enemy attack. The British Royal Navy, on the other hand, saw the two as different problems requiring different systems. While the RCN was committed to a digital system, the Royal Navy still pushed the more conventional analog approach to ASDIC technology. ASP was effective against submarines to the extent that it kept track of a relatively small number of targets, but the ASP analog approach was too complex and unwieldy when it came to handling a larger number of targets, as would be the case with air defense. The digital approach not only allowed the easy merger of air and submarine tactical data; it also allowed the use of a computer, which gave the system a flexibility that ASP could never have. Under DATAR, each ship's computer could extract, from a common pool of data, the information appropriate to its needs in a given context. The ASP system, on the other hand, presented information in a preset and rigid manner.

The Royal Navy also argued that, because of Canada's insistence on a comprehensive air defense and anti-submarine system, DATAR had become too large to fit on anything but a large battle cruiser. Many escort vessels, the British pointed out, were much smaller. The RCN countered that DATAR's design was modular, and that different sizes of DATAR could be installed on different ships. In today's terms, DATAR used a kind of open-system philosophy. The key was a standardized digital protocol for information exchange. DATAR was a decentralized network in which each ship used its own criteria in extracting information. Sinking one ship, along with its computer, would not disable DATAR. The British ASP system, in contrast, was a highly centralized approach to coordinating the protection of a convoy.

Despite DATAR's superiority, there was little likelihood that the British Navy would adopt it. But what mattered more for Canada was the US Navy's position on standardization. As the kingpin of the alliance, the United States could make or break any proposal to standardize naval tactical information systems. Because the US Navy had done very little to develop its own tactical, information-handling technology, Canada worried that the British would be able to persuade the US Navy to adapt ASP. After the success of the DATAR demonstration, there was little likelihood of this. The pivotal issue for the RCN then became the intentions of the US Navy. Would the US Navy, which still lagged considerably behind Canada in the area of automated tactical information-handling technology, embrace DATAR? Canada lobbied intensely to bring the United States on board. On 9 November, Minister of Defence Brooke Claxton, three rear admirals, the heads of various RCN directorates, the RCN's principal scientific advisers, and the RCN's officers in charge of DATAR gathered to sell the Canadian system to a high-ranking US Navy delegation headed by Rear Admirals W. G. Schindler and M. E. Miles and by Emanuel R. Piore, Deputy Chief of the Office of Naval Research.

After a demonstration of DATAR, the Canadians outlined the operational capacity of the full-scale version they intended to build. DATAR was to be the state of the art in automated naval warfare. Each ship was to be equipped with powerful air-search radar, high surface search radar, high-definition navigational radar, an airborne early-warning system, attack-and-scanning ASDIC, radio direction-finding equipment, and UHF radio communications. In a 16-ship escort fleet, DATAR would allow the simultaneous sharing of the input from all the above devices on all the ships. Each ship's computer was to have the speed and power to keep track of 128 targets simultaneously, to process an additional 1000 possible messages for each target, to compute the course and speed of any target automatically on demand, and to produce specialized plots for tactical control and target designation.[82] To overcome the performance, size, and reliability limitations arising from dependence on a memory drum and from the widespread use of vacuum tubes, the memory in the final version of DATAR was going to be all ferrite core.

The US delegation was impressed by the Canadian achievement, particularly by the Canadian claim that the equipment cost per ship for DATAR would be only $400,000. Piore had expected it to be closer to $1 million per ship. Canada had much riding on the American reaction to DATAR's techno-economic merits. If the Americans did not standardize on DATAR, Canada would have a very difficult time financing the further development and production engineering and the cost of procurement. Worse, if the US Navy did not agree to the Canadian data-exchange protocol, the RCN would be forced to drop DATAR whether or not it could pay for that system. The RCN realized that unless it retained the initiative and organized international talks on standardizing the contents and protocol of naval data exchange, DATAR technology would be useless to Canada regardless of its superiority.[83] Commodore Lay pressed the Americans to embark on a collaborative undertaking. Though US Rear Admiral Schindler endorsed the idea of drawing up common requirements, it was still too early to know how close to DATAR the US Navy was willing to move its weapons systems.

With the US Department of Defense reluctant to buy a foreign weapons system (particularly one as critical as automated electronic naval warfare), and with the powerful US defense industry lobby anxious to block foreign equipment programs, the only possible hope Canada had to sell the Americans on DATAR was to build and test a full-scale version before the US Navy had a chance to initiate a serious program of its own. Yet without some form of standardization assurances from the US Navy, investing heavily in a large-scale production version of DATAR could prove impoverishing. Caught in a chicken-and-egg dilemma, the RCN fell back again into its slow and cautious approach.

Time was of the essence, yet nearly a year elapsed after Canada's high-level meeting with US Navy representatives before the Electrical Engineer-in-Chief's Directorate felt it was ready to issue contracts for the construction of the first full-scale prototype of DATAR. Through most of 1954, the RCN's DATAR group reexamined, in great detail, the scope of the operational specifications it had defined in 1952 and then again in 1953. In part, this process was driven, as are many military development projects, by the tendency to seek every possible strategic advantage by pushing the state of the art to its limits. Increasing DATAR's marketability was another important factor in the expanded operational specifications. In 1953, the RCN had told the US Navy delegation that DATAR would eventually be able to handle 128 targets simultaneously. However, by November 1954 the specifications for DATAR had risen to include tracking 500 aircraft, plus ships and submarines. But no escort role could justify a 500-target air defense information system. Only a major battle group could. DATAR's increased capacity to handle many aircraft reflected the RCN's desire to make the technology attractive to the Royal Canadian Air Force and to the US Navy.

Air defense had become the dominant concern of the Canadian military. Driving this concern was the knowledge that Americans saw the defense of Canada's northern air space as essential for their survival. Canadians feared that unless they took a more active role in continental air defense they would lose effective control of their northern territory to the US military. It was within this political climate that Canada agreed to participate in the construction of various early-warning systems, and that the Royal Canadian Air Force received the lion's share of defense spending. The pressure for Canada to assume a stronger role in continental air defense also favored efforts by the A. V. Roe Company (AVRO) and the RCAF to design the now-legendary CF-105 Arrow all-weather fighter and interceptor. Though early-warning radar and a new generation of fighter planes were essential factors in countering long-range Soviet bomber attacks, sound air defense also required a sophisticated real-time tactical aerial information system to track enemy bombers, analyze their flight paths, suggest probable targets, and coordinate countermeasures. In the United States, the Air Force was working on a computerized information system called the Semi-Automated Ground Environment (SAGE) system.

The RCN wanted the RCAF to adopt DATAR technology for its air defense system. In the event that the US Navy did not buy into the DATAR development, then the burden of the high development and procurement costs could be shared with the RCAF. When the RCAF proposed to develop its own automated data processing, data transmission, and data display system, the RCN objected. The RCN argued that the RCAF's intended program would be a wasteful duplication of R&D already successfully completed by the navy.[84] Despite inter-service differences in tactical information requirements, the RCN felt there was sufficient overlap to warrant close collaboration along the framework of DATAR. To coax the RCAF closer to DATAR, the RCN offered it the equipment that had been used in the 1953 demonstration. To what extent the RCAF was willing to cooperate with the RCN is not clear. In 1954, the RCAF did contract Ferranti-Canada to do a preliminary study on how high-speed digital data processing techniques could be applied to Canadian air defense.[85] Ferranti-Canada's study proposed a decentralized system and emphasized the importance of rapid automatic dissemination of all data rather than the automation of the control and guidance systems. Ferranti-Canada had actually offered up a version of DATAR called Canadian Air Defence Automatic Reporting (CADAR). But the US Air Force's plans to press ahead with the full-scale production of its own system forced the RCAF to contemplate purchasing American equipment.[86] In the end, trans-national standardization with the USAF proved more of an imperative than national harmonization of R&D with the RCN.

An awareness of US Navy interest also explains the expansion of the initial 1951 specifications of DATAR into a generic naval tactical information system. How else can one explain the jump from a 128-aircraft tracking capacity to 500? Although the integration of air defense and anti-submarine tactical information seemed reasonable for a convoy-escort navy, the need to track 500 aircraft made no sense—unless of course, one envisioned DATAR for an American aircraft carrier fleet.

The early success of the DATAR project permitted Canada to play an important role in defining trans-alliance standardization. The RCN managed to convince the US Navy that binary was a better format than decimal for exchanging naval data.[87] Canada's insistence on error detection in the data-communication process, as well as the use of the UHF bandwidth for greater throughput, was slowly being accepted by Canada's allies.[88] Canada's espousal of an open, decentralized network architecture also appealed to the US Navy. The British approach, now known as the Comprehensive Display System (CDS), called for a centralized architecture in which all the processing was done on one ship and the required tactical information was then broadcast to each ship. In DATAR, although each ship contributed to common pool of tactical data, information extraction was a local process determined by each ship's own computational capacity and needs. Whereas the British approach reinforced the military tradition of hierarchical and centralized systems, the Canadian approach highlighted the advantages of a fluid, flat, and less vulnerable organizational structure.[89] By empowering each ship, the

DATAR philosophy maximized the survival of the entire system. If any one ship was destroyed, the ability of other ships to share and extract information would not be jeopardized, as would be the case in the centralized British system. When the United States had finally committed itself to the Canadian position that air defense and anti-submarine concerns form one integrated information system, the RCN knew it had finally triumphed over the British Royal Navy. Members of the RCN's Research Control Committee were no doubt pleased with the news that "the USN wanted to adopt a modified version of DATAR for the next phase of its own development program" because it was convinced that the Canadian approach was right for the 1960s.[90]

The RCN's victory, however, proved an empty one. The slow pace of DATAR's development not only jeopardized any possible export sales; it also made the project more vulnerable to the flagging Canadian political will to pay for costly defense R&D programs. Nearly 18 months passed between the successful demonstration of the experimental version of DATAR and the approval of contracts to start the design, development, and manufacture of the large-scale prototype.[91] The RCN had divided the DATAR prototype program into two contracts: one for data processing and display, one for the radio data link subsystem. When the specifications for the full-scale version of DATAR had finally been set, in November 1954, the plan had been to have the prototype ready in 2 years. Now more than a year had gone by and the basic components of the system had not even been designed.

A small part of the blame no doubt rests with bureaucratic inertia. It took about 4 months to push the contract request through the Department of National Defence to the Department of Defence Production and then to the Treasury Board for approval. Ferranti-Canada, which was to carry out the first contract for $750,000, started around April 1955. It took a year to select the company to carry out the second contract.[92] However, the slowness of the DATAR project's pace was due more to fundamental technical limitations than to bureaucracy lethargy.

As with the UTEC story, issues of design complexity turned the vacuum tube into a critical "reverse salient." The sheer size and power requirements of large-scale circuits made it increasingly difficult to fit vacuum-tube-based electronic equipment onto ships and airplanes.[93] Even more serious was the growing problem of reliability. As the number of tubes in military electronics grew into the thousands, the mean time between failure (MTBF) became unacceptably short.[94] The Ferranti-Canada team had experienced firsthand the urgency to miniaturize large-scale electronic digital systems and still improve reliability. The DATAR system tested on Lake Ontario had about 15,000 vacuum tubes. Whenever DATAR was being demonstrated, it was not unusual to find Ferranti-Canada's young engineers, stripped down to the waist because of the heat and armed with cartridge belts of vacuum tubes, racing around the ship's interior locating and replacing faulty tubes.[95]

Even though the full-scale version of DATAR was to have proportionately fewer tubes, the questions of reliability, space, and power consumption haunted the Electrical Engineer-in-Chief's Directorate. Throughout most of the 1950s, a considerable amount of money and effort was being directed to "ruggedizing" or miniaturizing the vacuum tube across the alliance. The RCN had even looked into dramatic redesigns of tubes, such as Josef Kate's "binary adder tubes," as a way of miniaturizing and improving reliability by minimizing complexity. But efforts to squeeze more tubes into smaller spaces only exacerbated the reliability problem. One senior RCAF officer had already warned in 1954 that "the rapid rise in the failure rate of electronic equipment coincident with the increasing complexity and miniaturization of that equipment is a cause for alarm" and that "the perfection of components would reduce failures in a ratio of only two to one whereas reduction in a ratio of fifty to one is required for acceptable reliability."[96] Unless the vacuum tube problem was addressed "the reliability problem [would] reach crippling proportions."[97]

The solution to the complexity problem resided in finding radically new electronic switching devices, not in incremental improvements to existing components. The military looked to the transistor as a potential replacement for the vacuum tube. Still early in its development, the transistor was itself plagued with serious performance and quality-control problems.[98] Its relatively slow switching also limited the transistor's utility for data processing and communications. Furthermore, the absence of tightly controlled transistor characteristics made it difficult to design high-performance digital circuits. Despite these problems, the transistor's potential for far greater miniaturization offered the only hope of getting through the complexity impasse. The RCN was convinced that unless DATAR could be transistorized it would never go into service.

Philco's 1954 surface-barrier innovation seemed to offer the RCN the answer to DATAR's miniaturization imperative. The SB-100 transistor offered reliable, fast switching. Over the next few years, Philco used its SB-100 transistor as the basis of the world's first all-solid-state computer. As soon as news of the SB-100 became public, the RCN arranged for a team of Ferranti-Canada engineers to visit Philco's Philadelphia plant with a mandate to explore the utility of the SB-100 transistor for DATAR.[99]

Ferranti-Canada immediately began experimenting with transistors and figuring out how to recast DATAR's circuitry into solid state. Though this course of action seemed straightforward, its execution proved difficult and slow. The young Ferranti-Canada design team had shown remarkable ingenuity in devising circuit techniques to minimize the number of tubes DATAR needed and to improve reliability. In the process, they came to understand the subtleties of the vacuum tube's performance and how to use them to their advantage. Switching to transistors meant abandoning all this tube expertise and reinventing digital circuit design know-how. Because transistors represented a paradigm shift in electronic design, miniaturizing DATAR went far beyond the mere replacement of tubes with transistors in the circuit diagrams. New principles were needed to exploit the strengths of transistors while minimizing their weaknesses.[100] Because no one in Canada (at a university, in government, or in industry) had mastered digital solid-state circuit design, Ferranti-Canada had to create the learning curve it was going to climb up.

Time had now become the DATAR project's biggest enemy. For nearly 7 years Canada had held an unquestioned lead in the formulation and technical pursuit of automated naval tactical information systems, but by 1956 a shift in the Canadian political will to finance military self-reliance and the sudden awakening of the US Navy to the "urgency" of automated tactical information systems were about to ravage this lead. These two factors combined to bring about the collapse of the RCN's DATAR project.

Canada's small economy could support only modest levels of military procurement. In such a limited internal market for military equipment, a balance had to be struck between spending to nurture strategic domestic design and manufacturing capacity and importing less costly off-the-shelf foreign weapons systems. The Korean War encouraged much higher defense spending than otherwise would have been the case. In the process, the balance tilted more toward military and technological self-reliance. With the end of the Korean War, the political will to maintain high levels of defense spending weakened. Thereafter, every military expenditure was put under the microscope to reduce any duplication between Canada's development programs and those of Canada's allies.

In June 1955, Minister of Defence R. O. Campney announced Canada's plan to modernize the Canadian Navy's aging escort fleet with 14 modern destroyer escorts.[101] At $23 million apiece, these escorts were to be, according to Campney, the most formidable "submarine killers" in the world. With the cost of each ship's armaments and electronics at $6 million, DATAR was no doubt an important technological component in the RCN's proposed state-of-the-art escort fleet. This proposed expenditure of $322 million for ships in the post-Korean era was challenged by the Conservative opposition. "Canada is not a country," argued one Conservative member of the House Defence Committee, "that can afford much higher defence costs."[102] Careful not to appear to compromise national security, the Conservatives did not attack defense spending per se; they attacked wasteful overspending. Whether the Canadian economy was capable of supporting the high pre-1954 levels of defense spending requires further study, but in 1956 Louis Stephen St. Laurent's government concluded that "it would soon be impossible to maintain the existing defence establishment within the present budgetary limits unless strict economies were made and a list of priorities established and adhered to."[103] The government was increasingly unwilling to bear the high costs of large independent development programs, particularly in electronics, missiles, and aircraft. The government ordered the Department of National Defence to cut costs and reduce duplication, particularly in the development of new equipment.

Since 1950, the Treasury Board had usually rubber stamped the RCN's development spending requests. But it approved only half of the RCN's $21.3 million R&D budget for 1956–57.[104] Though the refusal targeted the RCN's costly missile and anti-submarine helicopter programs, it nevertheless forced the RCN to rethink the funding of DATAR. The RCN had long justified the costs of the DATAR program on the ground that neither the United States nor the United Kingdom had any comparable system, but recent US Navy decisions had put this argument on very slippery footing.

Until 1955, the US Navy had done relatively very little work of its own on developing an automated, tactical, information system. That year, the US Navy concluded that this technology was of paramount importance to its future effectiveness. Driven by a sense of urgency, the US Navy reached down into the deep pockets of the US military and committed $10 million to a two-year crash program to build its own supercharged version of DATAR, to be called the Naval Tactical Data System (NTDS).[105] The chief of US Naval Operations saw NTDS as "absolutely essential for the survival of a task force in the post-1960 era." American naval officials took the attitude that "in the future, if a ship did not have NTDS, then it should not leave the port."[106]

In view of the RCN's 7 years of pioneering work and of the considerable financial and industrial R&D resources of the United States, there was little doubt that the US Navy's proposed NTDS would triumph over DATAR. Whereas the RCN had allocated $750,00 to Ferranti-Canada to design and build a transistorized information-processing and display system, the US Navy gave Remington Rand $4 million and Hughes Aircraft $3 million to undertake essentially the same challenge.[107] In view of the scope and the magnitude of NTDS, the advocates of DATAR knew that their program was extremely vulnerable to the cost-cutting imperative to eliminate duplication. If NTDS subsumed or at least was congruent to DATAR, why throw away any more money on a large-scale prototype when the Americans were already working on one? The only strategy open to the Electrical Engineer-in-Chief's Directorate was to try to weave DATAR into the NTDS project in a complementary manner.

From the outset, NTDS was designed to serve the needs of a large, carrier-based task force. Protecting destroyers, large battle cruisers, and aircraft carriers from enemy fighter planes was the prime function of NTDS. In designing NTDS, the US Navy had little interest in anti-submarine duty. Accepting NTDS as a fait accompli, the RCN argued that the US Navy system, as it was being developed, was not appropriate to Canada's anti-submarine mandate. The DATAR Sub-Committee recommended that the RCN use its expertise to embark on a development program to adapt NTDS to the needs of an anti-submarine navy.[108]

With Cabinet Committee on Research for Defence pushing harder for more cost-saving cooperative ventures, the RCN tried to keep DATAR alive by proposing a joint Canadian-American effort to design an anti-submarine version of NTDS. Offering Canada complete access to all the technical information on NTDS as it developed, the US Navy encouraged the RCN to continue on DATAR.[109] But the US Navy, with such a large portion of its financial resources devoted to NTDS, had little money to spare for the RCN proposal

With spending restrictions, and with little political support for an independent program, DATAR ran out of steam. The Canadian government's decision in 1956 to retreat from its earlier commitment to build the 14 destroyer escorts also contributed to the collapse of DATAR. These destroyer escorts were to be the state of the art in anti-submarine warfare. Not only was the design of the destroyer itself a dramatic improvement in speed and maneuverability, but these ships were to be equipped with the most advanced sonar, radar, weapons, and fire control. Individually, and collectively, the proposed anti-submarine destroyer escorts were truly a large complex technological system. DATAR was supposed to be the electronic nervous system and brain that made all the parts of the system work together harmoniously. With the collapse of the program to build a new fleet of advanced destroyer escorts, enthusiasm for any further costly work on an automated electronic naval battle system also waned. Canada did propose a three-year, $450,000 study to come up with suggestions for a Tactical Data System based on whatever NTDS developed. After 7 years of daring pioneering work in real-time digital information processing, communication, and display, DATAR was abandoned. Unlike the AVRO Arrow, which went out with a bang, DATAR died with not even a whimper, and along with it went the RCN's flirtation with military self-reliance.

In the Wake of DATAR

The DATAR story calls for a reexamination of the historiography of the role of the Royal Canadian Navy in the immediate postwar years. The Canadian historian Marc Milner has written extensively on the Royal Canadian Navy's role in the World War II battle for the North Atlantic.[110] Like others before him who studied the subject, Milner portrays how the RCN stumbled into this wartime role. At the start of the war, the RCN's officer corps dreamed of leading a fleet of aircraft carriers, destroyers, and battle cruisers into battle. The RCN looked down its nose at the idea of protecting convoys and hunting submarines. Canada's professional navy felt that this lackluster job was better left to reservists. Where was the glory, they asked, in serving on small, cramped corvettes that bobbed like corks in the water, and having to escort motley collection of freighters? Where were the career opportunities in serving on corvettes? While the RCN's officer corps trained for a fleet they would never have, reservists with little sea experience went to sea to guard the convoys. As the importance of anti-submarine warfare became recognized in Ottawa, Canada's professional navy was forced to take a greater interest in it. However, to the end of the war, Canada's anti-submarine contingency remained a navy apart.[111]

With the end of the war, Milner concludes, the RCN reverted to the prewar mentality of how a professional navy should look. "The post war RCN," writes Milner, "saw little of itself in the [anti-submarine] and escort war in the Atlantic."[112] According to Milner, Canada's professional navy, rejecting any postwar anti-submarine role, continued to harbor ambitions of building a conventional naval fleet. Milner argues that anti-submarine warfare had never been a part of the RCN's long-term plans, during or after the war. Milner adds that the RCN's postwar rejection of anti-submarine warfare was reinforced by a desire to distance itself from a wartime role that it considered an embarrassment.

The DATAR experience, however, contradicts Milner's conclusions about the RCN's postwar ambitions. As this chapter has shown, anti-submarine warfare was the axis around which the RCN's entire pursuit of self-reliance revolved. Anti-submarine warfare dominated the RCN's postwar R&D strategy. Milner's conclusions are extrapolations of the RCN's attitudes and actions up to the closing days of the war. Because of the problems it had encountered in trying to create a scientific and technological base to support its wartime weapons development program, the RCN realized in the last days of World War II that future peacetime military preparedness would demand the creation and maintenance of a military-scientific-industrial complex. David Zimmerman observes that the RCN's politically naive ambitions to build a large, well-balanced naval fleet and maintain a moderate but respectable R&D program ran up against Prime Minister Mackenzie King's desire to dismantle the military. "Much of the accumulated wartime scientific and technical expertise," Zimmerman contends, "was lost as result of [Mackenzie King's] cutbacks and the return of most personnel to civilian life."[113] Zimmerman goes on to suggest that small pockets of R&D remained. Though Zimmerman points to bathythermographic and acoustical research for ASDIC development, from this thesis it is evident that another, more ambitious pocket emerged in the Electrical-Engineer-in-Chief's Directorate in Ottawa. With the Mackenzie King government's refusal to pay for the construction of a full-scale naval task force of aircraft carriers, battleships, cruisers, and destroyers, the Naval Staff faced the dilemma of how it could define a self-reliant role for the RCN within the postwar North Atlantic Triangle.

The potential of DATAR sparked senior officers to reappraise the value of anti-submarine warfare to the RCN's postwar naval ambitions. The advances in submarine technology in the immediate postwar years called for radical innovation in anti-submarine warfare. The DATAR concept offered a dramatic technological leap in anti-submarine warfare. The idea of DATAR did not arise from the senior levels of the RCN. Rather, it was driven by the mid-level technical officer corps of the Electrical Engineer-in-Chief Directorate. Once the feasibility of digital tactical information exchange and processing became apparent, by the end of 1949, DATAR began to win acceptance at higher levels because it represented a technological redefinition of anti-submarine warfare that promised to catapult Canada to the forefront of modern naval warfare.

The RCN's extensive R&D program in DATAR, the digital electronic A/S Tactical Trainer, the hydrofoil, missile development, the program to build the world's most advanced frigates all demonstrate that the RCN's drive for power and self-respect now ran through the anti-submarine escort. A technologically advanced anti-submarine fleet designed and built with Canadian know-how held out the prospect of a unique and important niche within the Western alliance. Ironically, at the same time, power in the US Navy was coming to reside in Hyman Rickover's program to introduce nuclear technology into the lowly submarine.

The failure to sell the Americans and the British on DATAR underscores some of the contradictions the Canadian military faced in trying to advance Canada's technological and industrial capacity, even in specialized areas. This advancement hinged critically on the level of procurement that the Department of National Defence could guarantee. But the costs associated with the design, prototype development, production engineering, and deployment of very advanced large-scale defense systems were more than Canadian procurement alone could support. With Canada's relatively small peacetime Armed Forces, foreign sales were essential to underwrite any military flirtation with technological and industrial self-reliance. Commodore Lay knew that any efforts by Canada to fit the RCN with DATAR "would be largely nullified if the ships of other nations [its allies] were not so fitted also."[114]

The need to give Canadian industry access to the military markets of Canada's allies was one of the driving forces behind Canada's push for transnational standardization, first with Great Britain and the United States and later within NATO. Canada's pursuit of component standardization and interchangeability across the entire alliance was a reasonable response for a small nation trying to export to Great Britain and to the United States. The standardization of screw threads, for example, was a remarkable Canadian-led initiative to ensure, among other things, that Canada's manufacturers could supply both the British and the American military forces, without wasteful duplication of machine tools.[115] Though by the end of World War II the orientation of Canada's economy had shifted from transatlantic to continental, Canada's military still had strong roots in British military practice and traditions. Consequently, in the early postwar years the Canadian military still believed that in any future war Britain would remain an important destination for Canadian goods.

Standardization may have advanced the cause of Canadian mass-production manufacturers, but it did little to promote the sale of large, advanced military systems such as DATAR. For example, any early interest the Canadian Army may have had in DATAR-like technology was tempered by the realities of tripartite standardization. Of the three services, the Army was the most committed to international standardization. Consequently, the Army was reluctant to undertake any R&D in this area until the details of the standardization of signal equipment had been worked with the United States and the United Kingdom.[116] Similarly, the Royal Canadian Air Force backed away from CADAR and standardized around the SAGE system because the R&D and production costs of the former could never be covered by Canadian procurement alone. Though the Air Force never gambled on CADAR, it did wager a lot on the AVRO Arrow intercept fighter. The unwillingness of Canada's allies to standardize around CADAR made the AVRO Arrow economically unfeasible. The "not invented here" syndrome that served to block acceptance of DATAR and the AVRO Arrow by Canada's allies was a reflection of more than mere national pride. It was also a manifestation of the fact that global military power and economic power had engendered an international struggle to control and appropriate the science and technology that were sustaining the long-term global expansion of the postwar period.

The story of DATAR is as much about Ferranti-Canada's emergence as Canada's leading computer R&D group as it is about the RCN's quest for military self-reliance through technology. In 1948, the prospect of DATAR attracted the technology-driven Ferranti UK to establish an electronics R&D group in its Canadian subsidiary. This group was not to be a stereotypical "branch plant." Instead, the parent firm left the technical, administrative, and financial modalities of this enterprise to its Canadian subsidiary. The Ferranti corporate structure resembled a collection of independent trading houses. Ferranti-Canada did not fall within some international, multi-divisional, hierarchical management structure. As a result, the Canadian subsidiary was given considerable autonomy.

In the end, Ferranti-Canada developed the expertise to build DATAR on its own. Kenyon Taylor (a hands-on inventor) and Arthur Porter (an engineer with a theoretical bent) aggressively sought out young Canadian scientific and engineering talent. At the time DATAR was demonstrated, the professional staff for this project had swelled to 17 members; many of the engineers had master's degrees, and there were three physicists with doctorates.[117] By 1955, Ferranti-Canada's electronics group had grown to 30 engineers and scientists.[118] DATAR thus allowed Ferranti-Canada to grow into Canada's leading industrial R&D center in digital electronic information technology in the 1950s and the 1960s. Military funding also fostered a technology-driven ethos within Ferranti-Canada's Research Department—an ethos that was reinforced by the parent firm's corporate culture.

With the collapse of DATAR, Ferranti-Canada had to either pursue civilian markets seriously or dismantle the computer group. For the company to enter the computer business and survive, more than technical excellence would be required. How would the small Canadian subsidiary capitalize such an undertaking? What strategy would it have to adopt in order to compete in a North American market dominated by IBM, UNIVAC, and the Control Data Corporation? What would the markets be, and how would Ferranti-Canada address their needs? How would the subsidiary's aspirations to build and sell computers be reconciled with the parent firm's ambitions?

Becoming a computer manufacturer also raised important internal management issues. How would Ferranti-Canada's upper management, which was completely absorbed in surviving in the North American electrical capital goods sector, respond to the corporate needs of its small digital electronics group? Shaped by the exigencies of the electric power industry, could Ferranti-Canada's technological and management culture successfully embrace the engineering and business ambitions of the electronics group? During the DATAR project, the digital electronics group had worked in isolation, and few demands had been placed on the company's management. The loss of military funding and the subsequent need to find commercial outlets for the company's expertise in digital electronics brought the electronics group under closer scrutiny from the electrical power group.

The difficulty Ferranti-Canada encountered in miniaturizing DATAR was indicative of how little the transistor had diffused into Canadian industrial engineering circles. The transistor was of strategic importance in breaking through the complexity-vs.-reliability problem that had hitherto undermined digital electronic design in both military and commercial systems. The military understood all too well that a nation's capacity to design complex digital electronics had become inextricably entwined with the capacity to miniaturize.

3

Complexity and the Military Imperative to Miniaturize

The histories of digital electronics, computers, and transistor technology are inextricably bound together.[1] Each important advance in semiconductor devices has engendered a dramatic change in computer technology, which in turn has spurred further innovation in transistor technology. At the heart of this dynamic is the remarkable ongoing miniaturization of the transistor, which has led to a 30-year-long exponential decrease in the cost per bit of electronic memory. During the 1970s, this phenomenon led to the invention of the microprocessor and the birth of the microelectronics revolution.[2] The diffusion of the microprocessor in the late 1980s and the 1990s turned electronic digital computation into a mass consumer phenomenon with far-reaching societal implications.

The powerful civilian commercial dynamism that has characterized the mutually contingent growths of the computer and semiconductor industries conceals the important role of the military in nurturing this interrelationship, particularly in the early stages when there was not much of an incentive from the civilian marketplace to do so. As Ernst Braun and Stuart Macdonald have shown, the early development of the transistor was not propelled forward simply by obvious commercially driven opportunities or some internal, preordained autonomous technical logic.[3] Rather, it was the interplay of technical, corporate, economic, and social factors that moved the transistor from being a curious invention to being the substrate of modern industrial development. In pointing to the military as the dominant force in this interplay, Braun and MacDonald reinforce Richard Levin's earlier comprehensive synthesis of the development of the US semiconductor industry. The US military services, according to Levin, were willing to support R&D, to subsidize the engineering effort required to install production capacity, and to pay premium prices for procurement of new devices when the civilian market was incapable of doing so.[4] Similarly, Alfonso Molina reminds us of the following:

> ... the technologies and the industries behind today's microrevolution were shaped by the interests of a social constituency most heavily influenced by the military.... A broad range of advances in electronic systems, between 1952 and 1966, originated from an equally broad range of military projects.... The influence of the government-military constituents was wide-ranging, affecting all areas of electronics. Nowhere was this influence more decisive than on the development of the emerging technologies and industries of computers, industrial control, and above all, semiconductors, the technical base for the convergence of electronic systems.[5]

Any account of Canada's efforts to foster self-reliance in digital electronics would be incomplete if it did not also weave in the military's efforts to promote a corresponding national capacity in transistors. The preceding chapter dealt with how the Royal Canadian Navy tried to initiate this synergy by financially supporting Ferranti-Canada's efforts to miniaturize DATAR's design. The central theme of this chapter is the military's use of government laboratories as a source of innovation and as an agent of technological diffusion to industry.

In recognizing the early strategic importance of the transistor, Canadian military enterprise faced the dilemma of how much transistor self-reliance it could afford. Was it strategically necessary and financially conceivable to promote a costly domestic transistor manufacturing capacity? Was it more advantageous to concentrate time and money in nurturing a national core of expertise capable of designing military circuits using the transistor? The military's response to these questions was to have one of its own laboratories, the Defence Research Telecommunications Establishment (DRTE), embark on a 5-year, $400,000 program to design and build an electronic digital computer, even though the military had already concluded, with its cancellation of UTEC and the purchase of FERUT, that the commercial market offered acceptable access to computational power. Put into service in 1960, the DRTE machine was the first transistorized general-purpose computer designed and built in Canada. This chapter examines the circumstances that led a government laboratory with no experience in computers to undertake such a project and the role of DRTE in fostering digital electronics in Canada.

Complexity and Military Self-Reliance

Fear that the very proliferation of electronics that was revolutionizing warfare also threatened to compromise effectiveness of weapons systems underlay the military's interest in the transistor. As the scope and diversity of electronic applications grew, so did the size of these systems. And as the number of electronic components in weapons systems grew well into the thousands, the specter that greater design complexity would seriously undermine system reliability started to haunt the military mind. "I hope we have not created such complex machines," observed a former Assistant Secretary of the US Navy, "that we are in danger of contributing to our own ultimate destruction, for we are becoming so dependent upon automatic electronic brains for the conduct of modern warfare that their failure during combat could very well mean that difference between victory and defeat in an engagement."[6] The fear of falling victim to electronic complexity led one US Navy officer to allude to the lessons of Mary Shelley's novel: "Like the creator of Frankenstein, we have produced devices which, in the hands of the operating forces, are so unreliable that they could lead to our ultimate destruction."[7]

The reliability issue spanned the entire spectrum of electronic components, such as vacuum tubes, resistors, capacitors, and batteries. However, because of their size, their power consumption, their complex internal structure, and their more complicated circuit mountings, vacuum tubes became the focus of the reliability debate. One way to try to counter the problem was to manufacture much better vacuum tubes and circuit assemblies. Because such rigorous performance specifications far exceeded anything the commercial civilian market demanded, it was the military that had to finance the development of these more advanced and costly production methods. Project Tinkertoy and the US Army's secret "Automatic Component Assembly System" thus strove to increase electronic reliability through automation.[8] Just as the US Army had promoted new levels of precision machining in the nineteenth century with its insistence on machine-made interchangeable parts, military enterprise's obsession with stringent operational standards had again, in the case of electronic equipment, pushed precision and quality of manufacture to new levels.[9]

The complexity issue pervaded all aspects of military electronic design, but its threat was particularly harmful to the future growth of digitally based weapons systems. Unlike analog electronics, digital circuits supported extremely large structures, which, when coupled to the military's singular preoccupation with greater performance regardless of cost, found expression in the pursuit of very complex circuits.[10] Even as a small experimental prototype, DATAR had tens of thousands of electronic components. With such numbers, even very reliable individual components did not guarantee the overall system's reliability. The problem arises from the statistics of large numbers. For example, suppose a manufacturer of an electronic component promised a MTBF (mean time between failure) of 87,660 hours. On average, that component would fail in 10 years' time. If an electronic system employed 10,000 of these components, then the system could, on average, fail every 8–9 hours.

Even if the reliability challenge were met, "packaging" problems posed at least a great a barrier to the deployment of more advanced weapons systems. Vacuum tubes' large size and high power consumption placed severe limitations on how much performance and capacity could be designed into digital electronic systems. Only through an aggressive miniaturization program could the military hope to exploit the potential strategic advantages of digital technology. If (as DATAR had revealed) the task of deploying large digital electronic systems in ships was going to be difficult, it looked intimidating for aircraft and impossible for missiles. Despite considerable efforts,[11] the vacuum tube paradigm offered no dramatic gains in miniaturization that could realize the military potential of large-scale digital techniques.

The transistor offered the possibility of dramatic miniaturization that was inherently impossible with the vacuum tube. But there were considerable barriers to the transistor's industrial development. On the demand side, the transistor's high costs—8 times that of a comparable vacuum tube—offered little incentive for the civilian electronics market to incorporate it into their products.[12] Furthermore, in the civilian electronics market, which was dominated by radio and television manufacturers, there was no technical imperative to justify the high cost of solid-state miniaturization. Since a radio or a television used only a few tubes, complexity was not an issue. On the supply side, very costly development and production engineering programs were needed before reliable and standardized transistors could be manufactured in large numbers. Until demand reached a sufficiently high level, there was little economic rationale for components manufacturers to rush into expensive R&D programs to improve the reliability and performance of transistors. Furthermore, as Braun and MacDonald noted, many electronics manufacturers were reluctant to undermine their large capital investments in vacuum tube production by pursuing the replacement of tubes with semiconductors. And on both the demand and supply sides there was social resistance to any broad diffusion of transistor technology. Whereas the vacuum tube worked according to the laws of classical physics, the transistor was a product of the quantum-mechanical revolution. Engineers needed a whole new set of paradigms to deal with the transistor. The tremendous intellectual and experiential investment that electronics engineers had in the vacuum tube paradigm thus became an important source of attitudinal resistance to the transistor.[13]

In a vicious circle, low demand impeded the supply-side innovation required to make the transistor more attractive. No doubt the transistor would have eventually triumphed in the marketplace, but it may easily have taken 10 years longer to do so.[14] American military spending, however, boosted the weak, civilian demand-supply relationship and, as a result, greatly accelerated the rise of the powerful American semiconductor industry. The subsequent improvement in the transistor's performance-to-price ratio then initiated the synergy between digital electronics and semiconductor technologies.

Through much of the 1950s, Canadian military enterprise, like its US counterpart, grappled with the problems of reliability and size arising from the increasing complexity and scale of military electronic equipment.[15] "The rapid rise in the failure rate of electronic equipment, coincident with the increasing complexity and miniaturization of that equipment, is cause for alarm," warned Wing Commander K. P. Likeness of the Canadian Air Material Command.[16] "It is apparent," he added, "that the severity of the reliability problem may reach crippling proportions unless it is given immediate attention." Again DATAR underscored that electronic miniaturization had become critical to Canada's pursuit of military self-reliance. Although the Royal Canadian Navy's inability to convert DATAR's vacuum tube design quickly to a solid-state design was not the cause of the project's collapse, it was an important factor in the closing of an already small window of opportunity. Canada's efforts to develop a short-range missile and its plans to design one of the world's most sophisticated all-weather interceptors further underscored the importance of miniaturization to military self-reliance. With the transistor holding the strategic key to reliable miniaturization, electronics experts in the Canadian defense establishment worried about what could be done to get an ill-prepared national electronics industry to jump quickly onto the semiconductor learning curve. The barriers to accelerating the diffusion of transistor technology were far more formidable in Canada than in the United States. The disincentives to the commercialization of the transistor found in the United States were even more exaggerated in Canada. Foreign-owned subsidiaries, which dominated the Canadian electronics industry, were even more reluctant to embark on expensive product development programs than their parent firms.[17]

84   Chapter 3

Canada's considerably smaller civilian market further reinforced this reluctance.[18] Furthermore, the Canadian military did not have the procurement levels of its American counterpart as a reason to intervene as broadly and deeply in the promotion of the transistor as was done in the United States.[19] As a result, the Canadian approach to the role of transistor technology in military self-reliance became a financially driven trade-off between the capacity to produce transistors and the capacity to use them.

From the moment in 1948 when the Canadian military first took an interest in the transistor for possible use in the harsh Arctic climate, the military debated the importance of being able to manufacture the transistor in Canada.[20] The experience of World War II had convinced Canadian defense planners that undue dependence on foreign sources for electronic components compromised Canada's military sovereignty. Even trusted allies would, in the event of a war, look after their own needs before responding to Canada's. Canadian military planners believed that in the event of war there would not be enough time to develop, from the ground up, a national capacity to produce transistors.[21] Thus, from the Canadian defense perspective, dependence on foreign suppliers for transistors was unacceptable.[22]

The Department of Defence Production (DDP), created in 1951, almost immediately "expressed anxiety over the greatly increased use of electronic components in weapon development, while there [had] been no corresponding increase in Canadian development or application of more reliable components for increasingly severe operational load."[23] DDP felt that "Canada [could not] benefit fully from the energetic component programmes of the UK and US unless Canada actively sponsor[ed] similar programmes to ease increasing dependence on foreign sources of supply of these critical items."[24] In 1954, in an effort to stimulate domestic development and production of the new types of electronic components required by military technology, the Department of National Defence committed $660,000 to help underwrite the risky investment associated with such R&D.[25] The plan was to devote 20 percent of this investment to research, 30 percent to development, and 50 percent to establishing Canadian sourcing.[26]

The efforts of Canadian military enterprise to create a domestic capacity to develop and manufacture transistors proved a failure throughout the 1950s.[27] By 1956, when the need to miniaturize the exploding complexity of military electronic systems had become urgent, electronics experts in the Department of National Defence realized how little progress had been made in producing this revolutionary electronic device in Canada.[28] Northern Electric and Canadian General Electric, two of Canada's most prominent electronics companies, had shown little if any interest in committing their resources to the transistor.[29] Canadian Marconi and Rogers Majestic showed interest in producing transistors, but they had not even articulated an R&D plan yet. The only bright light in the otherwise bleak Canadian transistor R&D landscape was RCA Victor Ltd. in Montreal. This company had already produced, at its own expense, prototype junction transistors for the Canadian military. This Canadian subsidiary's unique willingness to invest seriously in R&D reflected the desire of its Semiconductor Department's manager to "break away from the general policy of following the American parent firm."[30] But RCA was the exception. Even efforts to get US manufacturers to set up transistor branch plants in Canada failed. Philco declined to manufacture its innovative surface-barrier transistor in Canada, arguing that the limited market made it uneconomical to invest $1 million in a manufacturing plant.[31]

By 1956, budgetary constraints and the changing nature of warfare had created divisions among Canada's defense planners over the issue of domestic sourcing of electronic components. The Department of Defence Production and the Electronic Component Development Committee saw promotion of "Buy Canadian" as an essential part of their mandates. But L. M. Chesley, one of the Assistant Deputy Ministers in the Department of National Defence, questioned the extent to which defense money should be used to foster industrial development: "Is it appropriate for ECDC to make a decision to spend public funds for the development of a source of supply in Canada when one might exist for instance in Detroit or elsewhere on this continent?"[32] In the years that followed the Korean War spending boom, the armed services came under constant pressure to reduce their budgets. They became increasingly unwilling to subsidize any domestic production of electronic components. Their definition of military self-reliance shifted accordingly. Concern about Canadian self-reliance was narrowed from worrying about manufacturing components to worrying about Canada's ability to fulfill, within the alliance, a role of designing, testing, and constructing transistor-based equipment. Furthermore, concern about being able to mobilize production of essential electronic components at the outbreak of war made less and less sense in view of the possibility of nuclear war. What counted was the capacity to design and deploy advanced weapons systems before the outbreak of war.

Self-Reliance and the Ability to Miniaturize

Although there was considerable debate within the defense establishment over whether to pour money into the domestic manufacture of the transistor, there was unanimity over the importance of fostering a national capacity to use this new electronic paradigm in original ways. By 1956 the DATAR experience had shown clearly that the challenge of miniaturizing large-scale digital circuitry had less to do with having access to transistors than with knowing how to use them. One could not simply pull a tube from a circuit and, with a little redesigning, plug in a transistor. Ferranti-Canada's short-lived effort to recast DATAR's complex circuitry into solid-state electronics was a rare event in the Canadian industry of the day not because of the difficulties it faced but because it was attempted. Not only did the Canadian electronic industry have little design experience in large-scale digital circuitry; as Frank T. Davies of the Defence Research Board noted, Canadian industry still lacked the knowledge and experience to design complex circuits based on the solid-state amplifier.[33] As late as 1958, Norman F. Moody, who played a central role in the Defence Research Board's efforts to advance transistorized digital circuit design, warned an audience of electronics engineers that the transistor was "different from the vacuum tube in all important respects." "It is true," Moody continued, "that both devices have one trivial property in common, for both give power amplification. To designers who are familiar with vacuum tubes, this trivial resemblance between the devices has been a trap. Many of their failures are strewn along the path of progress. The transistor is a nuisance because you must learn an entirely new circuit technique to use it satisfactorily. Even then you must avoid a tube-by-tube replacement by transistors."[34]

As early as 1954, DRB had decided that, rather than wait for industry to explore the use of transistors in military electronic design, it would create a special section within the Defence Research Telecommunications Establishment Electronic Laboratory (known as DRTE(EL)) to accelerate the process. DRTE(EL) was itself a product of the belief that, if Canada was to have an efficient program for the development of military equipment, scientists and development engineers would have to be immersed in the military ethos. From the services' perspective, "[electronic] experimental development, to be successful, should not be divorced from the applied research which is always incidental to it. . . . Experimental development for Service use requires the continuous absorption of Service experience . . . particularly when a new solution to an operational requirement is being sought."[35]

In creating a new Transistor Section within DRTE(EL), the Defence Research Board argued that Canada's armed services needed a source of new ideas on how to exploit the transistor in military electronics. Within a year of its creation, the Transistor Section, with Norman Moody at its head, staged a dazzling demonstration in the foyer of Ottawa's elegant Chateau Laurier hotel to raise awareness of the transistor within Canada's defense establishment. For most who attended the event, this was their first exposure to the new generation of semiconductor electronics. The circus-like unveiling of the transistor may have been a good piece of public relations for the Transistor Section, but it concealed the lack of a well-articulated program in transistor circuit innovation. However, the issue of digital electronic miniaturization soon brought the Transistor Section's mission into focus.

In 1955 the failure of Canadian industry to champion military use of the transistor prompted DRB and NRC to call for the government to take a more aggressive role in the creation and diffusion of transistor circuit design know-how to industry. Projects such as DATAR, the Velvet Glove missile, and the AVRO Arrow underscored the urgency of improving Canada's capacity to miniaturize electronic design. The Arrow promised to be the flagship of Canada's aircraft industry. For this new-generation fighter-interceptor, the Royal Canadian Air Force needed a new generation of sophisticated electronics: a comprehensive system that would integrate radar data, navigation, communications, the fire-control computer, and flight control.[36] Because of the drive to maximize electronic sophistication while minimizing payload, miniaturization was a central factor in the Arrow's avionics, of which digital circuits were a major component. In the early stages of the project, the Department of Defence Production (DDP) anguished over the prospect that the electronic "brains and nervous system" of what was to be an all-Canadian plane would have to be made in the United States. The DDP had estimated that if Canada tried to singlehandedly develop the Arrow's on-board electronic system the effort would easily absorb 25 percent of the country's electronics engineers.[37] But the problem was even worse than this estimate suggested. Simply counting the number of engineers needed to design and build the Arrow's electronic subsystems did not take into account that most of these engineers had not even started up the transistor learning curve.[38]

In 1955, DRB reacted to the problem with a more vigorous transistor R&D program that it hoped "would [in 2 years] provide a sufficiently broad basis of knowledge from which to guide industry."[39] In this plan, the Transistor Section bore the specific responsibility for developing, testing, and disseminating the fundamental transistor circuit techniques that would be needed to support the military's equipment development program.⁴⁰ Though this mandate admitted numerous options, it was the interplay of the electronic complexity crisis and Moody's past inventive talents in atomic energy research that shaped the contents of the Transistor Section's R&D mission, which by the end of 1955 had become "to master the techniques of applying semi-conductors to military problems involving digital computation."⁴¹ This program was to unfold in three stages: studying digital logic with particular reference to its implementation in terms of semiconductor circuits, developing circuits suitable for digital computation, and finally developing a digital computer as a test bed for logic and circuits.⁴²

Moody's experience with digital electronics had roots in Britain's World War II radar program. Under the direction of F. C. Williams, Moody had participated in the development of the circuits needed to generate and detect the rapid sequences of pulses used in radar systems. Moody later used this experience with electronic pulse techniques in devising better atomic energy instrumentation circuits. Finding his participation in Britain's postwar atomic bomb program distasteful, Moody wanted to be transferred to the Canadian program at Chalk River, which had a more acceptable mandate: exploration of peaceful and commercial uses of atomic energy. Moody's bridge from Great Britain to Chalk River was the physicist W. B. Lewis, Director of Research of Canada's nuclear program at Chalk River. During the war, Lewis had headed up the radar project at the Telecommunications Research Centre in Malvern, England, where Moody had worked in the basic circuits group. Lewis not only knew of Moody; as the only principal British scientist who had refused to become involved in the Manhattan Project, he understood Moody's qualms about the British atomic bomb project.⁴³

While at Chalk River (1949–1954), Moody led his team in a program intended to push the performance of trigger circuits to the very limits of vacuum tube technology. Before their use in computers, scientists used electronic digital counting circuits to measure the fluxes of radioactive particles emanating from nuclear reactions. Being able to quantify the number and energy of each component of radioactive emission was not only crucial to understanding the physics of nuclear fission; it was essential to designing a bomb or a reactor. The accuracy of binary digital counters depended on how well they could discern the passing of a stream of radioactive particles separated by extremely short intervals of time. If the counter circuits could not react faster than once every thousandth of a second, then they would miss all the particles in the radiation beam that were separated by only a ten-thousandth of a second. The switching time of the counter's many trigger circuits set the limits of the radiation counters. Trigger circuits with very fast switching meant faster and hence better nuclear instrumentation.

Moody faced two technical problems in speeding up the performance of trigger circuits. The first was how to devise a circuit that could provide strong, well-defined pulses to drive subsequent stages when operating at very high switching speeds. The solution was to incorporate a feedback element, a "trigger tube," in each binary counting stage. This technique, Moody wrote in his 1952 patent application, provided "the large, steep-fronted surge current" needed to ensure fast, reliable triggering of each successive stage within the counter.[44] The second problem Moody faced in his effort to increase the speed of trigger circuits was the vacuum tube's poor reliability. Every vacuum tube was characterized by a fixed family of plate-current-versus-grid-voltage curves that were crucial to determining the details of any circuit design. Ideally, these characteristic curves were exactly the same for all tubes of a particular type and remained constant over the tube's lifetime. But, despite the manufacturer's claims, vacuum tubes were not produced with this kind of standardization and reliability. Tubes were not manufactured with exactly the same characteristics. Furthermore, as a tube aged, its characteristics changed in ways that were not quantitatively predictable. No two tubes of the same type displayed exactly the same variations with age. Though these variations from the ideal were well within acceptable limits for the conventional analog electronics found in radio and television, they posed serious problems when designers tried to speed up the switching of digital circuits. Small variations in tube characteristics could easily throw off digital switching circuits that had been "tweaked" for maximum performance. In various patents filed while he was at Chalk River, Moody expressly tried to design high-speed trigger and binary counting circuits whose performance was resilient to variations in tube characteristics.[45]

Early pioneering work in electronic digital computers had drawn sustenance from the digital counters first developed for atomic energy research. As one technical study of digital circuit design explains, "a few scattered references began to appear on the use of flip-flops in counter circuits, where the applications in most cases was the counting of pulses from a Geiger-Muller tube."[46] John Mauchly, one of the two principal designers of the ENIAC, later recounted the technology transfer he had acquired from digital radiation counters: "Binary counters were already known to me; they existed in cosmic-ray labs [like my father's]. So I thought, see, if they could count at something like a million per second with vacuum tubes, why it's sort of silly to use the these [IBM] punch card machines, which can only do maybe a hundred cards, a minute and don't seem to do much at that."[47]

Though Moody had little direct knowledge of electronic computers, the strong technical affinity between radiation counters and computer circuit design led him to speculate, as early as 1950, that his techniques would offer important benefits to the designers of "high speed computers and multi-channel pulse communication systems."[48] Moody felt that his digital switches offered computer designers much faster and more reliable digital circuitry. No one in the computer world adopted Moody's vacuum tube switches, and Moody made no effort to promote them. Nevertheless, the idea that computers offered an excellent technological showcase for his techniques continued to percolate in Moody's mind. Five years later, as head of DRTE's Transistor Section, Moody saw another opportunity to establish the value of his trigger-circuit philosophy in the design of computers—this time, however, through transistor switches.

It is not likely that Moody would have suggested an ambitious computer design program as a way to advance expertise with transistor circuits, and it is even less likely that DRB would have funded such a proposal, had it not been for the complexity crisis that threatened to paralyze the deployment of new electronic weapons systems. Nowhere was the urgency of this crisis felt more acutely than in large-scale digital circuitry. Once again, it is important to recall that by the end of 1955 DRB had seen firsthand how DATAR's full-scale technical elaboration was being handicapped by the very slow pace of miniaturization. DRB's ambitions to define for Canada an independent missile development role within the North Atlantic Triangle alliance hinged in part on Canada's ability to design complex, robust miniature electronic systems for the Velvet Glove missile. The Canadian Air Force's development of airborne, digitally based, automated fire control systems and data communications systems also called for a solid-state solution to the power-consumption, size, and reliability problems that arose from greater complexity. What better place to hone miniaturization skills than on the scale of computers? The fundamental transistor circuit knowledge gained from building the computer could then be transferred to any digitally driven weapons system.

Another compelling reason for constructing a computer a DRTE was that a large and controlled statistical sample set could be used to measure and test the performance of transistors. Though it was clear by 1956 that the transistor offered the only hope of substantial miniaturization, the Canadian military was still very cautious (as was its American counterpart) about any sudden replacement of all its vacuum tube equipment. Because of germanium-junction technology, transistor reliability had come a long way from the unreliable point-contact technology. Nevertheless, there were still limits to reliability. The transistor's transition from an experimental device produced in small runs to a standardized, mass-produced commodity was still not assured, particularly for the newer generation of silicon transistors. According to Braun and MacDonald, "it was difficult to design a transistor to give even an approximation of the characteristics required, and even harder to produce two transistors with the same characteristics."[49] By 1956, the advantages of transistor reliability had still not been proved. "It is my suspicion," observed one senior US Air Force officer in a speech to a conference on electronic components, "that the chief reason we have not been able to advance the production date for this highly important equipment is due to the erratic and undependable behaviour of transistors."[50]

There seemed to be little statistical justification for the transistor manufacturers' assertions that they were producing a more reliable product. "There is no large body of experimental evidence to prove the [reliability] claims," the Head of Computing Devices of Canada's Semiconductor Engineering Department pointed out to DRB's chairman, "though all indications tend to support them. In consequence, many designers hesitate to introduce transistors into military equipments until they see adequate proof of this greater reliability than tubes."[51] Because of its inherently large scale, Moody saw the digital computer as an excellent framework within which to establish the statistical basis of the transistor's performance. More than a testing platform, the computer project also served as a valuable training ground in which Canadian engineers could move up the learning curve of solid-state digital electronics.

Today, silicon is synonymous with the digital revolution and is the self-evident material basis for the commercial computer-transistor synergy. Levin, however, shows very convincingly that the abandonment of germanium and the push to develop reliable silicon transistors had little to do with any self-evident civilian commercial opportunities and everything to do with the US military's pursuit of reliability and performance regardless of cost.[52] The semiconductor silicon was far more resistant than germanium to the high temperatures that were expected to be encountered in missiles and in high-performance fighters. It was also less susceptible to radiation damage. Considerable military spending on direct R&D and on future procurement was needed to get industry to move onto the silicon trajectory.⁵³ But, unable even to kick-start domestic manufacture of germanium transistors, the Canadian military could not even contemplate the far more expensive prospect of producing silicon transistors domestically. Even so, it was felt, the Canadian defense industry could become an early adopter and adapter of this technology. The Transistor Section's early and complete commitment to designing transistors around silicon illustrates that military needs, not civilian spinoffs, infused this group's exploration of new computer circuits.

The success of DRTE's plan to serve as a pole for the diffusion of innovations depended on the effectiveness of the mechanisms used to move information from a government laboratory to industry. In Japan, the government's Electro-Technical Laboratory worked in close and continual collaboration with industry.⁵⁴ DRTE's method, however, was to "broadcast" to industry the basic circuit innovations needed to apply the transistor to military needs. In this regard, DRTE's Transistor Section intended to publish and disseminate its work to all potential defense contractors by means of technical reports and papers. But how could the military, using only this method, ensure that the techniques developed in DRTE's laboratory would become part of industrial practice? Furthermore, technical reports could never convey the wealth of subtle and unwritten design know-how generated in a problem-solving environment in which engineers and scientists worked in daily close proximity and shared insights and experiences in countless informal discussions.

Rather than rely entirely on printed information as a mechanism of technology transfer, DRTE decided to move people around. In 1955, the Transistor Section set up a training program in which engineers from defense-related industries worked in Norman Moody's laboratory for a minimum of 6 months.⁵⁵ The hope was that these engineers would, like good disciples, go back to their respective companies and teach the secrets of transistor circuit design to others. Equally important, the trainees could be counted on to champion the Transistor Section's techniques once back in the company. The training program offered no government subsidies. Each company had to bear all its trainees' expenses. Even so, the response from Canada's electronics industry was overwhelming. Canadian Westinghouse, Northern Electric, Canadian Marconi, Computing Devices of Canada, Canadian Aviation Electronics, AVCO of Canada, and Rogers Majestic Electronics all expressed interest in sending one or more engineers for training.⁵⁶ Clearly, the initiative responded to a real need. Without extensive military procurement, most Canadian electronics firms were either unwilling or unable to embark on a costly transistor R&D program. In the absence of any contract, DRTE's proposal offered these companies the opportunity to participate in a large, sophisticated R&D environment. No doubt the electronics firms also felt that by participating they increased their chances of winning future contracts.

Moody's New Transistor Switch Drives Computer Innovation

Ready to exploit the transistor on its own terms, Norman Moody nevertheless still made creative use of his innovative work with vacuum tube switches. As with tubes, Moody set out to push the limits of the speed and the load capacity of transistor switching circuits.[57] Again, as with tubes, Moody tried to design high-performance, transistor circuits that would be fairly insensitive to the unwanted variations in component characteristics that were a mark of contemporary transistor manufacture. Positive feedback had been an important design element in Moody's earlier tube-based trigger circuits. Moody's answer to feedback in the semiconductor realm was the regenerative "PNPN" complementary structure. The tube-based flip-flops (i.e., the basic binary-valued circuit elements) used in computers in the 1950s were based on the Eccles-Jordan symmetric arrangement of two vacuum tubes. When the flip-flop was transistorized, circuit designers simply replaced the two identical tubes with identical transistors, thus maintaining the symmetry. Looking at transistor circuits as vacuum tube equivalents missed a fundamental difference. There was only one basic vacuum tube structure, but there were two kinds of transistor structures: npn and pnp. It was this asymmetry that Moody exploited in his regenerative PNPN complementary structure.[58]

The construction of a computer became a showcase for the Transistor Section's rather unique approach to high-speed transistorized digital circuits: the "controlled saturation" technique embodied in the PNPN structure.[59] Moody's contention that the PNPN trigger circuit's good performance "in the face of wide transistor and component variations" meshed well with the military's search for reliable high performance. His claim that the PNPN approach allowed for the development of switching circuits capable of driving many other stages was also attractive to the military because it implied greater digital miniaturization without any sacrifice in performance. The design and construction of a computer was to be the test of Moody's claim that the PNPN concept offered the military electronics designer significant reduction in digital circuit complexity without any sacrifice in performance.[60]

Beyond his invention of the PNPN structure, Moody played no further part in the design of the DRTE computer. He gave his most senior engineer, David Florida, the responsibility of being the project's manager and the computer's chief architect. In Florida's hands, the project's philosophy quickly took on an entirely new dimension. In early 1956, Florida suggested to a DRB audience that military preparedness demanded a deeper understanding of the design of electronic information-processing technology.[61] Whereas Moody had seen the computer as a vehicle for exploring the potential of the PNPN flip-flop, Florida was inclined to see the latter as a means of expanding Canada's capacity to build and use data processing systems. In 1956, with only a handful of electronic digital computers in operation in Canada and with Ferranti-Canada struggling to be Canada's first computer hardware designer, Florida recognized there was a need for more computer design research in Canada.

Florida transformed the mandate to showcase the PNPN invention into an opportunity to explore the complexity of computer hardware. After a year of experimentation, Florida and Moody had shown, using a sample of 40 transistors, the validity of "controlled saturation."[62] According to Moody and Florida, the ability to achieve high-pulse currents with very fast switching made the PNPN ideal for computers. By exploiting the PNPN's high fan-out capabilities, it was possible to transfer much of the computer's digital circuitry to passive diode logic without losing any overall speed.[63] Thus, for a given level of performance, the PNPN approach promised greater economy of transistors and hence even greater miniaturization. Ironically, the PNPN flip-flop's ability to maximize the performance of digital circuits reliably with a minimum number of transistors encouraged Florida to increase the number of features he wanted to design into hardware. Florida's fascination with hardware sophistication was stimulated by the desire to build an improved computational tool for scientists.

The design decisions made in the period 1957–1959 contain many examples of how, in resorting to greater circuit complexity, Florida wanted to take the computer from a mere test bed to a computational tool for scientists. For example, at a time when fixed-point arithmetic was the standard feature in computers, Florida added floating-point capability as a way to ease scientists' programming burden.[64] Another way to simplify the scientific programming environment was to eliminate doing arithmetic against an accumulator. Rather than use the then-standard one-address or two-address command word structures, Florida proposed a three-address architecture.[65] This method made an arithmetic operation resemble pencil-and-paper computation more closely.[66] The two addresses specified the memory location of each operand; the third address gave the location to which one wanted to transfer the result. Greater hardware sophistication thus created a environment in which the scientist programmed arithmetic operations directly against the entire memory, rather than always passing through intermediary accumulators. Florida reasoned that this would make programs shorter, and that they would hence take less time to write. Shorter programs also meant faster computational times.

In other areas, Florida tried to reduce the complexity of programming by relegating to hardware tasks that would normally have been left to software subroutines. Decimal-to-binary and binary-to-decimal conversion, division, and even square root were to be all hard-wired into the DRTE computer.[67] The decimal-binary converter alone contained 128 PNPN trigger circuits.[68]

From the military's perspective, the value of the DRTE computer project lay in the know-how in transistor circuit design that was to be gained. Florida, however, wanted to advance the state of the art of computer design by producing a machine that would significantly reduce the scientists' programming burden. His design philosophy was nevertheless shaped by Moody's view that the computer was a vehicle for demonstrating the merits of his PNPN trigger circuit. In the end, Florida's efforts to improve the utility of the computer as a tool for scientists meant creating greater circuit complexity to showcase the PNPN trigger.[69]

While the military funded the DRTE computer project as a means of advancing Canadian transistor circuit-design expertise, those working on the computer quickly came to see it as a end in itself. Florida's ambitions would have gotten nowhere had there not been a need for another military computer. To sell the construction of a sophisticated general-purpose data processing machine, Florida played on the growing computational bottleneck in Canada's defense laboratories.

From Hardware Development to Computation Centre

In 1954, Canada had only two computers: the British-made FERUT at the University of Toronto and the American-made Computer Research Corporation CRC102A computer at the Royal Canadian Air Force Base in Cold Lake, Alberta.[70]. Each owed its origin to military needs. These two computers met most of the computational requirements of Canada's defense research until 1955.

In the fall of 1954, the Chief Superintendent of the Canadian Armament Research and Development Establishment (CARDE) argued that his department's research simply could no longer rely on the FERUT computer.[71] Since the purchase of FERUT, it was observed, "work in ballistics has shown that heavy computational programmes will become more frequent as Canada's dependence on UK and US technical work is reduced, and as Canadian work in armaments matures."[72] Rather than expand the facilities at the University of Toronto, CARDE wanted to acquire its own computer. Direct access to more computer time was not CARDE's only concern, however; the scarcity of programmers was another.[73] "The principal value of a local digital computer, in the long run," it was argued, "will be the indoctrination of CARDE staff and the training of programmers. It is now obvious that one of the main obstacles to the application of computers to scientific and technical work is that scientists do not think in terms of computers except in general as mysterious devices used far away by specialists. Only the availability of a machine directly accessible to the individual scientist will overcome this difficulty."[74]

In the spring of 1956, CARDE acquired the American ALWAC III computer. Within a few months after its installation, a DRB memorandum noted that "even with this increased potential for solution of scientific problems by digital machine methods, facilities are being used at almost full capacity while the computation program has been kept of manageable size only by the exercise of restraint on the part of the larger users."[75] CARDE had been the largest user of computer time on FERUT. After the purchase of ALWAC III, CARDE's demands on FERUT were reduced considerably.

In CARDE's place, DRTE became the principal military consumer of computational time on FERUT.[76] Much of this had to do with DRTE's intensive research into the problems of military radio communications in the Arctic. Modeling the propagation of radio waves in the ionosphere (which was crucial to Arctic communications) required considerable scientific computation. At the same time, other government defense research laboratories, because of the time-consuming inconvenience of having to deal with the geographically distant University of Toronto, began to ask for their own computational facilities. The Naval Research Establishment (NRE) in Dartmouth, Nova Scotia, and the Pacific Research Laboratory (PRL) in Esquimalt, British Columbia, wanted their own computers. As with CARDE, the call for local computational facilities by other defense laboratories was not only a matter of fast and convenient access but also an expression of the need to expose scientists to this new technology.

DRTE Superintendent James Scott suggested to DRB Chairman Omond Solandt that "considerable annual savings might be made by establishing computer operating staff in Ottawa to utilize the [DRTE] computer [when built] rather than expand DRB support of the Computation Centre in Toronto."[77] Scott boasted not only that the DRTE computer would represent a significant technical advance over the American Philco transistor computer, but also that it would provide more computational power than FERUT. He also recommended that the proposed DRTE computer play an integral role in formulating DRB's future computational needs.[78] And David Florida even lobbied to have his machine become the mold from which other medium-size computers could be produced.[79]

After nearly 5 years of work and $400,000 of R&D investment, the DRTE computer went into service in the fall of 1960. Despite Florida's quest for a high-performance scientific computational tool, the DRTE computer still had serious deficiencies. Its fast parallel memory was linked to a slower serial arithmetic unit, and there were crippling input/output bottlenecks. Then as now, fast computation was fine, but if getting data in and out of the computer was extremely slow then the computer's utility to scientists was seriously compromised. Even more critical was the fact that little manpower had been devoted to developing any software that would encourage use of the computer.

From 1962 to 1964, the DRTE's computer group designed and installed a new parallel arithmetic unit to replace the serial one, expanded the memory to 4000 words, built a new magnetic-tape system, and developed the circuits and logical design for new punched input-output devices.[80] Work started on the development of an easily accessible scientific programming language. These enhancements were still funded under the terms of the project's 1955 mandate. However, from the point of view of "mastering the techniques of applying semiconductors to military problems involving digital computation,"[81] the improvements to the DRTE computer were of marginal utility. The continued availability of military funds stemmed instead from the immediate need for a machine to handle DRTE's ever-expanding computational requirements. Research in military communications systems, radar systems, and satellite technology called for an adequate in-house computation center.[82] Canada's Alouette satellite project in particular called for enormous data processing capabilities.

From 1961 to 1964, use of the DRTE computer increased by 400 percent.[83] The capacity of the machine had reached its limit. DRTE found itself having to rent additional time on IBM 7090 and 7074 machines, which were (as the statistics showed) far superior to the DRTE computer. (In 1964, 1300 hours of processing time on the DRTE computer were used, 60 hours on two large IBM computers.) But because of the superior throughput of these large IBM installations, the amount of actual processing done by these machines surpassed that done on the DRTE computer. The latter, scarcely capable of handling the data generated by the Alouette I project, clearly would not be able to cope with the new ISIS satellite program, to which Canada had also committed itself.[84] A new computer was needed.

George Lake, who had helped design and build the DRTE computer, was put in charge of the DRTE computation center when the computer went into service in 1960. In 1965, Lake estimated that by 1966 DRTE's computational requirements would increase twelvefold. Convinced that the existing DRTE computer would soon outlive its usefulness, he supported the purchase of a large IBM computer. The original DRTE computer continued to play an ever-decreasing computational role until about 1967. In 1968, this military exercise in self-reliance in digital electronics was sent to its final resting place: the Computer Hall of Canada's National Museum of Science and Technology in Ottawa.

Did the DRTE Computer Serve a Useful Purpose?

For many years, the DRTE computer sat in eerie silence on display under a spotlight in the dimly lit Computer Hall. On occasion, tour guides would characterize it as "Canada's earliest solid-state general-purpose computer." Looking back on this exhibit, one is tempted to ask whether there was any point to this machine.

Linda Petiot speculates that the DRTE computer project may have been "motivated by attempts to create a high-speed, code-breaking machine."[85] "Was information learned at DRTE being passed back to a 'more secret' group doing decoding and encoding work?" she asks.[86] Though Petiot offers little evidence to support her suspicion, the code-breaking scenario does, at first glance, offer a fascinating hypothesis to explain why this computer was designed and built within a military laboratory—particularly when, with the availability of commercial machines, the Canadian military had decided that the capacity to compute no longer demanded a capacity to build computers. Furthermore, Petiot's hypothesis is reinforced by the computer's close historic ties to code breaking.[87] But, as this chapter has shown, the origins of the DRTE computer project were more prosaic than Petiot's cloak-and-dagger scenario.[88]

In a broader technical, industrial, political, and military context, the DRTE computer takes on more historical significance as a material artifact of Canada's unprecedented flirtation with military self-reliance and of the ancillary pursuit of a capacity to design on the leading edge of technological change. Simply put, the DRTE computer's purpose was to move Canada's electronic miniaturization know-how rapidly up the global innovation wave. In the process, the DRTE computer contributed to Canada's early emergence as an international leader in satellite communications technology.

The imperative to deal with the rapidly increasing complexity of electronics used in weapons systems, which was sustained by ambitions of self-reliance, led the military to think of ways in which it could promote a domestic capacity to miniaturize. Complexity, a widespread problem in military design, promised to be particularly crippling to the development and deployment of large-scale digital systems. As the DATAR story reveals, the Canadian military's abandonment of domestic computer R&D in 1952 in the hope of ensuring access to data processing equipment did not extend to the broader area of digital electronics. But without reliable miniaturization, Canada's future ability to apply digital electronics in weapons development would be nonexistent.

Offering potentially dramatic advances in miniaturization, the nascent transistor became a new element in the Canadian military's notion of self-reliance. Faced with an electronics industry that was unable to champion the transistor, the military grappled with how to get Canada up the transistor learning curve more quickly. With the limitations of a small economy, self-reliance in all aspects of transistor innovation wave was impossible. Instead, the Canadian military tried to foster the capacity to design with transistors rather than the capacity to manufacture them. Norman Moody's Transistor Section became the pole of innovation diffusion. From DRB's perspective, the ultimate rational for building the DRTE computer was to create a national expertise in miniaturization of digital electronics that could bolster Canada's peacetime military preparedness.

Did the Transistor Section prove to be an effective source of innovation diffusion? From the point of view of computer technology, the answer is a categorical No. As a computational tool, the computer never found application beyond DRTE's walls in Ottawa. David Florida's predilection for hardware sophistication, reinforced by the military's experimental mandate, served to lengthen the project.[89] By the time the DRTE computer was ready, developments in the computer industry had removed any of the technical advantages it may have had. Perhaps if the computer had been completed by the end of 1957, as DRTE's Superintendent James Scott had confidently predicted to the chairman of DRB, the diffusion of this technology within government operations might have met with some success.[90] Once this window of opportunity was lost, the price tag for DRTE's transistor computer technology became too high to attract government users away from commercial systems.

Throughout the life of the DRTE computer project, there was little collaboration between the Florida team and other Canadian groups working in computer technology. There is no evidence that the DRTE group, during its design of the computer, ever interacted with the University of Toronto Computation Centre. In the period 1956–1960, when Ferranti-Canada had Canada's most advanced computer design group, no technology flowed from Moody's government laboratory to that company. While Norman Moody sang the praises of his novel PNPN techniques, Ferranti-Canada continued to develop DATAR with the more expedient Philco techniques based on the SB-100. Ferranti-Canada believed PNPN to be too complex, but Moody argued that Ferranti-Canada's Philco-based techniques were unreliable.

Having neither advanced the general state of the art of computer technology nor transferred any computer design expertise to industry, did the Transistor Section nonetheless succeed in some way in accelerating the ascent of Canada's industry up the transistor learning curve? Whether it successfully diffused fundamental circuit techniques to industry cannot be answered from the known public record, but there is no doubt that DRTE played an important role in raising the defense contractors' awareness of the transistor. The industrial trainee program that brought engineers from Canada's electronics industries into Moody's Transistor Section was a clear attempt to facilitate the transfer of technology from government laboratories to corporate product development. The success of this training initiative has yet to be evaluated. How long did this program last? How many engineers were involved? From which companies did they come? What was the nature of the transistor know-how taken back to the company after the training? Did these trainees become effective champions of the transistor within their respective companies? A great deal more evidence and research is needed at the corporate level before anyone can even start to answer these questions.

Yet, despite its defects, DRTE's experimentation with solid-state digital electronics and the construction of the DRTE computer were important to the rhythm and the direction of technological change in Canada. This digital transistor work contributed to Canada's ability to become the third nation in the world to design and deploy complex electronic satellite systems. From the outset, military enterprise drove Canada's interest in space and satellite technology. The Transistor Section supported the military's interest in space. In other words, this government laboratory bolstered public enterprise more than it did private enterprise.

The Canadian military's interest in space-related research arose from two distinct sources: remote sensing and the unique problems of radar and Hertzian communications in Canada's northern latitudes. The DRTE computer group's know-how first contributed to the study of these problems in 1958, when Canada's military decided to begin research on the effects of the ionosphere on the detection of missiles launched from the Soviet Union. For this project, a $2 million radar unit[91] borrowed from the US Air Force was refitted with specially designed equipment and installed in a new facility in Prince Albert, Saskatchewan.

Preprocessing the vast amounts of data that the Prince Albert Radio Laboratory (PARL) was to receive called for fast parallel core memory. Development of the digital electronics to drive this memory relied on exploiting the advantages of Moody's PNPN structure. Richard Cobbold, who was working on Florida's team to design this memory, faced the challenge of having transistor circuits deliver half-ampere currents at very high speeds (on the order of a microsecond). Here again was Moody's favorite inventive challenge: to develop a high-speed trigger circuit that could also reliably produce high currents. The solution of this problem led Cobbold and Moody to patent a new monostable trigger circuit based, again, on the complementary PNPN transistor pair.[92] In building the two 40,000-bit core memory units for PARL, Cobbold also patented innovations that simultaneously overcame limitations on the performance and the manufacture of core memories.[93]

On 6 June 1959, at the inauguration of PARL, Prime Minister John Diefenbaker received greetings from US President Dwight D. Eisenhower in the world's first satellite communication. The radio message was transmitted from the United States and bounced off the moon. The irony of the ceremony lay in the fact that Diefenbaker had killed the AVRO Arrow, an icon of Canadian technological self-reliance, but with the inauguration of the PARL had inadvertently participated in the creation of a new icon: Canada's entry into the space age and satellite communications.

The launching of Alouette I on 29 September 1962 made Canada the third nation into space, after the United States and the Soviet Union. The top-side sounding experiments carried out by Alouette I were an extension of PARL's earlier mandate: to understand how ionospheric disturbances, unique to Canada's far north, affected electromagnetic-based communications and detection systems.

Not only did the DRTE computer play a major role in processing the data from Alouette; knowledge and techniques gained in designing the computer's transistor circuits had helped to build the satellite. Minimizing the payload's size and weight was essential. And once a satellite was launched, little could be done about improperly operating circuits. Since miniaturization and reliability were critical in the design of satellites, the expertise in silicon transistor techniques developed by Moody's group proved timely and valuable for DRTE's subsequent work on satellite electronics.[94] Alouette's electronics designers had collaborated closely with Moody's group. For example, Colin Franklin, who spearheaded the design of Alouette I's electronic systems, had worked with Moody and others in the Transistor Section to develop miniaturized, self-contained navigational systems for use by Canadian fighter planes in the Arctic.[95]

Alouette I became the starting point of an ambitious Canadian space program. After leaving the DRTE computer project, David Florida led the ISIS I and II satellite projects. When the Department of Communications was formed, in 1969, it took most of DRTE's satellite R&D people, along with the electronics expertise that had grown out of the earlier transistor work. In 1975, under the auspices of the Department of Communications, Canada launched the Hermes satellite, used in one of the earliest experiments in direct broadcasting. Canada later led the way with MSAT (a mobile communications satellite) and RADARSAT (one of the world's most advanced remote sensing satellites).

Of course, it is possible to exaggerate the contribution to Canada's subsequent space program of the Transistor Section's early experimentation on silicon transistor digital circuit techniques. However, it is clear that the Transistor Section served as an important pole of technical knowledge diffusion within military enterprise, and that it helped Canada move up the transistor innovation wave. Among the principal heirs of technical knowledge acquired by DRTE's early transistor research were the Department of Communications and the civilian-based space program. DRTE's experimentation with the use of transistors in digital electronics was another important factor in military enterprise's promotion of the creation of a national core of experts, who in turn kept the digital electronics revolution within Canada's reach.

The legacy of the DRTE's digital transistor work also diffused into Canadian medical research. In 1959, just before the installation of the DRTE computer, Moody and Cobbold left DRTE and took their wealth of experience to the University of Saskatchewan. It was no coincidence that the Dean of Engineering at the time was Arthur Porter, the man who headed up Ferranti-Canada's work on DATAR. Porter had asked Moody to head up the Department of Electrical Engineering. Together, Moody and Cobbold turned their knowledge of transistor electronics to, among other things, the problems of medical research. Moody then went on to became director of the Institute of Biomedical Engineering at the University of Toronto. Cobbold again followed his mentor, and later he also was to become the director of the IBE. Moody and Cobbold became the nucleus around which research in the use of digital electronics in biomedical engineering developed at the IBE. By this time, microelectronics had replaced discrete transistors with integrated circuit silicon chips.

Although the DRTE computer project helped Canada to jump on the transistor innovation wave at a relatively early stage, it did little to help Ferranti-Canada develop and sell solid-state computers. No technology transfer between Moody's government laboratory and Ferranti-Canada's computer design team was possible when the two groups were entrenched in different transistor circuit philosophies. As the next chapter reveals, Ferranti-Canada, cut off from military funding by the collapse of DATAR, funded its way up the solid-state-computer learning curve by adroitly exploiting opportunities created by civilian public enterprise.

4

Civilian Public Enterprise Encourages Domestic R&D in Digital Electronics

The demise of DATAR symbolized the Canadian military's post-Korean War retreat from the high cost of technological self-reliance. As the Royal Canadian Navy's ambitions to develop its own digital electronic warfare systems stalled and then collapsed, Ferranti-Canada scrambled to keep its digital electronics group employed. One part of the group went into electronic nuclear instrumentation. The digital computer expertise built up during the DATAR project, however, perched precariously on the edge of a precipice. Unless Ferranti-Canada could find immediate civilian opportunities to underwrite further computer R&D, this critical mass of digital electronics engineers would disintegrate. This chapter explores how Ferranti-Canada exploited civilian public enterprise to sustain computer R&D at the critical moment when military enterprise pulled away. An important theme of this chapter is how innovation was shaped by the interplay of Ferranti-Canada's computer ambitions and the automation agenda of two public enterprises: Canada Post and Trans-Canada Air Lines.[1]

Moving the Mail

At the start of 1951, a climate of severe fiscal restraint in the federal government created an opportunity for innovation in digital electronics within the Canadian Post Office Department. As one of the largest government employers, the Post Office became a prime target in the Liberals' program to cut 10 percent of the operations of the public service.[2] The letter carriers were the first victims. Twice-a-day mail delivery became once-a-day, and a thousand jobs were eliminated.[3] The Conservatives, who had long attacked the government's inability to reduce expenditures, now criticized the government for trying to destroy this national institution. The public wanted cost cutting but was unhappy when twice-a-day delivery ended.

Criticism that the government's elimination of twice-a-day home delivery would slow the mail was not only blatant political opportunism but also misplaced. The real bottleneck in mail service did not arise from the reduced frequency of home delivery. Rather, delays in sorting the mail for the carriers were the primary source of slow service. The delays arose from the physical process of manual sorting (which consisted of men standing before an array of pigeonholes, reading the addresses, then placing each letter into the appropriate hole) and from the sorters' mental burden of having to memorize a very large number of geography-based sorting distinctions. These physical and mental obstacles placed important limits on the volume of mail a labor force of a certain size could process in a day.

Urbanization, suburbanization, and the constantly growing volume of mail were undermining the efficacy of manual sorting.[4] The continual redefinition of geographic sorting distinctions, arising from postwar growth, pushed the mental burden of sorting to its limits. In "forward sortation," the postal workers had to memorize the relationship between Canada's 13,000 agglomerations and the transportation system that served them. As the landscape of towns changed and new transportation routes were established, postal clerks struggled to keep pace with the sorting revisions. The rapidly evolving landscape within large cities, engendered by postwar growth, was even more problematic for sorters. In "city sortation," a physically distinct process from forward sortation, the sorter had to memorize the addresses served by each carrier. Deciding to which carrier's "walk" a letter should be assigned was not at all evident from the address. In Toronto, for example, there were 700 walks covering 10,000 distinct city sections. A walk could consist of a single building, a section of a primary street, or an entire secondary street. A postal worker underwent considerable training in the details of his particular sorting task. With every new office building or housing development came modifications to the boundaries of a carrier's walk. Relearning new geographical distinctions was time consuming and created bottlenecks in mail processing. Another important limitation to postal efficiency was the time-consuming process of having to learn and relearn sorting categories. Keeping up with change proved quite stressful for shift workers. "With the constant growth of population in Canada and the expansion of buildings," one Post Office Department expert observed in the early 1950s, "the strain on the sorters is increasing and officials fear an ultimate breakdown."[5]

One obvious solution would have been to employ more sorters, but in a climate of downsizing this option was not politically acceptable. Automation, if less costly than labor increases, was the preferred solution. Technology did exist to increase the physical reach of each sorter: purely mechanical devices allowed a sorter to place mail into several hundred pigeonholes, whereas the maximum number then being used for manual sorting was 108.[6] This technology did nothing, however, to alleviate the critical memorization bottleneck. If anything, it exacerbated it.

In December 1951, a bold automation plan that seemed to address the memory issue percolated up from an unlikely source within the Post Office bureaucracy. O. D. Lewis, a senior clerk within the postal bureaucracy in Ottawa, looked around and saw the great productivity gains and efficiencies that mass production had wrought in industry.[7] For Lewis, the road to more efficient and cheaper mail sorting was clear: Fordism.[8] But until the reliance on the sorters' memory skills could be overcome, the logic of the assembly line would be of little value to the Post Office. Lewis had no formal or informal technical experience, yet he was convinced that the miracle of electronics could eliminate the postal services' dependency on human memory and thus clear the bottleneck in the Post Office's sorting operations.

Lewis looked to IBM's card technology for inspiration. Why, Lewis asked, couldn't a technology capable of processing pencil-marked cards and questionnaires at very high speeds sort mail? Lewis suggested representing the address as "a code of vertical bars [using conductive ink] on the back of the letter." "Or if a virtually colourless conductive marking fluid could be developed," he continued, "then the front cover could be used."[9] In Lewis's mind, once the bar-coded translation of the address had been read electronically, the appropriate gates could be actuated directly to sort mail from a high-speed conveyor system into the appropriate destination bin. The government's push to reduce non-defense expenditures, the growing inability of postal workers to keep up with the volume of mail, and mounting criticism from the opposition made Deputy Postmaster General William J. Turnbull receptive to this clerk's adventurous idea.[10] Turnbull, like Lewis, knew nothing of technical matters. Yet Lewis's tale of moving greater volumes of mail at less cost resonated with ever-growing intensity in Turnbull's mind. Lewis's assertion that "a good half of the existing sortation staff would be displaced by . . . machines" surely made an impression on Turnbull. Exploiting the gender bias in wages, Lewis added that "[the sorters'] salaries would be replaced by a series of lesser and lower wages for typists."[11]

Investing heavily in new technology to achieve gains in productivity and savings in labor costs made economic sense only because of the existing economies of scale and geographic concentration in Canada's mail-sorting operations. Sixty percent of Canada's mail (which totaled 3 billion pieces per year) were sorted at just ten processing stations.[12] This concentration also had important political consequences. Despite the vociferous calls from the rural areas for equal postal services, more than 75 percent of complaints aimed at the Post Office came from large cities.[13] Urban needs drove the economics and politics of mail service. The most difficult, time-consuming, and costly task at each of these ten stations was the sorting of mail for local urban delivery. Investing in technology to reduce the time that mail lingered in these urban centers made good political sense.

Lewis and Turnbull quickly discovered the gulf between a vague technological desire and implementation. When they looked to Europe and the United States to see if anyone had developed an electronic mail-sorting system, they came up empty handed. Lewis's assumption that "the basic devices required for [an electronic sortation] machine are already in existence and patented" proved naive. Few if any of the existing patents directly addressed the specifics of designing fast and standardized methods of encoding postal addresses. Neither did they offer any useful designs for address-code printers and reliable readers capable of scanning the coded envelopes accurately at high speeds. The question of how the bar-coded address information, once read, would determine the appropriate sorting decision remained untouched. How would a machine know to which sorting category, from thousands upon thousands of choices, a particular coded address belonged?

In 1952 the world had no off-the-shelf solutions for the Post Office's automation needs. Unwilling to wait for the right technology to be created and proved elsewhere in the world, Turnbull had the bureaucratic astuteness and courage to gamble large sums of money on this risky enterprise.[14] But he needed someone capable to give this ambition the technical and managerial foundation it lacked. He turned first to Canada's most respected public R&D organization, the National Research Council, but he discovered that it had little to offer.[15]

Turnbull's search came to end when Maurice Moise Levy unexpectedly quit Canada's Defence Research Board to set up an R&D group at Federal Electric Manufacturing Co. (FEMCO), a Canadian affiliate of International Telephone and Telegraph (IT&T).

Though FEMCO was headquartered in Montreal, Levy's group was located in Ottawa, where it pursued federal contracts. In April 1952, upon meeting Levy,[16] Turnbull asked him point-blank: "Can you do electronic sorting of mail by putting code marks on envelopes?" Without hesitation Levy replied "Yes."[17] The extent of Levy's scientific achievements, IT&T's acknowledged experience in automated mail sorting, as well as the weight of IT&T's vast technical resources, gave this affirmation considerable credibility.[18] If Levy could put a sound and practical R&D proposal on the table, Turnbull was ready to push it through. During the next 3 or 4 months, Lewis and Levy fleshed out a detailed contract proposal. Lewis provided the know-how on postal operations and needs; Levy was the scientific and technical expert. By the end of May 1952, Levy had cast the notion of electronic mail sorting into a comprehensive design paradigm.

Levy was quick to have IT&T lawyers examine the potential for future patentable ideas from the Canadian postal proposal.[19] He discovered that the general idea of using coding marks to sort electronically was far from new. A dozen or more patents, some going back to the 1930s, had laid claim to many of the facets surrounding this broad concept.[20] Laying claim to a principle, however, was a lot easier than building a useful prototype. Few if any of these patents offered any concrete solutions to the array of formidable technical problems specific to mail sorting. Many of these patents presupposed very tightly controlled conditions for code reading that could not be realistically created in postal operations.[21] An efficient and economic sorting technology, adapted to the realities of postal operations, would require considerable R&D. Levy was convinced that there was still important patentable territory to claim.[22]

The most novel aspect of Levy's approach to mail sorting was to convert the scanned coded address into binary information and to use a digital memory to store every minute sorting distinction.[23] Previous approaches to electronic sorting had used the analog output signal from scanning to activate the appropriate mechanical sorting gates directly. In view of the complexity of postal sorting decisions and the need to alter constantly the geographic distinctions for sorting, the state of the art was limited, inflexible, and cumbersome. Flexible sorting, in Levy's mind, required that binary address information be compared against routing information in memory. Once a sorting decision was made, the memory could then know which mechanical switches to activate in order to route the letter to the right sorting bin. Memory had the marvelous advantage of being easily updated every time changes in sorting categories were needed.[24]

FEMCO pushed Turnbull and the government to sign a $100,000 contract for a detailed engineering study.[25] Convinced of the need for radical innovation, Turnbull nevertheless still needed more assurance that Levy's approach could produce a usable technology. When Turnbull turned to the National Research Council for guidance, its president, Edgar Steacie, claimed that "almost anything could be done with the aid of electronics."[26] B. G. Ballard, the director of NRC's Radio and Electrical Engineering Division, provided more useful assurance. After a thorough analysis of the Levy-FEMCO proposal, Ballard concluded that the proposal was sound and that such a machine could be developed.[27] Whether $100,000 would suffice to build a prototype depended, according to Ballard, on the sorting speeds required.[28]

Continued criticism of the postal service by the opposition pushed Turnbull to act. Conservative Member of Parliament Joseph Noseworthy emphasized the sorters' inability to keep up with the growing volume of mail.[29] Stress and uncompetitive wages were producing resignations at Canada's largest sorting station, in Toronto. These staff losses only made the problems of moving the mail worse.[30] To counter the growing criticism, Postmaster General Alcide Coté proclaimed to a Calgary audience in July 1952 that the Post Office was going to introduce "electronic marvels to sort the mail, instead of it having to be sorted by fallible hands."[31] The *Calgary Herald* was not impressed. Scientific marvels, the newspaper feared, could do little to "change the three or four days it sometimes takes Mr. Coté and his considerable staff to deliver a letter which is mailed within, and directed to an address within the Calgary city limits."[32] Still, with the electronic sorter now in the public and political arena, Turnbull had to move quickly. But before the Post Office Department could sign the deal with FEMCO, IT&T fired Levy.[33] Turnbull, having come to identify the electronic sorter with Levy rather than with IT&T, dropped FEMCO and immediately hired Levy to coordinate the electronic sorter project from within the Post Office.

Levy immediately pushed to create a research and development team. There was resistance within the postal bureaucracy to such a move, but Levy promised Turnbull that a stable in-house R&D team of four engineers and four technicians would have a prototype ready for commercial manufacture in 3 years.[34] Turnbull approved Levy's staffing requests.[35] To Levy's surprise, Turnbull also insisted that the Post Office immediately approach industry for technical assistance. Turnbull accepted the need for some sort of in-house research group, but in his mind the role of such a group was simply to demonstrate, in a laboratory setting, the technical feasibility of the contentious facets of the electronic sorter. Anxious for quick results, Turnbull wanted to turn industrial experts loose on the challenge of designing, building, and testing a full-scale sorter in Toronto or Montreal. New automation technology, he wrote to the Assistant Deputy Minister of Finance, had become urgent matter, as Canada faced a complete breakdown in the movement of mail.[36]

The central feature of Levy's novel approach to sorting was the use of a binary-based memory. The design of this memory had to be a compromise between capacity and speed. A great deal of memory was needed to hold all the details of forward and city sortation. And if letters were to move through the system at high speeds, the memory would have to make very fast sorting decisions.[37] Under pressure to demonstrate the feasibility of fluorescent or optical reading of encoded letters, Levy was ready to hand over the development of a memory to industry. Optical scanning was pivotal to the project and was the central reason for the creation of his new lab. If the mail-sorting "robot" could not see, memory would be a moot issue.

In early 1953, Levy made the rounds of Canada's electronics industry looking for the right company to design and build the digital memory and associated electronics for the small experimental model of the electronic mail sorter. When he met with Arthur Porter, the head of Ferranti-Canada's digital electronics R&D group, Levy's concept of memory implementation was still quite indeterminate. Porter, whom Levy described to Turnbull as an expert on memory, immediately recast the Post Office's automation needs into the framework of DATAR's technology.[38] Not only was Porter's move an astute transfer of technology; it also played on the high credibility that the DATAR project had conferred on Ferranti-Canada.

For reasons of reliability, real-time performance, and miniaturization, Ferranti-Canada and the Royal Canadian Navy had agreed in 1952 that DATAR's memory would be based on the core memory technology being developed by Jay Forrester at the Massachusetts Institute of Technology. When the RCN accelerated the DATAR program in response to the British Royal Navy's announcement of an early demonstration of its own anti-submarine tactical information system, short-term compromises had to be made. Core memory was still too experimental to use, so Ferranti went with the easier option: magnetic drum memory. From the RCN's perspective, magnetic drum memory had many disadvantages, but it was an expedient that ensured an early demonstration date. With his pitch to Levy, Porter turned this unwanted but necessary compromise into an asset.

Porter pushed Levy to see the proposed electronic sorter as a special-purpose computer built around magnetic drum memory. Like DATAR, the electronic letter sorter would make continual requests to a lookup table in memory. Comparisons with data in the lookup table required immediate decision and action. In the case of DATAR, it was a question of keeping track of many aircraft, ships, and submarines, then displaying their positions within the reference frame of each ship in the escort. The mail sorter had to keep track of each letter that flew by the scanner and then, at the precise moment, activate the appropriate switching gates to send each letter to its sorting track. In DATAR, radar and ASDIC were the "eyes and ears" sending input data to memory. In the mail sorter, a flying-spot scanner would be the "eyes" reading fluorescent bar codes. Porter no doubt argued that a high-speed magnetic drum offered a more than adequate memory for the Post Office. Not only did it offer an acceptable balance between memory capacity and real-time response; more important, the Post Office would be getting access to technology being developed for DATAR.

Postal Automation as an Opportunity for DATAR Technology

By April 1953, Ferranti-Canada seemed the perfect candidate to design and build the "brain" for the small demonstration model of Levy's mail sorter. There were no domestic rivals to Ferranti-Canada's technological approach, and DATAR had given the company's views immeasurable credibility.[39] Levy started by giving Ferranti-Canada a small contract to produce a small experimental drum to be used in the Post Office's own laboratory. Despite this contract, Levy's thinking on memory was still quite malleable. When Louis Ridenour, a highly respected American physicist and a vice-president of the International Telemeter Corporation, proposed the radical idea of photographic (or optical) memory, Levy started to vacillate on the Ferranti-Canada option.[40]

The long-term theoretical potential of optical memory captivated Levy. Optical memory promised potentially far greater storage capacity than Ferranti-Canada's magnetic drum. Levy was also convinced that optical memory would lead to a reduction in digital circuit complexity. If vacuum tube circuits could be made simpler, reliability would increase and maintenance costs would decrease. Ridenour's proposal was appealing but unproven. To pack a lot of information into a small space required that the spot of light be very small. Scanning with such a small light spot, however, raised serious technical problems: the reader would mistake minute scratches and pieces of dirt for data. Considerable R&D was needed to eliminate this problem.[41]

Levy wrote to Ridenour suggesting ways to tackle the problems of optical memory.[42] The length and the detail of Levy's written response attest to the extent to which optical memory had captured his imagination. Levy was inclined to give Ridenour's company a preliminary research contract to investigate the idea further, but Turnbull was "dubious about allotting $10,000 to develop a theory."[43] Turnbull had a more practical response to Ridenour's proposal: If he has such faith in his idea, Turnbull reasoned, then let International Telemeter Corporation pay to produce a experimental version that the Post Office could then test. But Levy persisted. At every occasion, Levy argued that optical memory, with its greater information storage capacity, would be needed for the large urban sorting centers.[44] Had it not been for Turnbull's desire to get a practical system into public view soon, Levy surely would have turned the project into a longer-term R&D effort in optical memory.

In the matter of memory, Levy was forced to reconcile his quest for the optimal system with Turnbull's impatience. Turnbull was anxious for a tangible sign that the idea of digital electronic sorting worked. Furthermore, the public-relations and political value of an early successful demonstration was not lost on Turnbull.[45] The success of a small prototype would announce to the public, to the opposition, and to Turnbull's political masters that the Post Office had innovative answers to the question of moving the mail efficiently and economically. In Turnbull's mind, Ferranti-Canada's magnetic drum approach presented the fastest route to demonstrating the overall effectiveness of electronic sorting.

On 14 July 1953 the Treasury Board approved the Postmaster General's request to "enter a contract involving the expenditure of $56,855.70 [with Ferranti-Canada] for the development of an Electronic Information Handling Equipment consisting of digital circuits and a digital storage system."[46] Despite this intended commitment, Levy continued to flirt with other memory options and manufacturers. Turnbull also expected Levy to nail down the most appropriate technical options for each component of the sorting system and let industry get on with the design and construction. Instead, Levy kept talking about the long-term significance of new developments on the horizon. In early July, Levy went to the United States to meet with representatives of RCA, Bell Laboratories,

IBM, and Remington Rand. One of the main topics of these meetings was "the latest progress in storage memories."[47] Levy's hesitation to freeze the design concept worried Turnbull. "I have quite early in the development," Turnbull reminded Levy, "stressed that after preliminary investigation we would have to decide the direction to take and then go ahead without being diverted by new developments or after thoughts."[48] Levy's persistent reference to the advantages of optical memory prompted an impatient Turnbull to ask "Are we committed to the magnetic drum or will it be photographic memory?"[49]

Levy, trying to reconcile his desire for the optimal technical solution with Turnbull's political and economic need for a speedy demonstration, pushed for two parallel memory efforts. While Ferranti-Canada worked on magnetic drum memory, Levy hoped, Ridenour's group would work on the longer-term problem of optical memory.[50] "By backing the development of this type of memory [i.e., optical]," Levy explained to Turnbull, "we are pioneering in the right direction. . . . Unless new progress is achieved in magnetic drums, I believe the photographic memory will be the best."[51] Levy underscored his request with this warning: "If we do not pioneer in backing this development of photographic memory, we risk to have to use a much more complicated design." Turnbull did not buy Levy's suggestion. On 10 August, Ferranti-Canada formally started work on the contract to develop an "Electronic Information Handling System."

In February 1955, the experimental prototype was unveiled to the world's electronics experts. At a national conference in the United States, Levy proudly announced that the Canadian Post Office Department had pioneered a digital electronic mail sorter that could process 200,000 letters per hour.[52] Based entirely on an extrapolation of the computer's processing speed, this assertion was more an act of hope than a statement about the sorter's actual operation. Levy's optimism concealed a project that was less a harmoniously integrated system than an assemblage of components whose design had not yet crystallized. The problem lay not with Ferranti-Canada's computer but with the scanning, encoding, code printing, mechanical, and sorting technologies—that is, the non-computer aspects of the project.[53] Despite these problems, Turnbull and Levy decided to push ahead with a full-scale prototype for Montreal or Toronto.

No sooner had Ferranti-Canada delivered its vacuum tube electronic brain for the small experimental electronic sorter system than it proposed jumping directly into a transistorized full-scale prototype. The potential technical benefits of building a solid-state computer did not come cheaply or easily, however. Though in 1955 transistors offered the most promising prospect for the reliable miniaturization for large-scale digital circuitry, only military enterprise could justify the high development costs associated with transistorization. Turnbull might well have asked whether a breakthrough in miniaturization was imperative to the success of the Post Office prototype. After all, vacuum tube technology was still the accepted electronic component in all the commercial computers of the day. Not until 1957 did Philco Corporation come out with first solid-state general-purpose computer. Higher costs were not the only drawback of transistorization. The technical problems of moving up the transistor learning curve were bound to introduce additional uncertainty into a project already fraught with technical problems of a non-computer nature.

In view of his concern over costs and his eagerness to get a working prototype out into the field for testing, it is surprising that Turnbull did not go with a vacuum tube machine, which in 1955 would have been the most cost-effective approach. Instead, convinced that the Post Office was getting a bargain, Turnbull, on Levy's advice, went with the Ferranti-Canada proposal. In early 1955, the RCN had just given Ferranti-Canada a contract to develop the circuit techniques needed to transistorize DATAR. The company argued that all the digital transistor know-how gained from the Navy project could simply be transferred to the mail-sorting computer. In August 1955, when the Post Office signed a letter of intent to pay only $65,000 for Canada's first solid-state computer, Levy and Turnbull no doubt believed that their project was getting a free technological ride on military spending.

When the RCN pulled back from DATAR, in late 1955, Ferranti-Canada faced the daunting problem of having to do all the costly transistor circuit development work for the Post Office computer, plus manufacture and test it, all for $65,000. To make matters worse, Ferranti-Canada's decision to base its computer design on the Philco surface-barrier transistor was proving problematic. Ferranti-Canada had overestimated the maturity of Philco's SB-100 and Transac circuits. Considerable variations in Philco's transistor made the Transac circuit unsuitable for reliable high-performance applications in computers.[54] Making the SB-100 transistor work proved far more difficult than had been expected. The difficulties were already manifest while the company was working on DATAR. However, Ferranti-Canada never revealed any of these difficulties when it pitched its transistor computer proposal to the Post Office. Fearful of losing a contract opportunity, Ferranti-Canada presented a confident face to Levy on the matter.[55] But the budget was less easy to gloss over. Ferranti-Canada's engineers quickly ran up against the financial limits of the contract's $65,000 budget. By early 1956, Ferranti-Canada had exhausted most of the $65,000 contract allocation and was still a long way from a final product as the summer deadline approached.

In the months that followed, Ferranti-Canada went back to the Post Office on several occasions to ask for more money. The Post Office had little option but accede. It had already invested a substantial amount of time and money on both the computer and the non-computer aspects of the mail sorter. Moreover, with the publicity surrounding the electronic mail sorter and the opposition's call for greater postal efficiency, there was political pressure on Turnbull to show that the Post Office could indeed modernize. The opposition insisted on knowing the state of Turnbull's electronic miracle. Without the computer there could be no sorter, and by 1956 it had become unrealistic to think that any other company could replace Ferranti-Canada.

Though the details of how Ferranti-Canada rationalized its requests for more money are not available, it is clear that the company had kept Maurice Levy in the dark about the problems it was having with the transistor circuits. But when, in August 1956, the cost of the contract had reached 300 percent of its original value and there was still no clear completion date in sight, the Post Office demanded an explanation. Ferranti-Canada's principal computer designer, Fred Longstaff, was forced to reveal that the company still could not get the necessary reliability and performance from its Transac circuit approach. Ferranti-Canada had therefore decided to embark on its own design approach to digital circuits, one that would overcome the manufacturing limitations associated with the SB-100. In addition to having to develop and test entirely new circuit techniques, abandoning the Transac circuit meant that Ferranti-Canada's previous circuit-board layout had to be rethought completely. The new circuits were bigger than the Transac, and difficulties encountered in trying to cram everything onto boards produced for smaller circuits.[56]

Delays and cost overruns, which plagued both the computer and the non-computer aspects of the Post Office's electronic mail sorter project, put Levy under a great deal of strain.[57] Levy's excuses only provoked Turnbull, who by now was exasperated at the slow pace of his project. "What I want," Turnbull wrote, "is action not criticism of other people who are trying to help us put of a situation which is becoming embarrassing."⁵⁸ Turnbull's commitment to the automation project was starting to hurt him politically. He realized that his Conservative adversaries would call for heads to roll unless the project was finished soon. "You have spent one million dollars and taken three years," he complained bitterly to Levy, "and so far we have nothing to show for this except a lot of bits and pieces, none of which seems to be finalized."⁵⁹

In August 1956, William Hamilton, a Conservative critic of postal operations, demanded that a parliamentary committee be allowed to inspect the progress of the electronic sorter project.⁶⁰ Hamilton reminded Hugue Lapointe, the new Postmaster General who had inherited Turnbull's project, that in the spring Lapointe had promised the House committee examining the Post Office's budget estimates of the Post Office that the project would be finished by the end of the summer. Hamilton explained that he was not against "modernizing many functions of the post office," but he wondered: "When is this million dollar monster going to get into operation so that we can see it in use?"⁶¹ Lapointe clumsily responded that the delays were due to the Post Office's burden of having to advance postal research singlehandedly in Canada.⁶² His prediction that "a machine will be installed in Ottawa by the end of the year" turned out to be wishful thinking, as Ferranti-Canada had to ask for additional money and time in November 1956. Levy warned the company that this was the last time the Post Office would go to the Treasury Board to amend the contract. Fortunately, Ferranti-Canada's "Route Reference Computer" was delivered in January 1957.⁶³

In the summer of 1957, the Post Office demonstrated its prototype electronic sorter at the annual meeting of the Universal Postal Union Congress in Ottawa. In its long history, the Universal Postal Union Congress had never before met in Canada. For Turnbull, the meeting represented a crowning moment in his career with the Post Office. The fact that a completed prototype of the electronic sorter could be demonstrated to this international gathering made the moment that much sweeter. But his taste of the electronic sorter soon turned bitter. The cost of the "million dollar monster" had now grown to $2.5 million. The Conservatives had come to power, and William Hamilton, the former Post Office critic, was now the Postmaster General. Unable to work with the Conservatives, Turnbull quit as Deputy Postmaster General within the year.

Canada's electronic mail-sorting system received international attention. Britain and Germany sent delegations to Ottawa to see the new technology in operation, and group of US Congressmen came to examine the Canadian project.⁶⁴ What the Americans saw no doubt facilitated

Congress's approval of $5 million for the establishment of a research laboratory to look into the computerized automation of postal sorting. But as the United States invested heavily in the computerizing of the movement of mail, enthusiasm for the idea seemed to wane in Canada. It is tempting to conclude, as the historian J. J. Brown did, that the demise of the Post Office's computerized sorter project was another example of "a melancholy procession of golden opportunities which we [Canadians] have let slip through our fingers . . . because we have not the vision to see their potential."[65]

Armed with a small amount of anecdotal evidence, Brown concluded that the newly elected Diefenbaker government, by killing this pioneering effort, had deprived Canada of some $300 million a year in exports. Despite his self-admitted lack of evidence, Brown puts the blame on the Conservatives. For Brown, the collapse of the electronic sorter, which happened at about the same time as the AVRO Arrow's demise, was another illustration of how Canadians "tend to be terrified little men, clutching frantically at what we have and afraid to take any risk whatsoever even for large rewards."[66] Brown's simplistic assessment naively assumed that invention is the better part of successful innovation and glossed over the roles of Turnbull and Levy in the failure.

The Conservatives did not intend to kill the project. The new Postmaster General, William Hamilton, wanted to implement the technology. He inspected the prototype in early August 1957. At that time, Levy again boasted that the prototype could be ready for commercial production in a little more than 6 months. With his expectations raised, Hamilton advised Turnbull: "I would like to see this in practical operation in Post Office work as soon as possible."[67] To accelerate the deployment of this technology, Hamilton was willing to increase the size of Levy's laboratory; he was even willing to consider higher contract fees.[68] But Levy's prediction of the completion date, which was no doubt an effort to garner support from the new Postmaster General, was ill conceived, because it concealed all the problems that still remained before all the components would be ready for commercial production.[69]

Though Hamilton had warned Turnbull in August 1957 that he would not tolerate the litany of postponements that had marked the project to date, work on the prototype dragged on until the project faded into obscurity and Levy's Electronics Laboratory was finally shut down.[70] The failure of Canada's computerized mail-sorting project was not due to some flaw in the Canadian psyche, as Brown intimated. There was no ingrained "pre-adolescent" Canadian fear of innovation. To the contrary, there was courage and daring. What the project lacked was good management and organization.

The electronic sorter was a costly and complex undertaking. Its potential impact on postal operations was far reaching. Yet Turnbull and Levy managed this project in "seat of the pants" fashion. There was little if any consultation across Post Office management to help guide the innovation process. No detailed cost-benefit analysis was undertaken. No one presented a cogent analysis of the estimated costs of seeing the development through to completion. No intra-departmental committee was set up to monitor the project's progress. The Post Office never evolved an explicit departmental policy or procedure to assess the merits of industrial contract bids. From the outset, Turnbull isolated Levy and his group from the rest of the Post Office.[71] Levy was to report to no one else but Turnbull. Did Turnbull believe that internal opposition would sabotage the project if given a chance? Did Turnbull's approach reflect an institutional structure and ethos in which consultation had little or no part? Was it simply a manifestation of Turnbull's personality? Whatever the reasons, Levy was left alone to navigate a bewildering number of competing interests.

In the end, Levy's and Turnbull's mismanagement reflected the absence in the Post Office of a corporate culture of innovation that could have provided the institutional structure to manage radical technological change. Though the electronic mail sorter project never translated into direct automation gains for the postal service, it served as a crucial opportunity for Canadian industry to expand its competence in digital electronics. The design and construction of the "Route Reference Computer" allowed Ferranti-Canada to be the first Canadian company to jump onto the wave of innovation in solid-state digital electronics. Whether this experience also provided competency that would help the Post Office to deal with the issues of automation over the next 20 years is an open question.

Trans-Canada Air Lines, another public enterprise, turned its ambitions in digital electronics R&D into an unqualified success. The key difference between the Post Office and TCA's patronage of R&D was the technological sophistication of the latter institution's senior management. Though TCA's management of radical technological change differed in significant ways from the Post Office's approach, both shared a common impetus in their pursuit of digital electronic solutions to automation: the political pressure to lower the cost and increase the quality of service. For the Post Office, it was to move the mail; for TCA it was to move people.

Moving People

State Monopoly and Competitiveness: Trans-Canada Air Lines and the Computerization of Reservations[72]

By 1937 Canada's regional airlines had linked metropolitan centers to their regional hinterlands, but Canada had yet to forge a nationally integrated air service. Clarence Decatur Howe, Canada's first Minister of Transport, was appalled that "many Canadians when travelling from one point to another in Canada find that they have to use airlines in the United States."[73] To the degree it siphoned the trans-Canadian flow of people, goods, and services, the emergence of American transcontinental air service weakened Canada's east-west economic and political axis. As one example, Howe suggested that the US air transportation network undermined the Canadian postal service. Air-mail stamps were sold in Canada, yet, Howe explained to Parliament, most domestic Canadian air mail was routed through the United States.[74] For Howe, the absence of a scheduled transcontinental air service clearly stood as a significant obstacle to national development. Public monopoly was the Canadian government's institutional response to the revolutionary change in transportation unleashed by aviation technology. In 1937, the Liberal government created Trans-Canada Air Lines as a wholly owned subsidiary of Canada's government-owned railway, Canadian National Railways.[75]

The period 1945–1965 was a period of remarkable growth for commercial aviation and for Trans-Canada Air Lines in particular. Advances in aviation technology sparked by World War II sustained high economic growth, and a vast land area helped to heighten the demand for air services. TCA responded to the soaring postwar demand by investing heavily in fleet modernization. This progressive response reflected the technology-driven brand of management that came to characterize TCA's management during the 1950s and the 1960s. A transcontinental service to span Canada's vast landscape needed new aircraft that could carry more passengers at higher speeds and with more comfort. TCA's older DC-3s were progressively replaced by 51 Viscounts, 10 Super Constellations, 20 Vanguards, 6 DC-8s, and some North Stars.[76] TCA's management bragged that the purchase made the airline one of the first in the world to use pressurized equipment.[77] TCA embarked on a program to convert its entire fleet to turbine power. In 1955, with the purchase of the British Viscounts, TCA became the first airline in North America to use turbo-propeller equipment.[78] In 1957, Gordon McGregor, the president who had overseen TCA's postwar growth, proudly told the Minister of Transport: "On the basis of orders now in the hands of manufacturers, TCA will be the first airline in the world to operate an all-turbine powered fleet."[79] Fleet modernization was a costly undertaking. For the period 1956–1961 alone, TCA had committed $100 million to it.[80] TCA's modernization program, which enabled it to multiply the number of flights, greatly expanded the airline's capacity. Equally important, TCA's capital investment gave it greater flexibility in maintaining a profitable mix of long-, medium-, and short-range routes. As the size and the flexibility of TCA's carrying capacity increased, so did the airline's ability to tap into the exploding market potential: in 1945 it had carried 183,121 passengers[81]; by 1960 it had carried 3,440,197.[82]

As is often the case, technological improvements in one sphere of an industry's operations accentuated the technological inadequacies in other spheres. TCA's growth concealed an important weakness in its sales operations that, in the long term, threatened to vitiate the airline's costly investment in carrying capacity. The problem was TCA's manual reservation system. Seats are the commodities that airlines produce and sell. The profitable clearance of this seat inventory rests on the speed, accuracy, and cost of the reservation system. Through the modernization of its "production machinery" (i.e., aircraft), TCA was able to "manufacture" seats that offered the consumer a greater variety of destinations and more flights between any two destinations. But could seats be sold quickly?

TCA's reservation system consisted of a long chain of human links between the ticket agent and the seat inventory. A passenger wanting to book a seat first called a local reservation office, where an agent would book the seat on a "sell and record" system. Agents were permitted to sell tickets on any flight until told otherwise. When a booking was made, the local ticket agent filled out a reservation card. One copy of this card was sent by conveyor belt to the teletype office, which then communicated the booking to the Central Space Control Office in Toronto. Another copy of the card was by another conveyor belt to the filing office. Clerks at the Central Space Control Office tallied the incoming bookings from all the TCA ticket offices. In addition to written records, the Central Office kept track of seat availability by means of a large Visual Space Indicator Board (VSIB) on which the availabilities of flight legs (open, closed, or short-haul closed) were indicated by disks of various colors. As soon as a flight was full, a red disk would go up in the appropriate spot on the Central Control Office's VSIB and a "Stop Sale" order would be sent by teletype to ticket offices across Canada. Local reservation offices had their own smaller VSIBs, where updates from the Central Office were also indicated by appropriately colored disks.

By 1953, TCA had concluded that its manual reservation system was increasingly unable to provide ticket agents with the information needed to sell seats in a timely manner. The dependence on so many human links (reservation agents, file clerks, teletype operators, dchart operators), and on relatively slow communications, created numerous errors. Often sales were not stopped or opened in time. If a flight was full, one cancellation could open up opportunities for bookings in many of the ticket offices across Canada; yet, by the time this cancellation reached the Central Office and was disseminated to all offices, the opportunity was easily lost. The converse was also true; agents took reservations before they learned of a "Stop Sale" order.

Half of the seat sales and most of the cancellations happened in the days immediately before flight departures. This high volume of booking requests over such a short period made the manual reservation system extremely unwieldy. This inflexibility translated into irrecoverable financial losses, for the seats in an airline's inventory were (and are) highly perishable and not easily interchangeable.[83] If there was an empty seat on a plane when it took off, that particular seat and the revenue it was supposed to generate were forever lost. "No-show" ticket holders further underscored the reservation system's increasingly sluggish response time as the volume and complexity of business grew. A study commissioned by TCA pointed out that the existing reservation system made it practically impossible to reallocate "no-show" seats in the short time just before s flight's departure.[84] Overbooking, an antidote to empty seats, created angry customers and fueled politically dangerous hostility to TCA's monopoly status in most markets.

During Gordon McGregor's tenure, TCA attempted to wrest more control of major corporate operational planning decisions from political interference. Until 1948, when McGregor had become TCA's first full-time president, Canada's national air carrier had been managed more as an appendage of the government and of Canadian National Railways (CN) than as a self-contained business entity with its own internal needs. "All major [management] decisions," according to one McGill University study, "were made by government or CN officials. [TCA's] presidents, until 1947, were, in fact, the presidents or directors of the already large CN, with only limited time available for the small, fledgling airline. The CN also provided many of the airline's corporate functions in these years."[85] Furthermore, from 1937 into the 1950s, all TCA's important management strategies were closely formulated by C. D. Howe.[86]

The new and all-encompassing pursuit of techno-economic efficiency became the foundation on which TCA built its autonomy strategy.[87] Shortly after McGregor was appointed president, Howe advised him: "Keep your hands out of the taxpayer's pocket, and I'll keep the politicians off your back."[88] McGregor understood the message well: As long as TCA was not a financial or a political liability, the government would provide the airline with a protective domestic market environment and would refrain from interfering in its strategic planning. A commitment to "engineer-driven" efficiency was the manner by which TCA's management sought to guarantee autonomy and all the privileges of monopoly.[89] By defining the good of TCA in terms of technology-driven efficiency, the technical experts, rather than the politicians, would be best placed to manage the public airline. McGregor and his executives argued that they, "the only ones to understand both markets and technologies relevant to the firm's business," could serve the public's interest best.[90] While TCA argued that "any interference from the [government] bureaucracy was dysfunctional because it lacked competence," TCA still looked to the government to maintain its monopoly advantage.

The reservation problem not only compromised efficiency; it also fueled the kind of bad publicity that threatened TCA's monopoly status because, as TCA executives understood, the "secret to success within the government [was] to avoid the attention of the politicians."[91] Frank Ross, who sat on TCA's board of directors, wrote in 1954 warning McGregor of the public's growing dissatisfaction with TCA's service and anger over deplanings arising from failure to reconfirm or from overbookings. "The greatest immediate danger of a public demand for competition," Ross asserted, "will come from dissatisfaction with our method of handling traffic."[92] On the issue of passenger anger over "deplanements" McGregor rationalized: "However much it is disliked, reconfirmation is a rule imposed throughout the continent to safeguard airlines as much as possible against empty seats resulting from 'no-shows.'"[93] Furthermore, McGregor downplayed the extent of customer dissatisfaction. Ross, a Vancouver resident, took exception to McGregor's suggestion that the customer dissatisfaction was an isolated phenomenon.[94] Ross warned that problems with service would only feed resentment that existed in western Canada over TCA's monopoly.

The seeds of western Canadian hostility toward TCA had been planted when western entrepreneurs were shut out by the creation of TCA. In 1937, when Howe contemplated the creation of TCA, western-based Canadian Airways was slowly positioning itself as a national service. This private carrier represented the formal merger of the successful Western Canada Airways with four struggling eastern airlines. The consolidation, however, had not yet produced an integrated east-west service when Howe called for the creation of a national airline. Despite Canadian Airways' willingness to participate in plans to create a transcontinental air service, W. L. Mackenzie King's Liberal government was determined from the outset that any national airline would have to be a public enterprise. The ensuing debate over private versus public air transport mirrored the struggle that pitted Canada's eastern-run, government-owned airlines against its western, private-owned airlines.[95]

From the time Canadian Pacific Railways established Canadian Pacific Air Lines (CPAL), it had tried to break the TCA's legislated eastern-headquartered monopoly over transcontinental service. At every turn, TCA blocked any CPAL route application that threatened TCA's interests. The Liberal cabinet's rejection in 1952 of CPAL's application for transcontinental service fueled western disaffection with the state's control over a significant portion of commercial air transport. McGregor dismissed anti-TCA sentiments as a conspiracy led by a group of disgruntled western "private enterprisers" who were angry at the government's recent rejection of CPAL's application. This vocal dissatisfaction, McGregor believed, was not representative of the public's satisfaction with TCA's service.[96] McGregor's dismissive attitude incensed westerners. The mayor of Calgary, D. H. MacKay, told the president of Canadian National Railways: "McGregor starts off with the promise that nobody has any business to complain and besides there is nothing to complain about. He simply will not make any investigation on the ground and is content with alibis. He doesn't seem to realize that the important thing is that people in this area are dissatisfied and fed-up and will do everything possible to break the monopoly so that we can get a competitive service to put [TCA] on its toes."[97]

If McGregor was dismissive of Ross's observations, surely he must have sat up and taken notice when his boss, Donald Gordon, chairman and president of Canadian National Railways, was caught completely off guard by the intensity of anti-TCA feelings among the western business and political elite.[98] What was to be a good-will luncheon for the president of CN, attended by the mayor of Calgary and some prominent members of city's business community, turned into a scathing condemnation of TCA. No one questioned the high quality of TCA's service in the air, but everyone objected to its service on the ground. Many of the complaints could be traced back to inaccurate bookings and frustrating delays in getting confirmations. Once again, westerners were charging that TCA's monopoly "produces indifference to the public needs and an absence of competitive spirit." Gordon went on to note that "allegations were made that many Canadians in Western Canada, and particularly on the B.C. coast, were turning to the United States because of exasperation with TCA ground service."

In 1956, the western-fueled criticism that TCA's monopoly was at the root of its poor service spilled onto the floor of the House of Commons. At every opportunity from March through August, the Conservatives attacked TCA's monopoly status during meetings of a parliamentary committee examining transportation policy. They characterized the airline's monopoly as a socialist evil that undermined efficient service to Canadians. For the Conservatives, the political weak points in TCA's monopoly were pricing and ground service. Not even the Conservatives could reproach TCA's engineering-driven use of the best aviation technology. Though TCA may have been the "best in the air," the Conservatives proclaimed TCA "worst on the ground."[99] Again, deplanements and problems with TCA's ability to monitor seat availability were the common complaints. "I cannot understand," exclaimed Diefenbaker, the leader of the Conservatives, "the dog-in-the-manger attitude of Trans-Canada Air Lines in continuing an attitude antipathetic to competition. . . . The time has come to take the hobbles off air development in Canada to remove the shackles of outgrown policy that denies competition, and to give Canadians those benefits that competitive enterprise will assure."[100]

While opponents bemoaned the lack of competition, McGregor saw government backed monopoly in aviation as an economic advantage for Canada. In his mind, TCA was "a protector of the public interest, rather than . . . a money-hungry business." The lack of competition, he argued, "actually benefited consumers."[101] McGregor's justification of monopoly rested on the notion that TCA was another form of public infrastructure. McGregor asked rhetorically: "Who would think of having a municipality with two sources of lighting supply, two sources of sewage disposal or two sources of garbage collection?" "It doesn't make good economic sense" was his reply.[102] To the Royal Commission on Canada's Economic Prospects, McGregor again repeated that under the umbrella of monopoly TCA had not only maintained the highest pace of fleet modernization in the world but also offered the consumer the widest possible network of flights.[103] Though McGregor could say that "no airline in the world [was] better equipped with aircraft for the type of service to be operated,"[104] he was well aware for several years of the growing inability of TCA's reservation system to keep pace with the airline's rapidly expanding business.

The modernization of the seat-inventory system had not kept up with that of the fleet.

At first it seemed that the only way TCA's existing reservation infrastructure could respond to the increasing demand for passenger service was through labor-intensive expansion. Throughout the early 1950s, TCA executives watched with "alarm" as the number of reservation personnel grew proportionately faster than the number of passengers carried and the cost of the entire reservation infrastructure grew faster than passenger revenues.[105] It became apparent to McGregor that the existing manual reservation system not only hampered current operations but would soon hobble the company's service and profitability.

To the extent that the reservation system underlay weaknesses in TCA's ground service, TCA's critics were mistaken to believe that poor service was a product of monopoly alone. The inadequacies in TCA's manual reservation systems typified the state of the art throughout the world's airlines.[106] By the mid 1950s, only American Airlines had actively sought a radical technological solution to the reservation problem.[107]

In 1949 American Airlines had started work on the "Magnetronic Reservisor," a small digital prototype that used a magnetic drum memory to handle seat confirmations. No sooner had American Airlines put its Reservisor system into limited operation (in 1952) than it realized that it needed a better solution. The absence of a general-purpose computer severely limited the scope of information processing, and better input/output devices were needed to make the ticket agent's interaction with the system much easier. Finally, the associated communications infrastructure did not provide a nationally integrated information-handling system. In 1953 American Airlines began extensive discussions with IBM to develop a far more advanced real-time seat-inventory system that could overcome these limitations.

Like American Airlines, TCA became convinced that incremental innovations could no longer alleviate the growing bottlenecks in its reservation system. TCA set out to avoid, with one technological leap, all the deficiencies of the Reservisor system. When TCA opted for radical innovation, two historical threads became intertwined: TCA's growth as Canada's national airline, and the emergence of a domestic industrial capacity to design digital information systems. While American Airlines and IBM started to collaborate on the design of the first computerized reservation system, TCA turned to Ferranti-Canada. What ensued was a race to develop the world's first fully computerized real-time passenger information system. American Airlines' role may be symbolic of the risk-taking virtues and foresight of competitive private enterprise, but TCA's initiatives reveals that Canadian public enterprise could be equally daring and visionary in the defense of monopoly.

Despite McGregor's contempt for his western critics, he knew that his argument about the advantages of monopolistic air service would ring hollow unless TCA met the public's rising expectations for comfortable, fast, flexible, cost-effective transcontinental service. To this end, TCA had to create a supply of seats to match the demand, had to run an efficient reservation infrastructure to convert a perishable inventory into profits, and had to develop an operations-research approach to long-term plans based on accurate forecasting methods. The fleet-modernization program addressed the supply issue. The second imperative dealt with sales and with timely liquidation of the seat inventory. The third imperative represented the "engineer's" systems approach to techno-economic efficiency. As with the aggressive aircraft-modernization program, the proposed daring computerization of its reservation system was an important weapon in TCA's efforts to protect its transcontinental route from its western rival.

Shaping Design to Meet Corporate Needs
The long path to radical technical change was characterized by TCA's methodical scrutiny of how the process of computerization should be structured to meet corporate expectations and priorities. At a meeting in early 1953, Gordon McGregor and his senior officials decided that the time had come to pursue a new approach to reservations. Because automation meant venturing into uncharted organizational and technological waters, McGregor agreed that the undertaking should proceed methodically in four very distinct phases. The first phase, to start immediately, was to be a preliminary internal analysis intended to define the problem and to formulate, in very general terms, the framework for automation. The second phase was to be a proper engineering evaluation by an independent consultant. Only after TCA had thoroughly digested the second phase did management plan to move to the third phase: discussing the technical and financial details of designing and constructing the new system with prospective computer manufacturers. The fourth phase entailed the system's installation and operation within TCA's organizational structure.[108]

In 1953, responsibility for the TCA internal study fell to Lyman Richardson, a young radio engineer in the airline's Telecommunications Division who had received his first training in radio technology in the

Royal Canadian Air Force. Hired by TCA in 1949 as a transatlantic radio operator, Richardson had risen to "communications engineer" by 1953. He found that the "majority of the manual operations in the reservation system were simple in nature and could be precisely defined by mathematical logic in such a way that they could be reduced to simple machine operations and done far more expeditiously with fewer personnel."[109] Although Richardson had never heard of computers when he started his preliminary analysis, in October 1953 he advocated the creation of a digital communications and computation network on a scale never before contemplated.[110] The proposed new electronic system was to encompass most of the simple manual operations needed to sell a ticket. Agents in TCA offices across North America, according to Richardson, would be linked interactively, in real time, to a centralized seat inventory kept in a computer.

In 20 TCA ticket offices across Canada and in the United States, agents would do all booking through a special input/output device, which Richardson simply called an "Agent Set."[111] "Agent Sets" in the same ticket office were to be routed into a "Local Store," a kind of buffer to keep and route information to and from the "Agent Sets."[112] All the geographically dispersed "Local Stores" would then pass on and receive seat information to and from a central computer that kept the entire seat inventory.[113]

Richardson's report served its purpose well. It sketched, in broad strokes, the nature of the reservation system TCA should pursue. The second phase called for more concrete evidence that a cost-effective approach to the Richardson recommendation actually existed. TCA solicited bids from seven companies. The objective of the contract was to elaborate, without going down to the level of circuit design, on the equipment configuration that would best meet Richardson's requirements. Only then, believed TCA executives, could they "definitely decide whether or not an automatic reservation system [was] justified from the viewpoints of efficiency and economy."[114]

Of the six companies TCA approached in the fall of 1953, only three bid for the contract to do the technical study: Ferranti-Canada (Toronto), Computing Devices of Canada (Ottawa), and Adalia Ltd. (Montreal).[115] Ferranti-Canada jumped at the opportunity to do the technical study.[116] To the digital electronics group at Ferranti-Canada, Richardson's proposal was tailor made to the achievements of DATAR. Richardson's proposal called for the design of a real-time data-exchange system that would link geographically dispersed users to a central information repository through a digital communications infrastructure. Like DATAR, TCA's search for an automated passenger service called for the fusion of computation, communication, and human interface considerations into one large information system.

After some reflection, senior management recommended to McGregor that it would be unwise to entrust this crucial technical study to one of the companies that would be bidding to develop and manufacture it. They feared that Ferranti-Canada or Computing Devices of Canada would "consciously, or unconsciously, promote [its] own preconceived solutions" instead of what best suited TCA's operational needs. Furthermore, TCA's management was concerned that, with a vested interested in building the system, these two companies would not do the "statistical analysis, informational analysis and operations research" TCA needed to help assess the cost-effectiveness of automation. In the end, the contract was given to Adalia Ltd. (a consulting firm, created after World War II by Sir Robert Watson-Watt, one of the pioneers of radar, that had no intention to enter the manufacturing field).[117]

In 1953, working knowledge of digital electronics, computers, and information processing was still a rare commodity in Canada, as elsewhere. Ironically, had it not been for the collapse of Canada's first effort to design and build a general-purpose digital electronic computer (at the University of Toronto), Adalia would never have had the know-how to tackle TCA's technical study. Recall from chapter 1 that, after plans to build a full-scale prototype of UTEC had been cancelled, the design team of Josef Kates, Leonard Casciato, and R. F. Johnston eventually went to Adalia.

The Adalia study by Kates, Casciato, and Johnston filled in the technical details of a system that could meet the requirements in Richardson's report.[118] The report analyzed the data flows associated with processing, storing, and updating information in the seat inventory. It also examined what communications capabilities were needed to ensure easy and reliable real-time access to the inventory. And, at a time when all aspects of computer design were still very much in flux, the report examined the various technical options for the design of the central computer. Adalia's study offered a technically informed corroboration of Richardson's model for automation. After their theoretical endorsement of electronic reservation, Kates, Casciato, and Johnston arranged to simulate the operation of an electronic seat-inventory system on the University of Toronto's FERUT, which at the time was one of only two general-purpose computers in Canada.[119]

None of TCA executives who gathered on the morning of 25 August 1954 to watch the demonstration had ever seen a computer before. They watched with fascination as FERUT stored the airline's entire seat inventory, updated the inventory continually as simulated bookings were made, and instantly revealed the status of every seat on every leg of TCA's flight schedule. After this demonstration, lengthy and meticulous internal analyses were carried out before TCA would commit money to the third phase: design and construction of the computer system. This detailed, systematic approach to innovation reflected the post-World War II emergence of TCA's new management ethos, which put technical efficiency and innovation at the core of the company's business strategy.

Although the Adalia study had confirmed the engineering feasibility of Richardson's concept in August 1954, it was not until the spring of 1959 that TCA solicited offers from industry to design and construct the computer. TCA needed 5 years to resolve two issues that were critical to the overall design philosophy of its computerization. The first, and perhaps the most immediate, was the human-machine interface. Could one design an input/output technology that made access to information a fast, reliable, transparently simple act for the ticket agent? Because of the large number (about 300) of Agent Sets that TCA wanted to install across the country, the price per unit for the human-machine technology became another important constraint. The second issue was the role of electronic automation in the broader context of TCA's future data processing needs. Should automation simply produce an electronic seat inventory that could be queried and updated in real time, or should it also serve to promote TCA's integrated system's approach to management? The prospect of building a computer provoked an examination of how this new technology could improve the quantity, the quality, and the speed of information flow among all TCA's interdependent operations so as to maximize efficiency, quality of service, and hence profits. The issue of introducing the computer into TCA operations engendered more than technical debate. It also touched on questions of power and control. Inter-departmental squabbles over who should manage this new technology had to be resolved. How all the above issues were resolved had direct and indirect effects on the detailed final design concept that TCA eventually put out to industry for development.

TCA turned the development process on its head. Rather than start with the biggest and most complex piece of hardware (the computer), TCA first focused on the human-machine interface. For TCA's management, the way the ticket agent communicated with the computer was the pivotal design issue.[120] An economical interface had to be created that would allow for easy and fast but economical and high-volume use of unscheduled interactions among numerous users dispersed over a vast landscape using one central computer. This was a truly ambitious goal in view of the fact that, with the exception of secret military projects such as Royal Canadian Navy's DATAR and the US Air Force's SAGE, real-time interactivity in an electronic digital information system with distributed users was still unexplored territory.[121]

The dominant paradigm for a human-machine interface during the 1950s was batch data processing. Batch-oriented card readers, punched tape, and magnetic tape did not offer users the opportunity to seek information from the system on an immediate feedback basis. Depending on the system load, the user would come back in an hour, in half a day, or the next morning to collect the results. Flexowriters (which resembled teletype machines) and printers offered more direct interaction between the user and the computer; however, usually the system had only a few of these input devices, and they resided in a centralized location to which only the system managers had direct access.

Uppermost among Richardson's priorities was the creation of an interface technology that would adapt to the needs of the ticket agent rather than vice versa. In his mind, the existing flexowriter-printer interface was not only an uneconomical option; it was also a cumbersome barrier between the ticket agent and the prospective customer. Richardson convinced his superiors that TCA could develop a relatively inexpensive interface that could be simply and naturally integrated into the ticket agent's existing patterns of work.[122] How? Richardson observed that ticket agents, as they talked to customers over the telephone, scribbled booking request information on pieces of paper. Why not somehow make the fast and simple act of scribbling the basis for a new human-machine interface? To further ensure the simplicity and speed of the interface, Richardson wanted the same piece of paper to convey the answer to the ticket agent's booking query. In 1956, TCA solicited proposals from the electronics industry to develop hardware that would embody Richardson's interface philosophy.

Of all the companies that bid, the director of TCA's Telecommunications Department noted, "Ferranti [Canada] of Toronto have made us by far the best proposal both from the technical and economic aspects."[123] Ferranti-Canada's proposal offered the "Transactor," a compact invention ingenious in its simplicity. All the information needed to update the seat inventory or to make any booking request was depicted on pre-printed, standard-size IBM-type cards. Every possible piece of information was laid out within a 300-element matrix. During the course of a telephone enquiry, the ticket agent, with great ease and speed, would translate the customer's reservation request into short pencil strokes that went through the appropriate pre-printed entry. For example, if the flight booking was to be on a Monday, a pencil mark was put through the word "Mon." Then, while still on the telephone with the customer, the ticket agent inserted the marked card into the Transactor through a neatly designed neck at the top. A movable plate would then, with 400 pounds of force, automatically press the card tightly against a reading plate. The reading plate consisted of a matrix of 300 sensing heads that fitted exactly over every possible entry on the pre-printed card. Each sensing head was thus responsible for detecting the presence of a pencil mark over only one spot on the card.[124] Within seconds, the information on the card was sent to the computer. And if the routing equipment and the computer were designed properly, the ticket agent would get an answer back in 4–5 seconds.

The Transactor's simplicity was twofold. First, the use of a static 330-head reading mechanism eliminated the need for complex and more costly designs that involved either cards moving past conducting brushes or optical scanners. Static sensing offered a cheaper approach to good reading reliability. This placed less stringent requirements on how well the ticket agent marked a precise area. As a result, data entry did not become an obstacle to efficient service while the ticket agent gathered information over the telephone from the customer.[125] The second facet of the Transactor's simplicity resided in how information was conveyed back to the ticket agent. The same input card served as the output. On the edge of the card, in pre-printed form, were all the answers the system could convey. To indicate the computer's response, the Transactor used one of nine pins to punch out a mark next to the appropriate pre-printed answer. In this way the need for complicated and expensive printing facilities in the interface was eliminated.[126]

The Transactor card represented a compromise between the costs of design and manufacture and optimal performance. The temptation to incorporate a large variety of input data had to be balanced by the cost of designing more expensive input systems. The economic and interface philosophy put a constraint on the diversity of information that could be transmitted from the ticket agent to the system. Richardson's challenge was to distill and then organize from all the operational information flows the data elements needed to run the entire reservation and sales operations efficiently and fit them onto the 300-element Transactor card.[127]

The Transactor's hardware simplicity was made possible by innovations in how operational information was organized and embodied on the card. The final Adalia report had suggested viewing automation as part of an integrated "Hybrid System."[128] "Consideration of the interdependency that exists between the reservations and scheduling structures," Lyman Richardson argued in a memorandum to the interdepartmental Automatic Reservations Committee, "[leads] to the conclusion that the most satisfactory automatic reservations system for use in TCA would be one incorporating a facility known as feedback."[129] Thus, the huge quantity of valuable data generated during the reservation process could be gathered easily, assimilated rapidly, and then used to improve forecasting of load demands, scheduling of traffic, long-term resource allocation, and evaluation of performance. Richardson's recommendation tapped into the TCA corporate ethos, which stressed efficiency.

On 24 October 1957, on the heels of its difficulties with the Canadian Post Office's electronic automation project, Ferranti-Canada demonstrated the effectiveness of the Transactor interface to TCA executives.[130] Ferranti-Canada designed and built six Transactors, the "distribution unit" that polled the Transactors, and the communications interface to the University of Toronto's FERUT computer, onto which TCA's entire flight schedule had been loaded. From the Toronto's Royal York Hotel, six ticket agents, each equipped with a Transactor, independently queried and updated TCA's seat inventory in real time. The performance of the Transactor convinced the Telecommunications and Sales Departments to prepare detailed design requirements, budget estimates, and cost-benefit analysis for a national system that would join TCA offices in 39 cities in Canada and the United States into an integrated digital electronic system for selling seats. Another year was needed before these two departments would complete their analyses.

The design requirements and advantages of an electronic reservation system had been studied thoroughly with respect to all the details of TCA's sales and scheduling operations.[131] Action on automation was now urgent. Not only had the manual reservation system become unwieldy and slow; it was now becoming too expensive to maintain. By eliminating most of the human links between the ticket agent and the seat inventory, an automatic system would cut the salary costs associated with reservations by at least fivefold. Even with the cost of the automated system factored in, the company would save $557,674 in the first year and more than $900,000 in the second year.[132]

The most expensive item in TCA's estimated $3.2 million automated reservation project was the computer. The new computer system for reservations represented a spectacular increase in TCA's computational facilities. TCA had a small vacuum tube IBM 650 computer, which handled the airline's administrative, financial data processing, and computations for the Operations Research Department. But it was the Accounting Department that administered the operation of the IBM 650. Would the computer be controlled by the Reservations Department and housed in Toronto, or would it be integrated into the existing data processing facility at Dorval, Quebec? Territorial conflict broke out over which department would oversee the proposed computer's operation.

The question over the administration of the new computer center was more than just a question of the prestige of controlling this postwar symbol of technological and corporate power. It was also a matter of influencing the design philosophy of the new computer. From the outset, the director of the Telecommunications Department, C. J. Campbell, who shepherded the automated reservation project from its inception, always visualized the computer as a tool for the Reservations Department.[133] At a meeting in McGregor's office with executives from other departments, Campbell argued that "if the Sales Department was going to be responsible for reservations, they should be responsible for the tool they use, which would be the computer."[134] As a result, the specifications and costing of the computer's design, he argued, should revolve strictly around the reservation task: real-time interaction. Hence, the capacity to do large-scale off-line data processing should play a minor part in the design philosophy behind the computer. Yet J. B. Reid, the Accounting Department's Manager of Electronic Data Processing and the company's acknowledged expert on computers, objected that reservations would consume only 6 percent of the proposed computer's capacity. With an excess capacity equal to that of 20 IBM 650s, Reid suggested that the proposed computer be reconfigured to solve all TCA's data processing needs.[135] The implication of Reid's analysis was that, with so much excess capacity for data processing, the new computer should naturally fall under the management of the Accounting Department. J. B. Reid proposed further studies to elaborate on those features in the computer's design that would be necessary to enhance the computer as a data processing tool. In support of Reid, W. S. Harvey, the comptroller, argued that only "one person should be responsible for all computers in the company."[136]

Campbell rejected Reid's estimates of excess capacity as unrealistically high. Director of Reservations W. G. Rathborne further argued that "the

Accounting and Reservations requirements could not be combined, according to the experience of every carrier which had tried it."[137] Pointing to the American Airlines–IBM SABRE project (a computerized reservation system that was being developed in the United States), Rathborne suggested that American Airlines had underestimated the difficulty of joining these two functions in one design. As a result, the SABRE effort had to compromise on the real-time performance of the reservation function.

Faced with an inter-departmental feud over the design specifications of the new computer, President Gordon McGregor intervened. He underscored the centrality of the reservation problem. However, he acknowledged that there could be some excess capacity during off-peak hours. He agreed that a second set of design specifications and cost estimates should be drawn up for a "multi-functional" computer.[138] Aware of the potential pitfalls of such an attempt, McGregor insisted that any multi-purpose design not compromise the speed of the reservation system in any way. He also cautioned against letting the data processing issue interfere with the scheduled appearance of the automated reservation system. "The company," he told his executives, "cannot face 1961 without this equipment."[139] On 3 April 1959, TCA solicited sealed bids to build the computer from 19 electronics firms. TCA's two main requirements were expandable real-time reservation response and reliability and the ability to handle batch data processing. The computer thus had to be a general-purpose—that is, programmable—machine.

The TCA contract was a pivotal economic and technological moment in the history of Ferranti-Canada's efforts to enter the computer business. After 10 years of important technological advances, Ferranti-Canada's digital electronics group had yet to establish a viable commercial business. DATAR had suffered a premature death, and with it had gone Ferranti-Canada's hopes of long-term, lucrative defense contracts. Next came the Canadian Post Office computer. The powerful technological momentum generated by DATAR carried Ferranti-Canada's ambitions in digital electronics into the civilian sector. Computer-based automation technology for postal service, however, proved to be a commercial dead end for Ferranti-Canada. Nevertheless, the project pushed Ferranti-Canada up the computer innovation wave by making it one of the first companies in Canada to gain design know-how in solid-state digital electronics and drum memory technology. There was also some hope that a profitable business in electronic check-sorting technology could be developed in the United States, but little had yet to materialize.[140] In bidding for the TCA contract, Ferranti-Canada found itself in competition with seven of the most prominent computer and electronics firms: Burroughs of Canada Ltd. (Montreal), Canadian Westinghouse Co. Ltd. (Hamilton, Ontario), E.M.I Electronics Ltd. (Middlesex, England), IBM (Montreal), Minneapolis Honeywell Co. (Montreal), and Philco (Philadelphia).[141] But none of Ferranti-Canada's competitors had computer design facilities in Canada.

TCA wanted a general-purpose computer, but all of Ferranti-Canada's experience up to that point in time had been in the design of hard-wired special-purpose computers. Ferranti-Canada had to make this technological leap quickly if it did not want to be completely frozen out of the rapidly growing commercial market for general-purpose computers. The TCA contract offered Ferranti its first serious opportunity in 10 years to develop general-purpose machines.

Ferranti-Canada's chances of beating out the likes of Burroughs, Honeywell, IBM, and Philco would have been slight had TCA placed equal emphasis on the computer's ability to handle reservation and batch data processing functions, as the latter companies had sophisticated off-the-shelf general-purpose computer systems specifically designed for off-line, batch-oriented data processing. Ferranti-Canada's best hope was to exploit the real-time interactive aspects of the specifications for the reservation system. Of all the competitors, Ferranti-Canada was most in tune with TCA's thinking on this matter. After all, Ferranti-Canada's digital electronics engineers had, since 1956, worked very closely with TCA's Telecommunication Department in the development of the Transactors and the auxiliary communications technology.[142]

Most of the bidders adapted standard data processing machines for their proposals. The Honeywell, IBM, and Philco proposals tried to exploit the data processing potential of their machines. Though Philco proposed a fast machine, none of these three offered an attractive solution to the reservation problem.[143] On the other hand, TCA's Committee on the Assessment of Electronic Data Processing Equipment concluded that "the Ferranti-Packard and Canadian Westinghouse systems were well matched to the Reservations requirements, but they could not be expanded into such useful general-purpose systems as the Honeywell, IBM, and Philco computers."[144]

Phasing in the automated system entailed eliminating most of the human intermediaries in the processing of ticket agents' booking requests. Once in place, TCA would be utterly dependent on the computer. Expecting nearly 40,000 reservation-related transactions a day,

TCA could not afford to have the system go down. The Canadian Westinghouse proposal, like most of the others, did a poor job of designing adequate redundancy into the reservation system. Common to most of the proposed computer architectures was the use of two computers in series. The front-end computer, the faster of the two, handled the reservation function, while the back-end computer processed data and acted as an emergency backup.[145] If the first computer failed, the much slower second computer was a poor substitute when it came to handling reservations.

Ferranti-Canada proposed a "Gemini system," with two identical computers operating in parallel. The twin computers were called Castor and Pollux. During peak loads, both computers would share the work. Each computer would independently poll the eight incoming trunk lines. Gemini's hardware design made Castor and Pollux into siamese twins joined by special registers and sharing the same drum and tape mass storage. Each twin knew exactly which of the eight trunk lines the other was currently handling and which drum memory address the other was currently reading from or writing to. In this way, the twin computers could work simultaneously and independently and yet avoid performance-degrading conflicts. Through this parallel architecture, two smaller computers offered faster real-time performance than one large computer. Of course, if one failed the other could take over.

As far as TCA's Telecommunications and Sales Departments were concerned, Ferranti-Canada's original design best addressed the airline's requirements for fast, distributed, on-line access to the seat inventory and adequate emergency backup. J. B. Reid, on the other hand, felt that the proposal, as it stood, would not produce a computer system that could adequately handle the future data processing needs of TCA's financial and administrative services. As a concession to Reid and to the Accounting Department, the committee chosen to assess the bids recommended to President Gordon McGregor that TCA pay for an additional $400,286 in modifications in order to make the Ferranti-Canada machine a useful tool for data processing.[146] If there was sufficient excess capacity, then during periods of low on-line demand one of the parallel computers could be taken off line for data processing. Despite these modifications, no serious compromises were made in regard to data processing. From beginning to end, the pursuit of fast on-line interactivity with the 330 Transactors framed TCA's view of its computer needs. This was the only technical card that Ferranti-Canada could play against the other computer manufacturers. Ferranti-Canada played that card well.

In August 1961, nearly 2 years after being given the contract, Ferranti-Canada installed the Gemini computers at TCA's Toronto offices. But not until January 1963 did automated reservation go into full operation. Named ReserVec, TCA's $3.25 million computerized reservation system consisted of 10,000 miles of full-time, voice-quality communications lines linking some 330 Transactors in 39 cities to an electronic seat inventory in Toronto.[147] In less than 2 seconds after the Transactor read the input card, any ticket agent could confirm or alter a reservation. Within a few months, ReserVec was handling up to 80,000 transactions a day.[148] Almost immediately, management noted a dramatic improvement in efficiency and substantial cost savings. The number of bookings canceled because of failure to reconfirm dropped immediately by 34 percent. There was a 250 percent drop in space cancellations due to record discrepancies. Productivity (as measured by reservation person-hours per passenger boarded) improved by 177 percent. All these improvements ensured a much higher return on the airline's massive expenditures on new aircraft during the early 1960s.

ReserVec in Perspective

The need to increase productivity turned the Post Office and Trans-Canada Air Lines—both public enterprises—into patrons of leading-edge engineering development in digital electronics. The story of the Post Office's "million dollar monster" underscores how, despite Maurice Levy's technical brilliance and William Turnbull's bureaucratic daring, this government agency was ill-prepared to manage the introduction of radical technical change. In contrast, TCA was prepared to constrain and manage radical technological change to suit corporate ambitions. TCA's operations were suffused with the postwar spirit of technical progress. From radar and telecommunications to the latest advances in aviation, TCA's business called for technological sophistication. In tandem with this high-tech reality was TCA's "engineering-driven" strategic planning, a management style nurtured under Gordon McGregor's tenure as president. The overwhelming emphasis that TCA's senior executives placed on operating efficiency thrust this government-owned airline into a Schumpeterian role. For example, McGregor bragged how TCA was a world industry leader in the introduction of the most advanced commercial aircraft technology. TCA was one of the first Canadian corporations, private or public, to make Operations Research an integral part of strategic planning.[149] Though TCA was not innovative in the services it chose to sell, the airline nevertheless exhibited considerable technical innovation in how it delivered the services.[150]

TCA's carefully elaborated priorities framed Ferranti-Canada's pioneering engineering work. TCA saw the need to promote radical innovation, but it was careful not to let the process get lost in the pursuit of ultimate performance. The American Airlines–IBM SABRE system cost $30 million to develop; ReserVec cost only $3.25 million. IBM designed an $11,000 interface for the ticket agent; the Transactor cost only $2633.[151] Except for IBM's STRETCH supercomputer, the two IBM 7090 computers that made up SABRE were IBM's most powerful commercial data processing computers, and they cost 8 times as much as the pair of computers used in ReserVec I.[152] The IBM 7090 could do 229,000 arithmetic additions per second; each of the Gemini computers could do only 17,800.[153] Despite the IBM 7090's faster computational performance, the Gemini computers processed 3–4 times as many transactions per day, and with faster on-line response per transaction. There were two reasons for ReserVec's higher throughput: Ferranti-Canada's novel parallel architecture (which optimized the real-time reservation task) and TCA's careful efforts to minimize the amount of information that the reservation system needed to work.[154] ReserVec offered a cost-effective and relatively quick solution to a pressing operational problem.

The design of SABRE was an outgrowth of IBM's experience building the real-time equipment for the SAGE and DEW Line air defense systems.[155] Similarly, Ferranti's capacity to design the TCA system was, to a great extent, a legacy of the Canadian Navy's DATAR project. DATAR and SAGE were the first defense systems to integrate real-time electronic digital data processing, data communication, and human-machine interfaces. Despite the high price American Airlines paid for the IBM 7090, the computer was an off-the-shelf item whose development costs had been covered by the US Air Force.[156] Ferranti-Canada used the much shallower pockets of Canada's Post Office Department to develop the digital-transistor-circuit and drum-memory know-how needed to design and build the Gemini computers.

Whereas IBM used its reputation as a leading manufacturer of computers to get the American Airlines contract, Ferranti-Canada hoped to use the TCA contract to become Canada's first designer and manufacturer of commercial general-purpose computers. Like Ferranti-Canada, all the other bidders for the TCA computer were subsidiaries of foreign-owned companies.[157] Except for Ferranti-Canada, each proposal from the subsidiaries of US companies offered mere copies of a parent firm's know-how and products. Ferranti-Canada's parent firm, Ferranti UK, was a prominent builder of computers. And yet the Canadian subsidiary chose to push its own design. Within 7 months of signing the Gemini computer contract, Ferranti-Canada announced that it was marketing an off-the-shelf general-purpose computer.[158]

The $3.25 million TCA contract provided Ferranti-Canada's computer group with the cash, the technological opportunity, and the credibility it so desperately needed to bolster its ambitions to design and manufacture computers in Canada. Despite this milestone, building a sustainable computer business remained elusive. For all the remarkable innovative tradition of Ferranti-Canada's electronics group, it only managed to generate "one-off" sales. DATAR and the postal system computer were custom systems that never generated any repeat sales. Perhaps ReserVec would be different.

Who would be the next customer be? Efforts to market RESERVEC around the world proved frustrating. Systems for national transportation carriers were invariably bought domestically. National railways were no different, and Ferranti-Canada's efforts to sell on-line computer technology to Europe's various national railways also failed. Furthermore, Ferranti-Canada received little help from its parent firm. Rather than promote the proven technology of its Canadian subsidiary, Ferranti UK wanted to develop and sell its own computerized airline reservation system to British Airways. Ferranti-Canada's failure to get Ferranti UK's cooperation reflected the parent firm's inability to formulate a coherent corporation-wide strategy. This inability, however, also permitted the Canadian subsidiary to slip through the rather large cracks in the parent firm's control and follow its own R&D agenda.

ReserVec was the world's first fully operational on-line reservation system using a general-purpose (i.e., programmable) computer. Nearly 2 years later, the American Airlines–IBM SABRE system became the second such system. Though ReserVec was not as ambitious in scope as SABRE, the Canadian system did the job at a fraction of the cost. With some enhancements, ReserVec ran successfully for 7 years before it was replaced, in 1970, by ReserVec II.[159] In the late 1980s, Air Canada (formerly TCA) installed an even more comprehensive computerized reservation system. Called "Gemini," the new system provided an important source of revenue diversification. Air Canada marketed the Gemini service to the entire Canadian air travel industry. The choice of "Gemini" as the name for the new reservation system was a fitting reminder of the pioneering work that in the 1950s gave Air Canada the experience and technical competence to be an important player in the computerized reservation business.[160]

In the early 1990s, nearly 40 years after TCA and American Airlines started to build computerized reservation systems, the direct descendants of ReserVec I and SABRE were at the center of a fierce corporate struggle between Air Canada and Canadian Airlines (formerly Canadian Pacific Air Lines). At stake were the substantial investments in, and revenues generated by, these two systems. American Airlines would consider a merger with Canadian Airlines only if the latter switched to the SABRE system. Canadian Airlines' merger with American Airlines, however, was being blocked by the former's contractual obligation to use Air Canada's Gemini reservation system. The long corporate struggle between Air Canada and Canadian Airlines became a battle over computerized airline reservation technology, and SABRE won.

Hope that the success of ReserVec would open up the world's transportation market to Ferranti-Canada soon evaporated. Ferranti-Canada realized that to succeed in a maturing computer market it would have to develop and sell a standardized digital computer, rather than the highly customized one-of-a-kind products it had sold in the past.

5

The Effort to Create a Canadian Computer Industry

From its inception, Ferranti-Canada's computer group existed in near isolation from the rest of firm. Historically, the company had been, first and foremost, a manufacturer of electrical power products for Canada's utilities. In 1949, the parent firm, Ferranti UK, grafted an electronics business onto Ferranti-Canada's power operations. For years Ferranti-Canada's senior managers, who were all trained in the electrical power business, knew little if anything about digital electronics, but as long as the Electronics Division did not lose money Ferranti-Canada's senior executives were content to let this group set its own R&D priorities and its own commercial strategy.[1] The parent firm, whose own technology-driven ethos reinforced the Electronics Division's ambitious R&D objectives, was content to adopt a similar stance. With the independence that this arms-length policy provided came reluctance on the part of senior management to share the Electronic Division's commitment to making the commercialization of computers a cornerstone of the Canadian corporation's existence.

As the Post Office and Trans-Canada Air Lines stories reveal, Ferranti-Canada's Electronics Division aggressively pursued opportunities to convert its military-derived expertise in digital electronics into civilian products even before the demise of DATAR in 1956. Once DATAR collapsed, these efforts to build new civilian markets became a matter of life and death for the Electronics Division. New sources of revenue were needed if the division was going to keep intact all the expertise it had built up during the DATAR work. Not only did the Post Office and TCA projects help provide the Electronics Division with sufficient financial stability to maintain a strong R&D program; these projects also provided the division with an opportunity to keep moving up the technical learning curve. Such an opportunity was essential to the Electronics Division's ambition of entering the commercial market for general-purpose computers.

The mail-sorting computer project for the Post Office expanded the division's competence in transistor circuit design and also allowed it to perfect its mass-storage technology. In the ReserVec I project for TCA, the division perfected the basic logic circuitry needed to build commercial computers and also developed experience in designing programmable computer architectures. Though Ferranti-Canada was not able to sell any turnkey computer reservation systems, the success of ReserVec I's Gemini computer intensified feelings within the Electronics Division that the time was ripe to commit greater corporate-wide energy and resources to establishing a niche in the commercial computer market. But there was a growing feeling of exasperation that the company's Canadian upper management had no real interest in building a computer business.

Faced with this increasing discontent and impatience, the manager of the Electronics Division assured his engineers that he was working on a plan to convince senior executives to let Ferranti-Canada manufacture the parent firm's existing advanced computers. With no computer-manufacturing facilities in Canada, the federal government was receptive to Ferranti-Canada's request for financial assistance to set up production facilities. All that remained to be done was to receive the parent firm's approval and to arrange for the necessary technology transfer. But Ferranti UK, overwhelmed by its own badly managed computer operations, rejected the proposal. Angered at that news and by the absence of any support from upper management, many of the senior computer engineers in Canada tendered their resignations. After a long meeting, Tom Edmondson, chief executive officer and chairman of the board of Ferranti-Canada, managed to convince the engineers to reconsider their resignations and persevere a little longer. However, the central issue remained unaddressed: How and when would the company, building on its Gemini expertise, get into the business of general-purpose computers?

Although the R&D ambitions of Canadian military and civilian public enterprise had served to move Ferranti-Canada up the digital electronics learning curve, it was American public enterprise that gave the company the first opportunity to create a niche in the general-purpose computer business. This chapter explores how the US Federal Reserve Banks' advocacy of digital automation as a way to keep up with the rising flood of financial paper transactions created Ferranti-Canada's first opportunity to sell general-purpose computers. The computer's design, as this chapter shows, was both a logical extension of the Trans-Canada Air Lines' ReserVec I system and Ferranti-Canada's search for a niche in a market dominated by large mainframe manufacturers such as IBM. Furthermore, the narrative will illuminate the circumstances that surrounded Ferranti-Canada's failure to turn the Federal Reserve computer into a commercial success and the subsequent demise of the company's computer design group. Finally, this chapter will show that the memory of the Canadian government's inability to act swiftly to support Ferranti-Canada's bid to commercialize general-purpose computers played an important but subtle role in the government's efforts during the 1970s to foster a domestic computer industry.

Technical Know-How Finds Commercial Opportunity: The Role of US Public Enterprise

By the early 1950s, faced with a mounting volume of paper checks, the US Federal Reserve System was searching for new technologies that might accelerate the "clearing" process. After a spate of bank failures, the US Congress had created the Federal Reserve System in 1913 to bolster the public's confidence in the solvency of the US monetary system.[2] Reliable and timely clearance of checks was essential to such confidence. For the thousands of small independent banks throughout the United States, the Federal Reserve System came to offer a fast, easy, convenient, and secure way to clear one another's checks.[3] The Great Depression reversed the preceding 10 years' steady climb in check transactions, but the volume of checks handled by the Federal Reserve System resumed its growth with the start of World War II and accelerated in the postwar years. From 700 million per year in 1939, the number of checks cleared annually by the Federal Reserve system had soared to 2.8 billion in 1956.[4] The increasing circulation of checks showed no signs of slowing down. One bank specialist in the Product Development Division of the National Cash Register Company suggested that the volume of checks cleared by US banks would climb from 3.5 billion in 1957 to 14 billion in 1960.[5]

Since check clearing was labor intensive, the considerable expansion of clerical staff needed to keep up with the number of checks in circulation troubled Federal Reserve officials. The Federal Reserve Banks (FRBs) found the prospect of large increases in its labor force economically unattractive. First, there was the high cost of salaries. Low birth rates during the 1920s and the 1930s and low rates of unemployment during the 1950s had raised the price of labor. There were also indirect but equally important costs associated with a labor-intensive solution to the problem of check clearing. A new army of clerical workers would require very large expenditures on new facilities. The FRBs preferred the path of increased labor productivity through technical innovation to the path of large-scale expansion of its labor force.

The FRBs had foreseen the long-term need for high-speed mechanization in the early 1950s. Under the leadership of H. H. Kimball, vice-president of the Federal Reserve Bank of New York, the FRBs set up a technical subcommittee to see if electronics could be applied to the check-clearing bottleneck.[6] Since the automation of check handling was a problem that faced US banks and clearing houses outside the Federal Reserve System, the American Bankers' Association had also set up its own Committee on the Mechanization of Check Handling. From 1953 to 1957, the FRBs were content to let the ABA committee take the lead on the automation issue. During this period, the ABA committee tried to raise manufacturers' awareness of the banks' technological needs. Central to the ABA committee's strategy was the quest for a consensus across the banking community on how to encode information on checks. Standardization, the ABA committee reasoned, would accelerate the industry's development of suitable check-clearing equipment.[7] As with electronic mail sorting, a major technical challenge in automated check clearance was the proper encoding of the essential information needed to process checks at high speeds and then to scan this code electronically and convert it to a binary digital format. The ABA committee advocated standardizing around a new technology called Magnetic Ink Character Recognition (MICR).[8] Despite the ABA's efforts, commercial equipment to mass process checks remained elusive. All the technological pieces for electronic check sorting were available, but no manufacturer had yet integrated them into a system.

As the nation's largest clearing houses, the FRBs could no longer stay in the background and wait for a solution. Pressed to find an answer to the worsening bottlenecks in their own operations, the FRBs decided in 1957 to take a more aggressive role in accelerating the development, manufacture, and implementation of electronic check-clearing equipment. In May 1957, after months of discussions, the FRBs engaged the Stanford Research Institute of Menlo Park, California, to study "the check handling requirements of the Federal Reserve Banks with a view of isolating those problem areas to which special attention should be given and from which mechanization can be expected to yield the greatest returns."[9] The SRI study set out "to make a preliminary translation of [FRB's] operating requirements into electronic devices and systems which incorporate currently known techniques; and then to make an initial study of the economic feasibility of such devices and systems."[10]

SRI's 4-month, $16,000 study examined the technological and economic issues of mechanization on two axes: (1) preparation of checks for sorting and actual sorting of checks and (2) settlement of accounts between banks. The first axis had to do with the problem of encoding a check with its amount, the second with the identity of the bank on which it was drawn. If the amount and the ABA number on each check were MICR-encoded, an electronic sorting system could "read" the check at very high speeds and sort them according to the banks on which they were drawn.[11] In the process, the electronic sorting system could keep a running total of the value of all checks drawn from each bank. Using these totals, the FRBs could then settle any outstanding differences between member banks by simply debiting or crediting their standing reserve accounts.

The economic feasibility of this electronic sorting depended on what percentage of the checks sent to the FRBs for clearing had already been encoded with an ABA number and an amount. A high percentage meant less manual encoding for the FRBs. If fewer than half of the checks received had an ABA number and an amount encoded on them, there would be no economic advantage to electronic sorting. Beyond the 50 percent level, savings became "significant." For example, with 75 percent of the checks pre-encoded, the FRBs would save 35 percent of the cost of clearing.[12] More important from the FRBs' perspective, these savings could, according to SRI, be traded for a 100 percent increase in throughput.[13] Thus, for mechanization to succeed, it was imperative that the FRBs promote the development and diffusion of amount-encoding equipment based on the MICR concept. Technologically, there were no obstacles to designing and building suitable amount encoders. The manufacture of MICR printing and reading equipment remained stalled, the SRI study noted, because standards had not yet crystallized. Unless simple aspects of standardization (such as where to locate encoded information on the check and what font to use) had unanimous acceptance in the banking community, equipment manufacturers would not risk any product development. SRI recommended that the FRBs throw their considerable weight behind the standards proposed by the American Bankers' Association.

Developing and manufacturing MICR printing and scanning equipment were straightforward matters. The same was not true of the digital electronic system that would control the sorting process and keep track of the running total values of checks drawn on each bank. SRI proposed an electronic digital dictionary lookup approach. Operating one check-scanning-and-sorting operation at 10–20 checks per second with drum memory was itself a challenge. And, for economic reasons, SRI recommended having at least three scanning and sorting processes running simultaneously. Thus, the challenge of designing this special-purpose computer was to allow several concurrent processes to access the memory independently and still keep each process moving at its top speed. In technical terms, the ability to field a cost-effective high-speed electronic check system required that real-time, time-sharing, or multiplexing features be designed into the system's computer.

In view of "little or no emphasis placed on research and production of [such systems]," SRI underscored the need for "the Federal Reserve System to take immediate action in the direction of playing an active role in directing business machine manufacturers toward such a development."[14] Either the Federal Reserve System would have to invest directly in R&D or it would have to incite industry to undertake this work by promising the whole prize to one company. The Federal Reserve System chose the second option. The FRB's electronics subcommittee asked SRI to have more detailed specifications of the Federal Reserve System's equipment needs ready for widespread public distribution by 1 March 1958. According to H. H. Kimball, "the subcommittee felt that such an announcement would induce competition in the development of equipment, not only among the recognized office equipment manufacturers, but also among manufacturers of other precision equipment."[15]

Six months after SRI made its system specifications public, Ferranti-Canada submitted its proposal to the Federal Reserve System.[16] Thanks to the Canadian Post Office's mail-sorting project (which had just ended) and its ongoing work with Trans-Canada Air Lines, Ferranti-Canada had acquired the blend of know-how needed to compete head-on with US computer and electronic manufacturers. Sorting mail and sorting checks were similar tasks. The crucial difference was the need to add accumulators to the special-purpose computer to keep running credit and debit tallies on all checks drawn from any one bank and deposited to any other. The special-purpose computer built for the Post Office showed the transistorized digital-circuit design expertise and the high-speed drum-memory technology that were required of a check-sorting computer.[17] But, equally important, Ferranti-Canada's work on DATAR and its 1957 contract with TCA to design, build, and multiplex six transactors for FERUT gave the Canadian company a keen appreciation of how to get many concurrently running processes to share a computer's resources in real time.[18]

After all the proposals had been received and analyzed, the Federal Reserve System decided in 1959 to launch an extensive pilot study to evaluate the technical, operational, and economic aspects of the most promising proposals.[19] Five Federal Reserve Banks were chosen as test sites. In the Federal Reserve Bank of New York test site, Ferranti-Canada teamed up with Pitney Bowes Inc. and with National Data Processing Inc. Pitney Bowes was to build the mechanical sorter equipment; National Data Processing was to build the MICR encoding equipment.[20] The contract drawn up between Ferranti-Canada and the FRB called for a rental fee of $7200 per month during a 6-month test period, after which time, if the system performed to specifications, the bank had to either buy Ferranti-Canada's Stored Reference Computer for $295,000 or continue to rent it for $7200 a month for an additional 4½ years.[21]

By September 1961, the field tests at the FRB of New York had been completed successfully. Estimating that its system would process some 500,000 checks daily, the FRB of New York received approval from the board of governors in Washington to purchase the Ferranti-Canada computer.[22] But before the New York bank had taken this decision, Ferranti-Canada tried to sell it an improved design. Having just installed the Gemini computers for TCA, Ferranti-Canada tried to transfer this technical experience back to its Stored Reference Computer. The company proposed a computer that considerably simplified the multiplexing hardware, improved the memory drum, offered input and output buffers, and provided new circuitry for self-testing, arithmetic, and control functions.[23]

If the Federal Reserve Bank of New York was interested, it said nothing. Ferranti-Canada meanwhile issued a press release announcing that it had sold the only Canadian-made computer for check sorting.[24] The delivery of the ReserVec general-purpose computer to TCA had already produced a heady atmosphere within Ferranti-Canada's computer group. The successful performance of Ferranti-Canada's check-sorting computer only intensified the feeling. Ferranti-Canada issued product literature and hoped to build a repeat business around check sorting.[25]

Although computerized check clearance was high on the American Bankers' Association's technology agenda during the 1950s, Canadian banks appeared uninterested. And yet the growth in the volume of circulating checks in Canada paralleled the US experience. A government report noted that "the annual fluctuations in checks cashed from 1924 to [1948] followed rather consistently the cyclical movements [of the periods 1924–1929, 1930–1937, and 1938–1948]."[26] The report also underscored that "the advance both in duration and magnitude in the cashing of checks from 1938 to 1948 was without precedent."[27] The postwar growth in the value of checks cleared was just as impressive: from $68 billion in 1945 to $212 billion in 1957. Though inflationary factors contributed to this growth rate, it nevertheless reflected a considerable increase in the volume of paper checks that Canadian banks had to handle. "During that period, the total number of checks was almost doubled," explained a Royal Bank publication. "[In 1964] the yearly total has reached one billion and continues to increase at the rate of around 10 percent a year."[28] But it was not before 1964 that the Royal Bank admitted that "if [it] retained its present methods, this volume would soon become unmanageable. . . . In short, existing methods, staff and premises are inadequate in the face of this increase in business."[29] Despite the claim that "the enormous increase in the number of checks handled by the Canadian banks in the decade after [World War II] prompted banks to search for faster methods of processing," it is difficult to find any reference in the literature of the 1950s to Canadian banks' trying to promote the development of electronic check clearing. One can only speculate as to why it took so long for computerization of check clearing to develop in Canada. Perhaps it has something to do with the way checks were cleared in a system dominated by handful of banks owning thousands of branches, where a higher percentage of check clearing entails inter-branch settling of accounts.

Moving Check Sorting from Special-Purpose to General-Purpose Computers

By 1962, technical advances had begun to undermine Ferranti-Canada's approach to check sorting. The special-purpose computer, despite its inflexibility, had been an attractive option for a certain class of data processing because of its lower cost and its higher speed. But as the price-to-performance ratio of general-purpose computers improved, they became, as Ferranti-Canada had earlier recognized, a competitive alternative to special-purpose computers, even in very restricted applications. Furthermore, programmable computers allowed check-sorting operations to be expanded to include a wider range of data processing tasks as circumstances warranted. Honeywell's efforts to introduce general-purpose computers into the check-sorting operations of banks made Ferranti-Canada realize how vulnerable its business plan was.

In January 1962, one of Ferranti-Canada's system engineers, Paul Dixon, proposed that the Federal Reserve Banks further improve their check-processing operations by purchasing a general-purpose computer from Ferranti-Canada.[30] "The use of the [Ferranti-Canada-designed] general purpose computer," Dixon argued, "marks an important step forward in the operating efficiency, reliability and flexibility, in the high-speed cheque sorting operation."[31] The use of a programmable computer, he explained, would allow the bank to determine the optimum sorting patterns and to change sorting patterns easily on demand, to prepare a greater number of more sophisticated control totals that would significantly reduce the manual work needed to maintain tight audit trails, and to make the system more adaptive to changing conditions. But the real advantage of the general-purpose computer proposed by Ferranti-Canada was its ability to run several real-time applications at the same time. The computer could run three independent sorting runs simultaneously, each with its own optimal sorting pattern and its own audit trail. At the same time, the computer controlled a variety of peripherals: a high-speed paper tape reader and punchers, high-speed line printers, card readers and punchers, and magnetic-tape units.

Real-time "time sharing" was an extension of the multiplexing features Ferranti-Canada had used in the special-purpose check-sorting computer. But in the new design, the "time-sharing" concept expanded beyond hardware to include software. "The time sharing feature," wrote Dixon, "enables the computer to operate several programs concurrently, while providing full-store and computational protection to each program handled. Hence, in the course of check sorting, the computer will be able to monitor its own operations and the accuracy of control accumulations which actually sorting."[32] In both the special-purpose and general-purpose computer designs, the economic need for the bank to have several sorting processes going on in parallel shaped how Ferranti-Canada came to define its competitive advantage in terms of innovation based on "time sharing."

The computer (named F.P.-6) had not yet been designed in any detail. But that did not stop Ferranti-Canada's computer engineers from boasting that the F.P.-6 was a "next generation general purpose computer" that "represented a genuine breakthrough in terms of performance."[33] This self-confidence, which rested on the successful transistorized logic circuit design that Ferranti-Canada had used in the TCA Gemini computers, was reinforced when the parent firm asked Ferranti-Canada computer engineers to participate in a crucial computer development program known as ORION II.

Acutely aware of its diminishing share in the world computer market, Ferranti UK had decided to launch a smaller machine called the ORION. In the design of the ORION, Ferranti UK gambled on a radically new circuit design. The ORION project was fraught with technical difficulties, cost overruns, and substantial delays. At one point in the project's life, Ferranti UK management worried that the technical and financial problems might prove insurmountable. As result, Ferranti UK initiated a backup project called ORION II. The parent firm decided to use Gemini's less sophisticated but proven transistor circuit techniques in the construction of ORION II. In 1961, engineers from Ferranti-Canada had gone to England to participate in the ORION II project. For a time two groups worked in parallel, one (in Manchester) building ORION and the other (in Bracknell) building ORION II. Orion II was never needed, but the experience later served to reaffirm the Canadian computer group's belief in its design abilities. When, in March 1962, the Federal Reserve Bank of New York agreed to buy the F.P.-6 computer, the ORION II experience further convinced the engineers and management in Ferranti-Canada's Electronics Division that they had the "right stuff" to build the "next generation" general-purpose computer.[34]

Ironically, the technical advantages Ferranti-Canada promised to deliver in its new computer were not the determining factor in the Federal Reserve Bank of New York's decision to buy the computer. The special-purpose check-sorting computer still worked quite well, but officials of the bank feared that when the system reached its expected processing output of 500,000 checks a day any breakdown in the computer would prove disastrous. Paul Dixon, a systems engineer and a salesman for Ferranti-Canada, had no doubt learned before he submitted his proposal that the bank had already planned to buy a second computer as a backup. The bank's decision to go with Ferranti-Canada was based in part on the success of the existing system, but it was equally important that the bank felt it advantageous to have one, rather than two, manufacturers responsible for the maintenance of its computer systems.[35] Dixon did not know that the bank had already allocated $260,000 for a second system.[36] The fact that Ferranti-Canada was asking only $152,350 for this "next generation" general-purpose computer turned necessity into a great deal for the bank.[37]

The news of Dixon's sales coup transformed the festering discontent within Ferranti-Canada's computer group into euphoria. At long last the company was in the commercial computer business. In a more detailed draft of the proposal, the F.P.-6 was replaced by the "FP6000 High Speed Check Sorting System."[38] This system represented an entire product line of components. (The computer itself was called the FP6001; however, by tradition this computer has come to be identified with the name FP6000.) Within a year, Ferranti-Canada had designed, built, and installed the FP6000.[39] Excitement was running high. In November 1962, while Ferranti-Canada was working on the Federal Reserve Bank's FP6000, it got an order for another from the Naval Research Establishment in Halifax.[40] Several months later, the Toronto Stock Exchange ordered another.[41]

The initial performance and cost considerations in the design of the FP6000 tried to exploit a specific opportunity with the Federal Reserve Bank of New York's check-sorting needs. At the same time, Ferranti-Canada's design philosophy was a reaction to the problems of its parent firm's computer business. Although Ferranti UK produced remarkable, powerful computers for scientific applications, the company had difficulties addressing the various levels of the burgeoning market for electronic data processing and office automation. Ferranti UK's computer business was unable to find the right technical and marketing response to this still-untapped market. Preoccupied with technical sophistication, the company lost sight of its customers and their needs. As Peter Drucker has pointed out, "while the technological leaders in the early computer days . . . were product-focused and technology-focused, the punch card salesman who ran IBM asked: 'Who is the customer? What is value for him? How does he buy? And what does he need?' As a result, IBM took over the market."[42] The loss of $450,000 in 1961 was clear indication of the commercial world's response to Ferranti UK's computer business strategy.[43]

Meanwhile, Ferranti-Canada's computer group felt that its best chance to survive against the likes of IBM, Univac, and CDC was to aim at a mid-level market and to stay clear of big systems. The challenge then became one of designing a lower-priced mid-size computer. To make the FP6000 a more attractive option for the mid-level data processing user, Ferranti-Canada migrated its "time-sharing" innovations for high-speed check clearing into the advanced feature of multiprogramming. In batch processing, all the user programs queued up and ran one after the other. The programs could range from complex scientific modeling to financial data processing. Each program made different demands on the computer system's central processing unit, core memory, magnetic tapes, magnetic drums and disks, printers, paper tape units, and other components. When only one program was being executed, there were always some portions of the computer's resources that sat idle. For example, during writing to mass storage devices or to the printer, the resources of the central processing unit were being wasted. In allowing many batch programs to run concurrently, multiprogramming improved the throughput and the economics of the entire computing center. By the end of 1960, attempts were underway to implement multiprogramming on very large systems, such as the IBM STRETCH computer.[44]

Although the advance of multiprogramming had become more widespread by 1963, it remained confined to very big and expensive machines. At a conference that had attracted North America's leading computer engineers, a Ferranti-Canada team explained how "time-sharing facilities normally only available on larger computer systems may be realized on a medium-size system [on the FP6000] by an intermarriage of hardware and software for minimum cost yet still retain all the safeguards necessary for an operable system."[45] For multiprogramming to be a success, not only did it have to optimize the integration of all the system's hardware; it also had to make the entire process transparent to the user. In other words, from the user's perspective, a program was written and run as if it were the only job on the system. According to one computer expert, E. F. Codd, the biggest challenge that the development of multiprogramming systems faced in 1962 was the design of flexible storage allocation.[46] The approach taken by Ferranti-Canada involved dynamic program and data relocatability. Each program was allocated a range in the core store specified by a starting address called the "datum" and an upper address called the "limit."[47] All instruction addresses in the program were assigned relative to each program's "datum." As programs terminated, blocks of free memory would be created. To prevent any fragmentation in the available core store, the FP6000's supervisory program would temporarily stop the execution of all the other programs each time a program terminated and recopy them hard against one another at the lowest end of the core store. The "datum" in each program would be changed accordingly, then the FP6000's supervisory program would set them all running again.[48] In this way, at any moment, all the system's available memory physically occupied one contiguous block at the top end of the core store.

The customer's ability to expand the system economically and flexibly was another important consideration in the design of the FP6000. The power and performance of the FP6000 could be expanded in a modular manner to accommodate a company's changing requirements and budget. The ability of the system's supervisory system (called the Executive,) to accommodate various core store and processor options with minimal additional software costs was an essential ingredient in Ferranti-Canada's modular strategy.

In his history of the British firm International Computers and Tabulators, Martin Campbell-Kelly suggests that the design of the FP6000 originated with Ferranti-Canada's parent firm.[49] This conclusion, however, assigns undue primacy to Ferranti UK's technical knowledge in shaping the development of the FP6000. This computer was not an isolated technical and corporate event in the life of Ferranti-Canada; neither was it a product of the parent firm's technological castoffs. Instead, the corporate capacity to design and build the FP6000 was the culmination of a process that started in 1949 with DATAR.

The transistor circuit techniques used in the FP6000 were first honed in the computerized mail-sorting system built for the Canadian Post Office. By 1956, when Ferranti UK installed its first vacuum tube Pegasus computer, the Canadian subsidiary had abandoned the vacuum tube paradigm and undertaken an independent program to develop transistor-based digital circuit techniques. Throughout the latter part of 1955 and the early part of 1956, the digital electronics group at Ferranti-Canada worked closely with Philco in the United States to explore the implications of Philco's new high-speed SB-100 transistor for the design of digital electronics. It was during this period that the Canadian subsidiary gained firsthand experience in the design and manufacturing issues associated with the construction of systems based on standardized transistor cards.

The Gemini computer's transistor logic circuitry, which formed the basis for the FP6000, was the culmination of Ferranti-Canada's early commitment to solid-state electronics. The expertise needed to develop the real-time communication's aspect of ReserVec's information-processing capabilities grew out of the DATAR experience. The Gemini computer's unique architecture was itself a direct consequence of Trans-Canada Air Lines' on-line-reservation imperative. The FP6000's multiprogramming marketing feature was the technical culmination of the Ferranti-Canada's long experience (first with DATAR, then with ReserVec, and finally with the check-sorting computer) in multiple, on-line, real-time access to computer resources.

Although it is important to recognize the important innovations that originated at Ferranti-Canada, it would be foolhardy to push this argument too far and dissociate the company from the internationally available pool of ideas and techniques. Ferranti UK's experience no doubt had some influence on how Ferranti-Canada approached the task of designing a multiprogramming environment. After all, it was Stanley Gill of Ferranti UK who in 1958 wrote the first comprehensive research paper on the question of multiprogramming.[50] Gill's work sparked widespread interest in multiprogramming. Ferranti UK had also incorporated multiprogramming features in the ORION. Moreover, multiprogramming had become a widely discussed topic within the international, computer engineering community when Ferranti-Canada took on the FP6000 contract for the Federal Reserve Bank of New York.[51] Ferranti-Canada engineers were exposed to these discussions, particularly those within North America, as much as they were exposed to Ferranti UK ideas.[52]

Innovative engineering involves balancing the realities of market and corporate constraints with the pursuit of technical artistry. The decision to design a mid-size computer was, in part, a consequence of the terms of Ferranti-Canada's contract with the Federal Reserve Bank of New York and of a conscious recognition that, as a small Canadian company in a field dominated by IBM, Ferranti-Canada would have to design a smaller, less costly computer for the mid-level user. In the early 1960s, the Digital Equipment Corporation (a US firm) was pursuing a similar strategy to produce a line of relatively cheap and simple computers for specific situations. These computers, named PDP, were intended to perform the functions of conventional electronic data processing systems. After the success of its early line of PDP computers, DEC created and dominated a market for more general-purpose data processing machines called minicomputers.[53] In contrast, Ferranti-Canada's approach, from the start, was to create a mid-size computer with the features of larger mainframes. The innovation in the FP6000 lay in finding the right "intermarriage of software and hardware" that would allow multiprogramming with a less costly mid-size machine.[54]

When Ferranti-Canada had first approached Ferranti UK with the idea for the FP6000 computer, the parent firm disagreed with the concept and wanted little to do with it. Technology-driven and immersed in technological elitism, the parent firm viewed the FP6000's design as too conventional and as tainted by overtly "commercial" motives. But in fact the FP6000 was a creative adaptation of leading-edge technology to the competitive challenges faced by a small company in an industry dominated by very large corporations. The fact that Ferranti UK let Ferranti-Canada pursue its FP6000 strategy nevertheless reflects the remarkable degree of R&D autonomy that the Canadian subsidiary enjoyed.

Within a year, Ferranti-Canada had received three orders for the FP6000 computer system.⁵⁵ That was a modest but promising beginning. Subsequent orders for the FP6000 proved elusive, however. Despite Ferranti-Canada's sale of the FP6000 to the Naval Research Establishment in Dartmouth, Nova Scotia, two other attempts to sell to the Canadian military failed. The sales team was particularly discouraged by what it perceived to be an unwillingness of local and provincial governments to "buy Canadian." However, it was in the federal government, which was the single largest user of computers in Canada that such unwillingness was most acute.

By the mid 1950s, interest in electronic digital computers was growing in many departments of the Canadian federal government. Few in government had any working knowledge of this technology, however. Procurement, staffing, and maintenance of computer centers represented considerable expenditures. In 1955, fearful that departments, left to their own devices, would make unsound or unnecessary purchases, the Treasury Board instituted the requirement that it had to appraise all government expenditures on electronic digital computers. That same year, the Treasury Board set up the Interdepartmental Committee on Computers to advise on proposals to install new computers, to eliminate duplication and overlap in departmental expenditures on computer centers, and to advise departments on the application of computers.⁵⁶

Concern over the most cost-effective way to introduce the computer into government operations became a central theme in the work of the Royal Commission on Government Organization. In the fall of 1960, the government gave this commission (chaired by John Grant Glassco, president of Brazilian Traction) the mandate to examine how to modernize and streamline the operation of the federal government along the best practices of private-sector management. The commission emphasized how computers were transforming the productivity, scope, and efficacy of white-collar work. "By making vastly increased resources of information readily available, in whatever form required," the commission reported, "the new technology continues to enlarge the boundaries of effective action as a rapidly accelerating pace."⁵⁷ No doubt impressed by the accomplishments of automation in the secondary sector of the economy, the commission looked to the computer to "create new administrative 'machine shops'—data-processing centers which can respond to the demands of administrators and operators throughout the organization with increasing versatility."⁵⁸ If the government was going to modernize its operations and be more productive, it would have to integrate the computer effectively into all its operations. Not only did the federal government have to readapt its administrative structures and practices to the logic of this new technology; it also had to be a lot smarter in how and where it established data processing centers. Despite the Treasury Board's initiatives, the commission underscored the "uncoordinated," "hit-and-miss" manner in which data processing centers were being established.

The Royal Commission on Government Organization was particularly scathing in its assessment of the Interdepartmental Committee on Electronic Computers: "The Committee has not provided leadership and management guidance, or given adequate technical assistance to the departments. . . . No formal reporting procedures have been set up to facilitate coordination of automatic data processing activities, and no attempt has been made to measure actual results against forecasts or to assess the impact on departments. Furthermore, the Committee has not taken the initiative in promoting and developing comprehensive plans and policies for the introduction of automatic data processing."[59]

While the government and the Royal Commission on Government Organization grappled with the management and financial issues associated with the computerization of government operations, little attention was given to whether the government should use its purchasing power to promote domestic manufacture of computers. However, the commission did venture into the manufacturing issue when it stressed the importance of defense procurement in advancing Canada's overall technological and industrial capacity: "Those responsible for defence procurement must . . . be more than skilled purchasing agents. The procurement effort must be guided by an adequate appreciation of the present and future potentials of the economy in fields of research, development, and technology which can, by proper stimulation and support, underpin the country's economic strength and potential economic growth."[60] Why the commission failed to extend this argument to all those in the civilian government who procured data processing systems remains a mystery.

Aside from the fact that the absence of a suitable "buy Canadian" policy limited opportunities for the FP6000, Canada's structure of import tariffs served as a disincentive to the manufacture of systems in Canada. Because the duty rate on components was higher than that on complete systems, Canadian firms seeking to manufacture entire systems were at a disadvantage. Not only did the actual rates work against domestic manufacture; the manner of their application exacerbated the disadvantage. Instead of applying the tariff rate on the fair market value of the data processing system being reported, Canada Customs applied the excise tax to the cost of manufacturing the system. This strange logic had arisen because US computer manufacturers had claimed that electronic data processing systems were normally leased, not sold, in the United States. "As cost of manufacture is generally 25% of sale price," the Department of Industry's Electrical and Electronics Branch noted, "the evaluation is correspondingly low, resulting in only nominal duty being paid with no incentives for subsidiaries to manufacture in Canada. Furthermore, Canadian subsidiary importers can take protection in their sale price of duty at arms length or sale price evaluation which a private importer would have to pay on individual imports."[61] Under a tariff structure that encouraged Canadian subsidiaries of US computer manufacturers to import entire systems at far lower prices than anyone in Canada could sell for, it was only natural that Ferranti-Canada's systems would always be seen as too expensive. Try as they might, Department of Industry officials found that there was "no way of overcoming the problem under the existing legislation and regulations."[62]

It is important to underscore that the "buy Canadian" issue is not a question of whether the large computer manufacturers had, or had not, a subsidiary in Canada. Most operated some sort of sales and service functions within Canada. However, during the 1950s and the 1960s, only Ferranti-Canada maintained any R&D program in Canada. To make matters worse, the other subsidiaries did not even operate as manufacturing branch plants. Instead, they merely imported, or at best reassembled, the computer systems produced by their parent firms. In describing the tariffs of Canada's nineteenth-century National Policy, Michael Bliss argued that they created a wall specifically over which US business was to supposed to jump.[63] Though the tariff discouraged imports of manufactured goods, it encouraged the creation of branch plants.[64] The tariff structure on data processing systems, however, eliminated the need for US computer firms to jump over any walls or to manufacture in Canada.

Ferranti-Canada's difficulties in selling the FP6000 were not only externally generated. There were also internal impediments. Though the FP6000 was an attempt to break away from its parent firm's technology-driven product development habits, Ferranti-Canada faced serious marketing obstacles. In its sales efforts, the Canadian computer group found it difficult to escape the dominant, technology-driven corporate culture that it had inherited from its parent firm. The Canadian firm sold computers to technically sophisticated users more on the basis of design achievements than on an appreciation of the customer's operations and needs.[65] The dominant mind-set seemed to be that technical achievement would somehow sell itself. Ferranti-Canada's previous sales had been to technology-driven users: the Royal Canadian Navy, the Canadian Post Office, Trans-Canada Air Lines, and the Federal Reserve Bank of New York.[66] But selling to the mass business market was an entirely new experience for Ferranti-Canada. Fred Longstaff, the FP6000's chief designer, put the problem quite succinctly: "Our marketing was zip. We were poles apart from IBM."[67]

As one OECD study later concluded, the ability of a nation's manufacturers to compete against IBM was less a matter of technology and more an issue of marketing and management.[68] To be successful, domestic computer manufacturers had to develop the management and marketing mindset that allowed them to design and build "the type of machines and services needed by the market, and in cases where the market is non existent or too small, to create and enlarge this market," which Ferranti-Canada did do in the case of TCA and the FRBNY.[69] The OECD Expert Group on Electronic Computers went on to emphasize that "manufacturers must sell a service, and not simply systems, however complex."[70] Designing and building the FP6000 proved much easier than selling it. Not only were new marketing skills required, but the company would also have had to invest a great deal money in expanding its marketing efforts if it were going to chip away at IBM's dominance of the Canadian market. Could any Canadian company have mounted an effective enough marketing effort? Probably not.

Proper marketing demanded considerable capital investments in an extensive sales and service network. The Ferranti corporate structure made it difficult for the Canadian subsidiary to mobilize large sums of money. From the time Sebastian Ziani de Ferranti regained control of Ferranti UK in 1923, he and his descendants were determined to keep the company under family control. Tradition played a powerful role in the family's refusal to go public as a way to raise capital. As a result, Ferranti UK relied only on profits or the British banks to finance long-term expansion. As the economies of scale and corporate concentration in both the electrical and computer-manufacturing sectors kept increasing, so did the capital investment needed to compete. Pursuing venture capital was not an option open to Ferranti-Canada. The Ferranti family's refusal to go public became an onerous financial handicap that crippled both parent and subsidiary.

Ferranti-Canada's own struggle to survive in the North American electric power market compounded the problems of raising capital. During the 1950s, offshore competition and dumping wreaked havoc on the entire Canadian electrical capital goods sector. In the early 1960s, excess electric generating capacity, which reduced the demand for new capital goods, further exacerbated the Canadian electrical industry's business hardships. The Canadian Electrical Manufacturing Association warned that unless the federal government acted to protect local manufacturers from unfair foreign competition, Canada would lose its industrial base. Other analysts not directly tied to the industry argued that Canada's problems arose from internal structural problems—too many producers in a small market.

From the point of view of its historical traditions, revenues, profits and losses, capital investments, and labor, Ferranti-Canada's corporate existence was very much rooted in the design and manufacture of electrical capital goods. Thus it was understandable that Ferranti-Canada's struggle to survive as an electrical manufacturer was preeminent in the minds of the company's senior executives and board of directors.[71] In an attempt to overcome the limitations of the small and fragmented Canadian electrical capital goods market, Ferranti-Canada committed much of its energy to going after the US market, an unprecedented move south for the Canadian electric power industry.[72] The deeply entrenched "buy American" attitude of the US electrical utilities had long impeded selling south of the border by Canadian firms. Ferranti-Canada hoped to break this stranglehold. The "Electrical Conspiracy" litigation of the late 1950s, which saw the US Department of Justice convict America's largest electrical manufacturers of price fixing, had shaken the utilities' faith in their suppliers.[73] US utilities, now more open to Ferranti-Canada's sales pitch, even launched their own $9 billion lawsuit against these companies. Ferranti-Canada's success in the US market, however, depended on how effectively and quickly it could improve its capacity to design and manufacture larger but more cost-effective power transformers. To achieve this improvement, Ferranti-Canada had to devote most of its human and financial resources to electrical power R&D. To underwrite any serious entry into the computer business would have entailed a considerable reallocation of capital away from Ferranti-Canada's unswerving plan to save its electrical business.

While Ferranti UK gave Ferranti-Canada relatively free rein to set its own R&D agenda, the parent firm balked at actively promoting the FP6000 in the UK. When a delegation from Ferranti-Canada went to England in the late fall of 1961 to get the head office's approval to enter the general-purpose computer business, Ferranti UK responded coldly.

Since Ferranti UK's senior computer engineers disagreed with the design philosophy being proposed, they wanted nothing to do with the undertaking. The parent firm gave reluctant consent for Ferranti-Canada to purse the new opportunity with the Federal Reserve Bank. Ferranti UK had originally agreed that, if the FP6000 project went well, it would help market the computer in Britain. However, in the 2 years that followed the first sale of the FP6000 the parent firm never honored this commitment. By 1964, unbeknownst to Ferranti-Canada, Ferranti UK's unwillingness to market the FP6000 in Britain and the rest of the Commonwealth reflected the parent firm's decision to cut its own heavy losses and get out of commercial electronic data processing. In the end, however, Ferranti UK's unwillingness to promote the FP6000 in Britain was not the pivotal business issue for Ferranti-Canada's Electronic Division. The real market challenge for the Canadian subsidiary lay in wresting a small piece of the domestic market for data processing equipment from the likes of IBM and Univac.

In 1964, Canada's Department of Industry began to examine the strategic economic importance of having a domestic computer industry. In 1963, shortly after regaining political power, the Liberals had created the Department of Industry. The new department's principal responsibilities were to improve the balance of payments, to foster the efficient development of the Canadian manufacturing industry, and to encourage industrial R&D. The very terms of the new department's creation were a call for more nationalism in Canadian industrial policy. "In order to maintain Canada as a political entity," senior officials adopted the policy that "Canadians had to be prepared to pay a certain economic premium" in order to foster domestic design and manufacture.[74] A political interest in technological and industrial self-reliance, promoted by the military in the early 1950s and abandoned by 1958, now reappeared under a civilian guise. Many of the more nationalistic electronic experts in the Department of Defence Production now transferred to the new department.

The Department of Industry perceived that computers had become a powerful engine of economic growth in the world economy. Anxious to push for a domestic capacity to manufacture computers, the department contemplated spending some $2 million per year to help Ferranti-Canada develop and market a commercial line of computers.[75] In seeking technical advice from the Defence Research Board's Chief of Computation and Analysis Branch, the Department of Industry explained that it wanted to focus on Ferranti-Canada because "it was the only company free of US backing with the experience and knowledge to enter the general purpose computer field."[76] Since 1962, the Canadian government had already been engaged in an extensive R&D cost-sharing program with Ferranti-Canada to develop data processing peripherals for export.[77] But now, with the FP6000, Ferranti-Canada had an entire system to sell. Before committing any funds to Ferranti Packard, the Department of Industry wanted to assure itself that the technical merits of the FP6000 warranted the investment. Knowing that DRB's Naval Research Laboratory in Dartmouth had recently bought an FP6000, E. A. Booth, chief of the department's Electronics Division, wrote to DRB on 15 July 1964 asking that it, as a user, evaluate the merits of the FP6000.[78] The contents of DRB's evaluation remain unknown; however, no report, no matter how glowing, could have helped. A corporate drama in the United Kingdom, over which Canada had no control, rendered the Department of Industry's intentions to help Ferranti-Canada become Canada's first successful manufacturer of commercial data processing systems irrelevant.

British Corporate Needs and the Death of Computers at Ferranti-Canada

In early 1963, Vincent Ziani de Ferranti approached International Computers and Tabulators Limited (ICT) with an offer to sell Ferranti UK's non-military computer operations. Behind this gesture was the visible hand of the British government trying to consolidate the British computer business. Great Britain, like most other industrialized nations of the day, was desperately seeking ways to strengthen its domestic computer industry against IBM. ICT expressed little interest in Sir Vincent's proposal. As technically sophisticated as Ferranti UK's computer product line was, Ferranti UK's product mix did not match ICT's perception of market needs.[79] Like IBM, ICT had roots in punched-card machines and other office equipment.

Despite its rejection of Sir Vincent's first offer to sell, ICT realized that it could greatly benefit from the R&D potential and manufacturing capability that Ferranti UK had to offer. The appearance of the FP6000 computer increased the attractiveness of Sir Vincent's offer to sell. ICT had already decided to develop its own mid-size computer. The FP6000 offered a readily available, proven technology. An internal Ferranti UK report acknowledged the usefulness of the Canadian computer to ICT's plans and "Were we to begin designing now a machine in the same price/performance range as the FP6000," the report noted, "we would have in some 18 months time a system that would not be significantly better—if indeed it were any better—than the FP6000."[80] Seeing the computer firsthand and speaking with Ferrari-Canada engineers confirmed Ferranti UK's evaluation. In April 1963, ICT's upper management concluded that the FP6000 should serve as the technological basis for ICT's entry into the market for mid-size computers.

With ownership of the design rights to the FP6000 central to the deal, ICT acquired Ferranti UK's non-military computer operations in June 1963. This deal was premised on the guarantee that the design rights to the FP6000 would be included. John Picken, a former technical director and member of the board of Ferranti UK, later recalled: "Without the FP6000 we would not have gotten the deal we wanted from ICT. The FP6000 was the golden brick in the sale of our operations."[81] In the words of Donald MacCallum, another Ferranti UK board member (and also the director of the company's Edinburgh operations), the FP6000 was "the jewel in the crown that ICT had bought."[82]

It remains unclear whether Sir Vincent consulted Tom Edmondson, Ferranti-Canada's president and CEO, before agreeing to sell the FP6000 rights to ICT. More than 25 years later, Edmondson contended that the deal was contingent on the Canadian subsidiary's consent.[83] With FP6000 crucial to the ICT deal, it is doubtful that Sir Vincent, as the owner of all the Ferranti-Canada shares, would have agreed to sacrifice the deal to keep the FP6000 in Canada. If Ferranti-Canada's senior executives did indeed have a say in the matter, then the sale of the FP6000 to ICT reflected in part a preoccupation with saving their troubled electric power business. The $1 million compensation offered to Ferranti-Canada must have seemed very attractive to executives faced with several years of financial losses. But equally important, any Ferranti-Canada consent also reflected the realization within the Canadian company's most senior executives, all of whom had backgrounds in the electrical power industry, that the company did not have the capital resources to back a serious fight for market share with the likes of IBM, particularly since it would surely mean sacrificing much of the electrical business. Immersed in the electrical power tradition, they still saw the company's mission in terms of the older technology. Knowledge of the deal did not reach the middle management or the engineers in the Electronics Division until 1964.

Although the deal permitted Ferranti-Canada to follow up sales leads that had been started in 1963, it effectively killed the company's computer business. The Ferranti-Canada computer group approached ICT with a proposal to design and manufacture ICT's proposed 1905 and 1906 computers in Canada. The group tried to convince ICT not only that it was technically well suited for the task but also that Ferranti-

Canada offered ICT a gateway to the North American market. Unfortunately, Ferranti-Canada was not part of ICT's corporate strategy. By the summer 1964, that had become apparent to the computer group at Ferranti-Canada.

The resignations of most members of the members of the hardware development team (including Gord Lang, Fred Longstaff, Don Ritchie, and Ted Strain) was quickly followed by the resignations of the entire software development team (David Butler, Brian Daly, Bob Johnston, Ted McDorman, Jim McSherry, Roger Moore, Audrey Sharp, Ian Sharp, and Don Smith). The hardware team then formed a digital electronics firm called ESE. When ESE was acquired by Motorola Information Systems, Lang, Longstaff, and Strain moved into that company's top management. The software team formed I. P. Sharp Associates, which went on to build a profitable on-line global database service and to become one of Canada's largest computer-services houses.[84] The men who had been involved with Ferranti-Canada's high-speed tape readers and drum memories (Cliff Bernard, Rod Coutts, Lawrie Craig, and Al Vandeberg) resigned in 1967 to set up a digital electronics engineering company called Teklogix.

The success of ICT's 1900 series, which strengthened the British computer industry, turned the FP6000 into another Canadian icon of lost technological and industrial opportunity. In 1972 it was asserted that the FP6000, "the first true time-sharing, multi-programmable computer on the world market back in 1961–1963," represented a lost opportunity to develop a domestic, general-purpose computer industry.[85] D. H. Wilde, who had been the manager of the largest FP6000 installation at the Saskatchewan Power Corporation, similarly claimed that the FP6000 was "certainly ahead of its time" and that after 8 years in operation the computer "still measure[d] up in many respects to current models."[86] In their book *Entering The Computer Age*, Beverly Bleackley and Jean LaPrairie repeated the argument that the FP6000's multiprogramming capability was "a significant technological breakthrough" that "preceded similar capabilities later introduced in the equipment of other manufacturers."[87] In more dramatic language, the journalist David Thomas has contended that the sale of the rights to Canada's FP6000 to ICT "annihilated Canada's capacity to build major computers."[88]

Yet the FP6000 did not mark any radical breakthroughs in the art of computer design. The story of the FP6000 is one of creative adaptation. Endowing a mid-size computer with multi-programmable capabilities became the technical solution to the problem of how a small Canadian company could carve out a niche for itself in a market dominated by IBM. No one can deny that the sale of the design rights to the FP6000 dealt a mortal blow to Ferranti-Canada's computer business. But the commercialization of the FP6000 faced many serious, if not insurmountable, problems even before its acquisition by ICT. Ferranti-Canada's inability to make a solid business of the FP6000 could not be simply blamed on a branch-plant economy that thwarted national interests. Neither was it, as J. J. Brown argued, due to timid Canadian entrepreneurs' unwillingness to run risks. The demise of the FP6000 was due to a constellation of factors: the growing corporate concentration in the North American and British computer industries, the very high costs of marketing computers, the different technological cultures within Ferranti-Canada's management, the inability to sell to business users, the difficulty of raising equity within a privately owned company, the precarious financial state of the parent firm, the British industrial policy, and the Canadian tariff structure.

Though it is doubtful that the Canadian government's intervention at the last moment could have reversed the events that led to the sale of the FP6000 to ICT, Ferranti-Canada' subsequent withdrawal from the computer business underscored the absence of a national industrial strategy in the area of computer technology. How long could the Canadian government continue to avoid having such a policy, particularly when most of the major industrialized nations were scrambling to strengthen their domestic computer industries? A mounting anxiety gripped the industrial policy makers of Japan and Western Europe. At issue was the devastating effect that the international expansion of the US computer industry was having on these nations' capacity to design, manufacture, and market computers successfully to their domestic markets.

International Panic over the Growing Technological Gap

Between 1952 (when the first commercial computer was sold) and the mid 1960s, the US computer industry grew so strong that it overwhelmed the domestic computer industries of other nations.[89] Not only was the United States by far the largest user of computers, both in absolute and relative terms; US firms accounted for 95 percent of the world's production of computers.[90] How did the US computer industry rise to such unrivaled supremacy? "First-entry advantage" is the usual response. Though it is correct, this answer conceals the entire matrix of conditions that sustained this advantage.

In the beginning the United States shared the technical stage with other countries, but soon a conjunction of socio-economic and political factors forced all the other countries off center stage. Not only did US military enterprise provoke first entry, but its spending dramatically amplified and sustained US technical competence.[91] Heavy civilian government procurement provided the first profitable commercial market for these technological advances. The capacity of the government to underwrite the early growth of the US computer industry reflected the United States' overall economic hegemony in the early postwar years. While much of Japan and Europe lay in ruins, the United States emerged from World War II with the most powerful industrial base the world had ever seen. The sheer size of the US market, coupled with prosperity, allowed local US computer manufacturers to convert the advantages of large public procurement to growth based on a strong industrial demand. Homogeneity, coherence, and standardization further strengthened the advantages that this large internal market conferred to the US computer industry. Europe's market was potentially larger; however, national differences—reflected in the diversity of social, economic, and political institutions—fragmented this potential.[92] As with much of the United States' industrial history, the internal market became a springboard to dominance in world markets.

As the war-torn economies of Japan and Western Europe regained their vitality, they were determined to narrow the enormous lead that first entry had conferred on the US computer industry. These governments increasingly believed that the widening gulf in computer technology that separated them from the United States undermined the socio-economic and political strength of their nations. At a Ministerial Meeting on Science held by the Organization for Economic Cooperation and Development, a group of experts underscored the strategic significance of the computer industry: "Its importance lies not only in its economic output, which is considerable, but in its far-reaching effects on the whole economic, industrial, social structure of a country."[93] The viability of a domestic computer industry had now become equated with the vitality of a nation's political economy.

European and Japanese policy debates about the United States' domination of the computer industry were directly aimed at countering the power of one US company: International Business Machines, which enjoyed near-monopoly status in both domestic and foreign markets.

IBM first asserted its dominance of the markets in 1954, with the introduction of the medium-size 650 computer. Within a few years, IBM went from no market presence to an overwhelming presence. For the next 10 years, IBM systems made up about 75 percent of the installed base of computers in the United States.[94] In 1965, IBM accounted for 74 percent of the annual revenue generated by the computer industry in the United States.[95] With world demand expected to grow very rapidly, the computer market attracted many competitors. Forced to share a small piece of the treasure, companies from around the world looked for weaknesses in IBM's technical and market strategies that they could exploit.

In the early 1960s, IBM, determined to keep its tight grip on the world computer market, undertook a bold technological and marketing offensive: the design and sale of the 360 series. The 360 concept represented a dramatic departure in product development strategy. Until the mid 1960s, every computer manufacturer made two or three different models. Varying in size and performance, the models were attuned to the needs of different customers. Each model had its own unique software environment, with little if any compatibility between models. Operating systems, software tools, and applications were inextricably tied to the specific architecture and logic design of each model. This incompatibility required customers to face the daunting and expensive process of migrating their costly investment in software applications to a new software environment every time they wanted to move up to a new system.[96] From the manufacturers' perspective, software incompatibility limited their ability to sell customers newer models. The 360 series was the first fully integrated line of small, medium, and large computers. From the users' perspective, there would be compatibility up and down the product line. Users could start with a small computer; then, as their needs increased, they could "buy up" to larger systems without sacrificing their existing investment in software applications.

Put on the market in 1964, the IBM 360 represented a truly ambitious undertaking, which one historian "compared with the massing of troops and equipment for the D-Day invasion of France during World War II."[97] No doubt this is how European and Japanese manufacturers must have felt about IBM's 360 campaign. Before the 360 program, IBM had never reported annual capital expenditures of more than $340 million. By 1966 (2 years after the 360 entered the commercial market), the figures had risen to an astounding $1.6 billion.[98] But, equally telling, from 1961 to 1967 IBM's long-term liabilities rose by $1 billion.[99] In view of the enormous R&D costs and risks, no other company had the deep pockets to take such a gamble. Furthermore, any company that tried to introduce an integrated line of computers had to contend with the costs and technical problems of migrating its existing client base to the new line. IBM's mystique, market dominance, and financial resources allowed it to impose the 360 without fear of a wholesale defection of its old customers to other computer makers. With the 360 line, IBM had increased the stakes considerably. Every competitor knew that it had to respond. The world's computer manufacturers had two options: invest heavily or fold.

The enormous success of the IBM 360 series had a profound impact on the industrial development strategies of governments around the world. Dissipating the first-entry advantages of US firms had just become harder. Though IBM faced some competition from other US companies (Sperry-Rand, Control Data, RCA, Honeywell, Burroughs, General Electric, and National Cash Register), it faced little competition from foreign firms. In the late 1960s, the dominance of the US computer industry showed no signs of diminishing.[100] Japan and the Western European countries argued that it was more imperative than ever before to establish strong domestic computer industries in order to regain lost revenues. To make matters worse for European and Japanese computer manufacturers, the growth of sales in the United States slowed down considerably from about 1969 to 1971. Aggressively seeking new markets, US firms took their competitive battle with IBM to the international arena.

Although Western Europe and Japan shared the policy goal of promoting domestic capacities to design and manufacture computers, they adopted very different strategies. While each country in Western Europe looked to a "national champion" to defend the domestic market, Japan favored pre-competitive cooperation among several candidates. International Computers Limited, Compagnie International de l'Informatique, Siemens, Philips, Olivetti, and Saab, respectively, were to be the favored instruments of British, French, German, Dutch, Italian, and Swedish industrial policy. Governments pushed small firms to merge with the state's chosen instrument. Competition in domestic markets was reduced as government policy chased after economies of scale and their perceived advantages.

The British government favored the rationalization of its fragmented computer industry into one large entity capable of mounting a credible defense against the American invasion. Ferranti-Canada and its FP6000 computer got swept up in this rationalization when the British government forced Ferranti UK to sell its data processing business to International Computers & Tabulators. Twenty years later, Harry Johnson, who had been a senior executive in Ferranti UK's computer development program, remarked: "It was becoming clear that the business, particularly in the burgeoning commercial applications sector, was one in which only a small number of suppliers prepared to invest on an enormous scale could compete successfully."[101] In 1968, the British government pushed for the merger of ICT with English Electric Computers, with the participation of Plessy, to form International Computers Limited. Not only did the British government become a holder of 3.5 million common shares; it also committed £13 million to supporting ICL's R&D over 3 years.[102]

In 1951, when computers were first coming on the market, the French company Bull produced the successful Gamma 3 model. Despite a series of more advanced models, Bull's share of the French computer market kept shrinking while IBM's kept increasing. By the early 1960s, Bull and IBM sales accounted for 30 percent and 60 percent of the French market, respectively. Bull was already suffering considerable losses; its future became hopeless after the appearance of the IBM 360 series. The value of Bull stock plummeted.[103] By 1965, 90 percent of the French computer market was in American hands and Charles DeGaulle was in power. For France as for Britain, a strong domestic source of computers was more than just a question of economic nationalism and balance of trade; military self-reliance and the broader issues of sovereignty were also involved. The initial refusal by the United States to allow France to buy US computer technology for its nuclear research program provoked the nationalist DeGaulle to act. Even after the US government capitulated and agreed to let Control Data to sell computers to France's atomic energy research program, DeGaulle wanted to make a clear political statement of French sovereignty. In December 1966, determined to put a significant portion of the French computer market back in French hands, the government forced two smaller computer manufacturers to merge to form the Compagnie International de l'Informatique. Under a new national strategy called Plan Calcul, CII was given 5 years to produce a full range of computers to compete against those of US manufacturers.[104] To this end, the French government also committed considerable money over 5 years in the form of direct aid for R&D and easy access to credit.

Germany also intervened directly to promote two national champions: Siemens and Telefunken. The first was to specialize in business data processing, the second in large-scale information systems and industrial automation. The goal was to help these two companies develop the technological competence needed to design and build their own lines of business and scientific computers by 1972. The German government committed $75 million (US) in matching grants for R&D and $87.5 million in loans to make computer manufacturing more competitive. As a further measure to ensure the viability of a domestic computer industry, subsidiaries of foreign firms were effectively eliminated from receiving any of this public money.[105] And in the Netherlands, the government looked to the manufacturing giant Philips as a domestic champion. Like the British and the French, the Germans and the Dutch advocated coordinated government procurement as an important tool to foster their domestic computer industries.

The preoccupation of foreign governments with the issue of a domestic computer industry went beyond the mere existence of local manufacture, of which US subsidiaries were doing more. By the late 1950s, IBM had dramatically expanded the scale and the scope of its international computer operations.[106] The most dominant computer manufacturer on the globe, IBM, was also the most international and decentralized firm. Under the organizational umbrella of IBM World Trade, IBM's international subsidiaries operated apart from the company's US operations.[107] The 360 project contributed to greater cooperation between IBM World Trade and the US operations. Spread across several countries, the development program of the IBM 360 provoked greater coordination and rationalization across the company's worldwide operations. By 1969, IBM had established ten manufacturing facilities, one basic research laboratory, and six development laboratories in Europe, where IBM produced computers locally and supported local product design. Employing some 65,000, these facilities were staffed and managed almost entirely by Europeans.

IBM thus played an important part in advancing the technological and managerial competence of the Western European nations. Yet these same governments portrayed IBM as the principal obstacle to European self-reliance in computer technology. At issue was the nationality of ownership. Identified with national well-being if not national survival, self-reliance in computer technology, like self-reliance in defense, became a geopolitical imperative. The ultimate decisions affecting the fate of a nation's computer industries could not be left in foreign hands. Even Thomas Watson Jr., CEO of IBM, understood the nationalistic pressures at work. "It is easy to see," Watson observed, "why an Englishman, Frenchman, or German would not want absolute control of an industry crucial to his security and survival or even to his country's economic future; why he would not want all its shots called by men of another country, thousands of miles away whatever the advantages of American management and technology can bring him and his fellow countrymen."[108] To what extent did direct foreign investment, and hence the nationality of ultimate corporate control, diminish a country's capacity to expand the breadth and depth of its technical and managerial competence? Regardless of the answer, the existence of a strong domestic computer industry was not only a strategic asset but also a matter of international prestige for the United Kingdom, France, Germany, and Japan.

In contrast with the Western European countries, Japans's attempt to catch up to the US computer industry was based on "cooperative research superimposed on a highly competitive national market."[109] Japan's remarkable success in catching up to (and some aspects surpassing) the US computer industry rested on four institutional cornerstones: the Ministry of Trade and Industry and its technical arm, the Electrotechnical Laboratory; the government-owned telecommunications giant Nippon Telephone & Telegraph; the Ministry of Education, which ran the national universities; and industry in the form of the Nippon Electric Company, Fujitsu, Hitachi, Toshiba, Mitsubishi Electric, and Oki.

In the early 1960s, the government first started to expand Japan's technical and managerial competence in computer technology through the use of high trade barriers. To get access to the important Japanese market, foreign firms had to enter into a joint venture with Japanese firms in which there would be an important transfer of know-how. The visible hand of MITI was always present in setting up these ventures. Like the rest of the world, Japan was also forced to reassess its policy on computer technology when IBM launched its 360 line. Hybrid integrated circuits were a new feature of the 360. With little competence in the emerging technology of integrated circuits, MITI created the Super High Performance Electronic Computer (SHPEC) program. Running from 1965 to 1971, the SHPEC program was an ambitious undertaking that brought together the government labs of Nippon Telephone & Telegraph and the Electrotechnical Laboratory and the industrial labs of the Nippon Electric Company (NEC), Fujitsu, Hitachi, Mitsubishi Electric, Toshiba, and Oki into a coordinated effort to advance Japan's competencies in microelectronics, a critical facet of designing and building large computers. The SHPEC program fostered a sophisticated pre-competitive network of users and suppliers that, through a cumulative process of learning by doing, moved the competence of the entire Japanese computer industry closer to that of its US competitors. Another cooperative network that encouraged industry-wide learning revolved around the Japan Electronic Computer Company (JECC). A government-sponsored joint venture, JECC was a "middleman" that bought computers from the six Japanese manufacturers and rented them to users. MITI also gave R&D grants to the members of JECC. Membership in JECC, however, was denied to foreign computer manufacturers. [110]

While the governments of the various European countries were encouraging the consolidation of their respective computer industries around national champions, Japan pursued diversity and competition in its domestic market. Joint ventures with US computer manufacturers helped sustain this diversity.[111] But in the early 1970s, many of IBM's American competitors were facing serious financial difficulties. Realizing that many of Japan's joint ventures with US firms were now in danger of collapse, MITI encouraged even greater pre-competitive technical cooperation within Japan's computer industry. Hitachi joined forces with Fujitsu, NEC with Toshiba, and Mitsubishi Electric with Oki. By the mid 1970s, the efforts of the Japanese government to foster a strong computer industry had begun to bear fruit. Japan was now producing mainframes that outperformed the IBM 370.[112]

Canada's Search for a Policy Framework: Limited Options for a Small Market Economy

Since the late 1940s, Canada's pursuit of military self-reliance had spawned a coherent and coordinated policy to build a national industrial competence in digital technology. In the late 1950s, this flirtation with self-reliance came to end. A civilian-oriented industrial policy was slow to fill the void left by the military's retreat. In 1963, an institutional framework for civilian-oriented industrial policy was established with the transformation of the Department of Trade and Commerce into the Department of Industry, Trade, and Commerce (DITC). With the department now organized along industrial lines, computer technology came under the Electrical & Electronics Branch. Despite the new institutional framework, a policy on the computer industry was not immediately forthcoming. And yet, the scramble by the governments of Western Europe and Japan to create domestic computer industries underscored the need for Canada to articulate a national strategy. With the forecasted market for computers in Canada ranking about sixth in the world in absolute terms[113] and per capita ranking second behind the United States, could Canada continue to risk not having a strategy? By 1968, the belief that industrial competence in digital technology was essential for nation's future well being had finally established a toehold in the politics of Canadian industrial policy.

In January 1968, Minister of Industry, Trade and Commerce Charles Drury underscored Canada's plight when he presented the cabinet with his department's first policy pronouncement on computer technology.[114] Although each of the US computer manufacturers established some form of marketing and sales subsidiary in Canada, none had production facilities in Canada (beyond perhaps some very rudimentary assembly operations). Even IBM, with its vast international network of R&D and manufacturing facilities, had little value-added activity underway in Canada. If allowed to continue, Drury argued, the absence of computer R&D and manufacturing would exacerbate Canada's negative balance of trade with the United States in sophisticated manufacturing, would promote the exodus of highly skilled labor from Canada, and would seriously undermine the innovative capacity of other key Canadian industries. A domestic competence in computer design and manufacture was essential to Canada's ability to seek appropriate applications of digital technology in other important spheres of economic activity. Drury warned that unless the government took action "Canada would be left in a unique and unenviable position among modern industrial nations." Drury continued: "The development of her computer industry and thus the most effective application of computers throughout the economy will be largely dependent on external forces possibly acting counter to the national interests."[115]

DITC's proposed policy response to the absence of a Canadian industrial competence in computer development and manufacturing had four dimensions. The first emphasized the role of the multinational corporation. Unlike Western Europe and Japanese computer policy, who were obsessed with reducing the relative participation of US manufacturers in their local markets, DITC welcomed greater US involvement. No nation had received more direct US investment than Canada.[116] In the late nineteenth century and most of the twentieth, countries erected tariff barriers to keep US firms out. Canada, however, set up tariff walls to encourage US firms to jump over them and locate in Canada. There was always the belief that multinational corporations and Canadian interests, though not congruent, could have sufficient overlap to benefit Canada's long-term development. Under the banner of "moral suasion," DITC was determined to prod, coax, and if need be even coerce US computer manufacturers to invest in R&D and high-level manufacturing activities in Canada to a level commensurate with their sales in Canada.

With 70 percent of the Canadian market, IBM was the prime target for this "moral suasion." Despite IBM Canada's legal membership in IBM World Trade, the Canadian subsidiary did not share the same opportunities as its European counterparts. IBM's Canadian activities had remained relatively static through the 1950s and the 1960s, while its European R&D and production activities had grown.[117] By 1970, IBM increasingly viewed the Canadian subsidiary as just another division of its US operations. DITC wanted IBM to give its Canadian subsidiary the same opportunities enjoyed by its subsidiaries in other advanced industrialized nations.[118] In a letter to the president of IBM Canada, Drury underscored that the government's request for greater Canadian content was not unreasonable in view of the existing "IBM policy of internationalization through the establishment of computer manufacturing and research and development facilities in all its major markets."[119] To ensure that IBM understood the importance of cooperation, Drury asked the cabinet to consider the threat of anti-IBM procurement policies as sanctions for non-compliance.

Specialization was the second dimension in the DITC's strategy. Although DITC did not expect multinationals to transfer entire product lines to Canada, it did set out to convince parent firms to give their Canadian subsidiaries global mandates to develop and manufacture products along narrower specializations. In effect, DITC was asking multinationals to rationalize their production on a global scale, with subsidiaries being given specific roles.[120] But what would a multinational gain by such a move? Could the financial inducements offered by DITC offset any of the organizational problems the parent firm would face from relocating its R&D and production activity? How would shifting more operations to Canada increase a multinational's overall global competitive advantage? The cabinet memorandum and the supporting documents were devoid of suggestions on how to sell the strategy of specialization within the multinational's own frame of reference.

The imperative to export was the third dimension. DITC was worried over the growing negative trade balance in computer related products. Like other nations, Canada was willing to accept the logic of comparative advantage over a variety of product groups, but it could not do so for computer products. In the name of national well-being, governments believed that they could not abandon their internal computer markets to foreign suppliers. Such was the perception of the computer industry's strategic importance. Yet DITC understood that Canada's internal market alone could never sustain a strong domestic computer industry. To thrive, this industry would have to export. The entire premise of specialization for foreign-owned subsidiaries was to sustain exports. DITC wanted parent firms to give each of their Canadian subsidiaries the mandate to be a center of excellence in one aspect of the multinational's global product development and manufacturing activities. Only on such a scale of production could a subsidiary sustain a strong R&D capacity over the long term.

When Drury went to the cabinet with his department's proposed strategy for promoting the design and manufacture of computers in Canada, he only had one concrete example of a subsidiary successfully exploiting a product mandate for export markets: the Sperry Gyroscope Co. of Canada, or Sperry Canada. Yet, ironically, DITC's response to Sperry Canada's needs was ambivalent if not confused. DITC's R&D assistance turned out to be too little too late. In the drive to break away from unreliable military procurement, Sperry Canada pursued Schumpeterian product strategies in numerical control and in computer numerical control. These initiatives were conceived, managed, and financed at the subsidiary level, with no parent-to-subsidiary transfers of managerial or technical competence. The subsidiary singlehandedly moved itself up the learning curve. But DITC, like most of the management scholars of the day, could conceive of the parent-subsidiary relationship only as a one-way, top-down decision-making process. Parent firms handed out mandates; subsidiaries were passive entities. But the Sperry Canada story offers a striking example of the possibility of locally driven strategic initiatives within large multinational corporations. Expecting the parent firm to fund Sperry Canada's R&D efforts, DITC never fully appreciated that Sperry Canada was on its own when it came to numerical control and computer numerical control. As result, DITC was more parsimonious in its support of Sperry Canada than it would have been had NC and CNC products been developed by a Canadian-owned company.

The fourth dimension of DITC policy was promotion of a Canadian-owned computer industry. Again DICT believed that Canada's interests would be better served if an important part of the computer industry was in Canadian hands. Again DITC stressed the pursuit of global niche markets. In DITC's view, only local ownership could sustain the high costs of innovation and attain the economies of scale needed to compete. Advocating local ownership was easy; devising a plan to achieve it was much harder. DITC had no strategy other than to encourage Canadian entrepreneurs, through incentives, to identify narrow niches. DITC wanted a local winner but did not know where to look. Software as one possible viable area of specialization for Canadian-owned companies. Although the market for off-the-shelf "shrink-wrapped" software was still many years away, the emerging Canadian computer services industry was one area where specialization in software could work. But since computer services did not manufacture "boxes" and did not export, DITC failed to make computer services an integral part of its strategy to support a locally owned computer industry.[121] Still unsure of specific specializations to encourage, DITC could only wait and react to whomever would step up to bat.

Although the memorandum sounded a clear alarm, its recommendations to the cabinet were remarkably cautious. With a small market and relatively open economy whose secondary sector was dominated by US foreign direct investment, the Canadian government did not have a great deal of latitude in the policy options it could reasonably pursue. DITC officials had carefully examined the approaches taken by other governments. But none seemed appropriate for the Canadian context. There were no calls for a domestic champion to do battle with US computer manufacturers; no suggestions of a "made in Canada" computer system. First, there were simply no Canadian-owned candidates to fill the role of "national champion." To promote the creation of a new company or consortium was also not a realistic option. Entry costs were extremely high. National champions needed a steady and rich diet of government support for R&D if they were to have any chance of survival. A great deal of money would have had to be spent before a startup company could gain the technical and the managerial competence needed to design, develop, test, manufacture, market, and support a competitive Canadian computer system. The government of the day had neither the will nor the capacity to finance such a risky venture. But, equally important, the Canadian government's procurement practices could not offer a national champion the guaranteed market needed for survival during the crucial early years of operation. The procurement of computers within the Canadian government was decentralized. Each department was free to chose the system it wanted. But since the purchase of a mainframe was such an important and costly decision, senior managers tended to fall back on the security of the IBM name, or even the Control Data or the Sperry Rand name.[122] Though DITC may have wished that government computer purchases be centralized and used as instrument of industrial policy, it knew that there was little hope of stripping departments of their procurement prerogatives.

Having rejected the European and Japanese policy models, DITC's Electrical & Electronics Branch found inspiration elsewhere. Three of the four important elements in the DITC strategy—multinationals, specialization, and exports—bore close resemblance to elements of a landmark trade agreement that Canada had signed with the United States for the automotive sector.

Subsidiaries of multinational corporations based in Detroit had long dominated the Canadian automobile industry. Historically, Canada and the United States had both imposed substantial tariffs on the entry of assembled vehicles and parts. Behind a tariff barrier, Canadian subsidiaries evolved into miniature manufacturing replicas of their US parents (Ford, General Motors, and Chrysler) and produced only for the domestic market. Although producing for a much smaller market, Canadian automotive manufacturers still had to offer the same variety of products as their US parents if they were to compete successfully. As a result, Canadian producers could not benefit from economies of scale. The Royal Commission on Canada's Economic Prospects had already underscored this problem as an systemic problem of small-market manufacturing.[123] Dependence on a small market, one study for the Royal Commission observed, prevented Canadian manufacturers from "obtaining maximum economies from mass production and specialized operations."[124] This report adds: "In comparing manufacturing plants in Canada and the United States, one is struck no so much by the relatively greater size of the United States production units but by the fact that the Canadian plant in practically all cases produces a much greater variety of products for its size."[125] One solution to this problem lay in getting competitive access to large foreign markets. In 1965, Canada and the United States signed the Automotive Products Trade Agreement (more commonly known as the Auto Pact).[126] This bilateral agreement established a kind of free-trade zone between the two countries' automotive industries.

Welcomed by Detroit's Big Three, the Auto Pact was also a boon for Canada. Under the duty-free umbrella, the large US automotive manufacturers treated their production facilities in Canada and those in the United States as one integrated system. Multinationals rationalized their production and inventories over Canada and the United States. In agreeing to an open border for automotive products, the Canadian government negotiated safeguards for its domestic industry. US manufacturers could ship duty free into Canada as long as a car was produced in Canada for every car shipped into Canada. For Canadian subsidiaries, rationalization led to specialization. Instead of producing all models of cars and a large catalogue of original equipment parts for just the domestic market, Canadian plants were given mandates to supply a smaller range of models for the entire continent. At long last the Canadian automotive industry had a market large enough to support the economies of scale and the efficiencies of its US counterparts.[127]

Although the full extent of DITC's participation in the shaping the Auto Pact has yet to be assessed by historians, it cannot be denied that the agreement was a major coup for DITC, particularly since the responsibility for industrial planning had been added only 2 years earlier. The success and the lessons of the Auto Pact surely reverberated throughout all the branches of DITC. The E&E Branch had no illusions of getting a parallel agreement with the United States on computers. The corporate and spatial structures of the two industries were very different, as was the international competitive scene for the two industries. In the automotive sector, each of the major industrialized nations had its own strong champions. Three US giants ruled the Canadian automobile industry. The computer industry, on the other hand, had only one giant: IBM. But of more significance was the long existence of a considerable manufacturing investment by the US automobile manufacturers in Canada. Not only did manufacture in Canada bypass tariff walls; it also offered a low-tariff gateway into the British Commonwealth. In other words, the fact that each of the Big Three had significant auto-production facilities in Canada was an essential precondition to the Auto Pact. In the case of computer industry, there were no manufacturing facilities in Canada. Nevertheless, it is clear that DITC did try to adapt certain elements of the Auto Pact: the central role of the foreign-owned multinational corporation, specialization, international rationalization of production, and the need to access large export markets.

A few months after submitting his memorandum "Canadian Computer Industry" to the cabinet, Charles Drury announced to the public his department's intention to champion the emergence of a strong Canadian capacity to design and manufacture computer technology.[128] Drury pointed to the design and manufacture of digital controls for machine tools as a clear example of where his department's efforts were heading. Though he mentioned no names, Drury was referring to the efforts of Sperry Canada, a US-owned subsidiary, to compete globally in this new market niche. The visible hand of public policy had little to do with Sperry Canada's decision to pioneer the use of computers to control batch manufacturing, but DITC's attempt to incorporate Sperry's strategic initiative into a government industrial policy proved to be an ambivalent if not a confused undertaking. DITC's policy had difficulty responding to the needs of highly independent subsidiaries.

There is another important dimension to the story of Sperry Canada. The development of electronic data processing equipment by Ferranti-Canada suggests a Schumpeterian vitality to the behavior of foreign-owned branch plants that goes against the dominant view, promulgated by "nationalist" political economists, that subsidiaries were agents of dependency and de-industrialization. To be sure, one can argue that Ferranti-Canada was an isolated case and that, as a British-owned firm operating under what Alfred Chandler called "personal capitalism," this company does not in anyway detract from the focus of the nationalist argument in which "arrested" development is identified with the domination of US-owned subsidiaries. As the next chapter will make clear, Sperry Canada, a subsidiary of a large US multinational, also acted as an agent of radical product innovation in digital electronics with its own ambitious global business agenda.

6
The Sperry Gyroscope Company of Canada and Computer Numerical Control

In 1962, while Ferranti-Canada was trying to define a technological and commercial niche in the electronic data processing market, a Montreal firm, the Sperry Gyroscope Company of Canada Limited,[1] set out to become the first company in the world to design and commercialize the fusion of the programmable computer and the machine tool. The development of computer numerical control (CNC) was not a case of simply taking an off-the-shelf computer and attaching it to a machine tool. CNC required significant adaptive redesign of existing mainframe-oriented hardware, a completely new software environment to reflect the real-time needs of the machining process, and the design of appropriate servo-control technology. The major questions in this adaptive innovation process were how to match design goals with the available corporate R&D resources and how to define performance specifications that would not make the new CNC technology prohibitively expensive for machine tool shops.

Despite the considerable adaptive innovation needed, none of the components of Sperry Canada's CNC system marked a dramatic advance in the state of the art. The technological and marketing boldness of Sperry Canada's 1964 innovation, which was called UMAC-5,[2] was to conceive a programmable computer and a machine tool as a single integrated production unit on the shop floor. Each machinist thus directly operated his own computer and machine tool center. This digital version of automation differs from the notion of a centralized mainframe, which sits in an air-conditioned office far removed from the harsh environment of the shop floor and sends orders to all the numerical-control-equipped machine tools there. Acting as an agent for the parent firm, Ferranti-Canada also competed against Sperry Canada in the numerical control business. And yet, despite the expertise of Ferranti-Canada and Ferranti UK in computer technology, neither of them attempted to make the conceptual jump to CNC in the early 1960s.

In 1968, 3 years after the demise of the FP6000, Sperry Canada's aspirations to be in the vanguard of computer numerical control technology faced a similar fate. Why and how did Sperry Canada, created in 1950 to serve primarily as a manufacturing branch plant and having little experience in computer technology, ever venture into this costly high-tech gamble? This chapter examines how, as part of a continual struggle to break away from its total dependence on military procurement, Sperry Canada drifted into computer design and CNC. In the process, the Canadian subsidiary created its own R&D agenda and aggressively tried to shape its own global product mandate. Sperry Canada's determination to develop and market CNC was the act of a classical Schumpeterian firm. Sperry Canada's entry into computer technology again shows, as the Ferranti-Canada case study did, that the nationalist historiography, which depicts subsidiaries in Canada as passive and subservient entities, has been rather one-sided. Another important element in this story will be the important role that the government played in assisting in the development of UMAC-5. The Department of Industry, Trade, and Commerce's support arose from a policy that sought not only to stimulate the use of this technology but also to create a national capacity to design and sell it. The collapse of Sperry Canada's CNC business is a story of how, within the Canadian political economy, a Schumpeterian company was unable to make the monopoly-like profits usually associated with being an early innovator in the fourth innovation wave.

Sperry Canada: A Child of Canada's Flirtation with Military Self-Reliance

On 8 November 1950, the Sperry Corporation established its Canadian subsidiary Sperry Canada. Created on the eve of the Korean War, Sperry Canada had a clear and focused business mission: to exploit the commercial opportunities of the new wave of defense spending. In 1950, 42 percent of Canada's defense procurement was devoted to aircraft.[3] This high figure reflected the growing strategic role that the Canadian government assigned to the Air Force in continental defense. For the American parent firm, Canada's defense spending offered an excellent business opportunity in its major technological strength: aircraft instrumentation, particularly sophisticated military navigational systems.

Sperry Canada's charter had a three-pronged mandate: "to do business in any aspect of gyroscopes or gyrocompasses; to design, manufacture, purchase, sell, lease, import, export and deal in any gunfire, bomb, torpedo and rocket control equipment appliances and devices, for use on or under the sea, on land, or in the air; and to export and deal in and with anti-aircraft guns."[4] Sperry Canada's creation was, first and foremost, driven by the mandate to sell the navigational technology developed by the parent firm. The second mission statement, however, held out the promise of some form of advanced R&D in Canada. Yet subsequent events revealed that the parent firm, the Sperry Corporation (later to become Sperry Rand), was unwilling to risk any of its own capital to further this mission statement.

Initially, Sperry Canada served simply as an sales agency. Working out of the International Aviation Building on Lagauchetiere Street West in Montreal, Burton Wensley ("Wence") King, the company's Managing Director, marketed the parent firm's aeronautical instruments.[5] King, a Queen's University graduate in civil engineering, had obtained his expertise in aviation technology during World War II. Upon joining the Royal Canadian Air Force in 1940, King had attended its Aeronautical Engineering School. From Chief Engineer at the RCAF Station in Trenton, King had been promoted to the rank of Wing Commander at the 63rd RCAF Base in Great Britain.[6] Immediately after the war, he had gained valuable experience in government procurement in the aviation industry when Canadair Ltd. had hired him as an Executive Engineer. When General Dynamics had acquired Canadair, King had been promoted to Manager of Procurement. In the winter of 1950, he had left Canadair to create Sperry Canada.

The scope of the government's announced rearmament program created a propitious climate for Sperry Canada to expand its sales infrastructure beyond the one-room sales agency. Sperry Canada's board of directors passed a resolution to borrow $643,000 from the parent firm to buy the Ottawa-based firm of Ontario Hughes-Owens Co. which had a long tradition in the precision engineering, architectural, and surveying instrument business[7] and which for many years had also represented two US aircraft instrument manufacturers: Sperry and Kollsmann.[8] Ontario Hughes-Owens provided an established sales infrastructure with offices in Halifax, Montreal, Toronto, Winnipeg, and Vancouver. This arrangement, however, did not yet give Sperry Canada a manufacturing presence.

The extant public and corporate records do not reveal if, when, or where the parent firm planned to set up design and manufacturing facilities in Canada. But whatever Sperry's plans, Minister of Defence Production C. D. Howe's determination to ensure a greater Canadian defense production capacity was what accelerated the timetable and expanded the scale of Sperry's manufacturing presence in Canada. In rearming, Canada risked a crippling outflow of currency to pay for imported military equipment. Fearful of the havoc that rearmament might wreak on Canada's balance of payments, the government wanted to ensure that military spending created as much business and as many employment opportunities for Canadians as possible. In a speech before the House of Commons on 8 February 1951, Howe argued that military preparedness called for expansion of Canada's productive capacity.[9]

The absence of a domestic capacity to manufacture sophisticated aircraft instrumentation stood out as a glaring weakness in Canada's pursuit of a technologically advanced aviation industry. By the spring of 1951, Sperry Canada and the Department of Defence Production (DDP) had entered into negotiations to establish a Canadian manufacturing facility for aircraft instruments. Determined to promote the transfer of an important portion of the parent Sperry Corporation's production engineering capacity to Canada, the government was prepared to underwrite the construction and equipment costs. In return, Sperry Canada was to lease the facility on favorable terms on the condition that it manufacture aircraft instruments for Canadian defense needs. Underlying this arrangement was the tacit assumption that the government would, at some future date, offer the plant and its equipment back to Sperry Canada at an attractively depreciated price.

Sperry Canada's board of directors authorized B. Wensley King to submit an offer to sell the land and the building at 45 Spencer Street, Ottawa, to the Canadian government for $159,000.[10] The building had been a part of the Ontario Hughes-Owens purchase. Whether this offer signified an attempt by Sperry Canada to get the government to pay for the construction of facilities for manufacturing aviation instrument in Ottawa is not known. "[But] the government," as King reported to the board of directors, "was not interested in buying the land and building."[11] Instead, on 14 June, DDP agreed to build a new facility in Montreal's Cote de Liesse area.[12] By the summer of 1952, when the 90,000-square-foot factory opened, DDP had poured $3.3 million dollars into the new facility: $63,815 for the land, $1,790,000 for the building, and $1,496,185 for the equipment.[13]

In a little over a year, Sperry Canada's staff swelled to more than 1100,[14] mostly production workers engaged in high-precision manufacturing.[15] Airspeed indicators, altimeters, machometers, gyrocompasses, and gyrohorizons were the company's bread and butter. The Sperry machometer was a sophisticated piece of equipment that gave a continuous indication of a plane's Mach number. Because a Mach number relates to shock waves and aerodynamic stability, the machometer was considered essential for the safety of high-speed fighters. The Sperry gyrocompass consisted of a gyroscopic unit and a "flux valve" mounted on a fighter's wingtip. The flux valve sensed the Earth's magnetic field and used this information to keep the plane on course during turbulence.[16] By 1955, when the defense-spending boom had passed, Sperry's Montreal plant had manufactured nearly 6000 aircraft instruments.[17] In addition to its Canadian market mandate, Sperry Canada also obtained the North American mandate to manufacture air-driven gyrocompasses and gyroscopic horizons.

Government procurement had successfully encouraged the diffusion of production engineering to Canada, but it had yet to provoke any transfer of design engineering or the creation of any significant R&D program. The parent firm designed all the aeronautical instrumentation manufactured by Sperry Canada.[18] In rocket and automatic fire control, for which Sperry Canada had an ostensible R&D mandate, there was no progress. One reason for that was the lack of any direct parent firm investment to support such a program[19]; another was the absence of government contracts in this domain. During the Korean War, the Canadian military viewed Sperry Canada as simply a local manufacturer of US instrument technology. As the government's resolve to finance the technological and industrial burden of "military self-reliance" waned in the years immediately after the rearmament boom, Sperry Canada abandoned all hope of developing its mandate in rocket and fire-control systems. Instead, the company set about to create a new product mandate for itself within the Sperry empire—a mandate that would break the Canadian subsidiary's total dependence on military contracts.

The Imperative to Define a New Technological and Business Strategy

Government procurement at Sperry Canada maintained its momentum through 1955, then dropped by more than 50 percent in 1956 and 1957.[20] The contraction of Canadian spending on military equipment in the mid 1950s underscored Sperry Canada's dependence on defense business. From a peak labor force of 1100 at the height of the Korean War, Sperry Canada had slashed its work force to 400 by 1956. In view of unpredictability in the timing and size of military contracts, it became very difficult for the company to maintain a stable labor force. Another problem, of course, was the underutilized productive capacity of the 90,000-square-foot plant during lulls in military procurement. The unreliability of defense procurement made year-round plant utilization impossible to maintain.

Having built an important Canadian manufacturing facility from the ground up, Wence King and his Canadian executives were not about to let their company die because the military well was running dry. Their minds turned to finding civilian applications to which Sperry Canada's technical expertise could be turned. Around 1956, the company set forth a long-term strategy to redirect half of its engineering, manufacturing, and sales efforts to the civilian market.[21] To make this transition, Sperry Canada had to face some hard questions. What should the new product be? How would the development of this product be financed? How could Sperry Canada's efforts to define a technological and commercial future articulate itself within the parent firm's long-term business strategy? The imperative to diversify into the civilian sector also forced Sperry Canada to confront its lack of an R&D capacity.

Changes within the parent firm created both an opportunity and an additional compelling reason for Sperry Canada to redefine its business activities. On 30 June 1955, Sperry—a large, multi-divisional corporation whose business activities spanned all manner of sophisticated weapons systems, aircraft, and marine instrumentation, electronic component manufacture, hydraulic servo controls, and farm machinery—merged with the Remington Rand Corporation. Remington Rand had its origins in gun production: E. Remington & Sons, having founded a business in 1856, had become famous for a rifle bearing their name. Over the next hundred years, the Remington organization had branched out, through numerous acquisitions and mergers, into the office equipment market.[22] Like its office-equipment competitor IBM, Remington Rand had moved into electronic data processing; in 1951, it had created the Univac Division after acquiring the Eckert-Mauchly Computer Corporation[23] and Engineering Research Associates.

By the time of its 1955 merger with the Sperry Corporation, Remington Rand had grown into a business with net sales exceeding $250 million.[24] It employed 13,000 people in 30 manufacturing facilities in 20 cities. It produced all manner of office equipment and supplies, civilian and military electronic computer systems, photographic supplies, and electric razors.[25] For its office equipment business, Remington Rand maintained a 6100-person sales-and-service network in 230 locations and a network of 10,000 authorized dealers.

Rather than drastic reorganization and rationalization, the merger of Sperry and Remington Rand led to the creation of two parallel lines of management authority. The coordination of these two powerful interests took place at the highest level, in the Executive Committee.[26] The Sperry portion of the new company consisted of four fairly independent units, which reflected different markets and technologies: Sperry Gyroscope ("Gyro Group"), Vickers, Ford Instruments, and New Holland Machinery.[27] Sperry Canada was one of eleven geographically dispersed divisions that constituted the Gyro Group.[28]

"Divisionalization and decentralization" became the prime directives soon after the merger.[29] Fundamental to the new strategy was the principle that each division should define its own product mandate. In effect, Sperry Rand became a federation of cooperating profit centers, each with its own distinctive, self-funded R&D programs and its own product lines. An essential feature of the cooperation was that the profit centers avoid competition among themselves. In the case of the Gyro Group, these ideas were enshrined in 1958 as corporate commandments: "EACH UNIT OF THE GYRO GROUP SHALL: 1) Develop, design, manufacture and sell the product lines prescribed by its charter; 2) Refrain from entering new product areas, or areas assigned to other divisions, without prior approval, or specific approval, by the Chairman of the Products Committee; and 3) Purchase its requirements from other units of the Gyro Group or of the Corporation to the greatest possible extent."[30] The last commandment attempted to reinforce complementarity, as well as to minimize cash outflows, by requiring each profit center to favor internal over external sourcing.

Divisionalization and decentralization not only offered Sperry Canada an opportunity to define its own product mandate area; they made it essential. The product-oriented rationale that defined the corporate and geographic organization of Sperry Rand under divisionalization also compartmentalized R&D.[31] The absence of a central corporate source of R&D thus opened the door for an independent Canadian design program—provided that this program did not impinge on some other division's territory. For Sperry Canada, being a profit center became a two-edged sword. With R&D autonomy came the burden of self-financing. In seeking to develop a new product area for itself, Sperry Canada had to juggle market opportunities, avoid territorial conflicts with other divisions, make the best use of a manufacturing facility originally designed for the production of instruments for fighter planes, and test its ability to finance the development and commercialization of all prototypes.

In 1955 Sperry Canada sent engineers to various industries to assess technological opportunities. Control technologies were the focal point of the company's industrial inquiries. This particular technological inclination was a legacy of the only (and as yet unrealized) R&D mandate Sperry Canada had in its founding charter. Unable to design, manufacture, and export rocket and fire-control systems, Sperry Canada switched to civilian applications that entailed similar principles of negative feedback and servo control. Inquiries to the pulp and paper industry to discover potential applications for control technology proved a dead end. Atomic Energy of Canada Limited, on the other hand, asked Sperry Canada to design power controls for its reactor technology. Despite the initial success, Sperry Canada eventually dropped this business in the early 1960s because it yielded only one-off sales, as Atomic Energy of Canada Limited took Sperry Canada's designs and manufactured them in cheaper shops.[32] In 1955, the Canadian subsidiary also started to consider the still-embryonic field of numerical control (the application of digital electronics to the operation of machine tools).

Canada's First Designer and Manufacturer of Numerical Control Technology

Numerical control was pioneered under the patronage of US military enterprise. The development of supersonic aircraft, which began in the late 1940s, created new manufacturing demands on the aviation industry. As the weight of on-board weapons systems climbed, it became imperative to minimize the weight of an aircraft's frame while at the same time maximizing its strength. Machining large sections of an aircraft from solid pieces of metal produced frames with a higher strength-to-weight ratio than the conventional practice of riveting an aircraft together from a large number of parts. To machine down the walls of the aircraft's frame to a minimum thickness without compromising structural integrity required cutting metal in an extremely precise manner along very complex paths. This high-precision machining was extremely time consuming. Furthermore, if, after hours of work one mistake was made, it was necessary to scrap the entire piece and start over. As fighter designers pushed the limits of performance, the precision and the complexity of machining demanded ever-higher precision. Machining time grew linearly with the top speed of the aircraft.[33] Similar machining demands were present in the construction of rocket, missile, and satellite systems.

Work at the Massachusetts Institute of Technology, under the patronage of the US Air Force, demonstrated that the use of electronic digital servo-control techniques could accelerate the production of these complex machined parts. The machining operations were encoded in a digital format: punched holes in a tape. The NC tape resembled the continuous loop of punched cards that Joseph Marie Jacquard had used in 1805 to automate certain aspects of weaving. In Jacquard's system, digital information on a particular weave pattern was conveyed from the cards to the loom through mechanical rods. In the NC system, the tape's control information was passed on electronically to the machine tool's actuators. The key to NC technology lay in the closed feedback loop that automatically, quickly, and accurately ensured that the cutting tool was doing precisely what it should be doing at precisely the right spot on the work piece.

In 1955, when Sperry Canada decided to experiment with numerical control, the technology was still esoteric, and only a handful of US companies were trying to commercialize it. No one in Canada had any real technical knowledge of it. Sperry Canada's experimentation with NC technology had to start from scratch. The company's first task was to find a development engineer with a good grasp of both mechanical and electronic design. Late in 1955, Sperry Canada hired Peter Herzl. Born in Vienna, Herzl had immigrated to Canada in 1939. He had received his Bachelor of Mechanical Engineering degree from McGill University. After his first job, at Northern Electric, Herzl had gone to RCA Victor in Montreal, where he had become involved in designing new ways to automate various manufacturing processes. Though this automation had not been based on numerical control, Herzl's experience in both mechanical engineering and electronics made him, in Sperry Canada's eyes, an excellent candidate to head up the NC experiment. This choice turned out to be a wise one. Herzl's considerable inventive talents became the cornerstone of Sperry Canada's successful entry into the NC business.

From the moment Peter Herzl's first prototype proved successful, in 1957, Sperry Canada set out to claim numerical control as one of its product mandates within the Sperry Rand Corporation. The Canadian subsidiary's NC ambitions, however, had to confront an immediate challenge posed by the head office of the Sperry Gyroscope Group ("Gyro Group") at Great Neck, New York. The need for high-precision small-batch machining of complex parts was an essential feature of the development work on weapons systems done by the all the US divisions of the Sperry Gyroscope Group. Around 1956, work started at the Great Neck headquarters of the Gyro Group to develop in-house NC technology to meet the group's own manufacturing needs. The Sperry Rand Corporation also hoped to spin off a commercial business from the in-house NC developments at its Great Neck facility. The parent firm's NC development program was focused on very large, sophisticated, and expensive prototypes.³⁴

With limited development funds of its own, Sperry Canada's NC design philosophy focused on simpler and less expensive solutions to the handling of less complex machine operations. The imperative to diversify required putting a new product on the Canadian market quickly. Modest design ambitions ensured a shorter innovation cycle. For similar reasons, Sperry Canada confined itself to point-to-point machine tool operations. Herzl's patents demonstrated that there was room for considerable product innovation. within these design constraints

Within the logic of divisionalization and the harsh realities of declining profits, it made no sense for the Gyro Group to have two divisions with NC product mandates. Although the group at Great Neck had succeeded in building several sophisticated NC systems for its own machine shops, no commercial sales had ever resulted. Meanwhile, Sperry Canada had begun to find customers for its simpler NC technology. By early 1958, the head office had decided to relinquish any commercial interest in NC, and the president of the Gyro Group had recognized Sperry Canada's claim to a global NC mandate. Out of the eleven divisions in the Gyro Group, Sperry Canada was the only one listed with a program to develop numerical control. If there were doubts at the head office about this decision, they were silenced after the Great Neck manufacturing facility put Sperry Canada's technology on several of its own machine tools.³⁵

Through a series of inventions, Peter Herzl addressed all the key technical components of an NC system. The precision of the digital machining data contained on punched tape would be of little utility if it could not be properly translated by the machine tool's positioning system, which was analog by nature. The electric or hydraulic actuators that moved the machine tool's table and cutting instruments also were analog devices, as were the systems used to measure position. Depending on the machine tool, the table on which the work piece sat could weigh several tons. Getting this table to move very quickly and precisely in response to digital NC instructions demanded very responsive and accurate analog actuators. Herzl's work produced positional and actuating systems of considerable accuracy and responsiveness.³⁶ Though NC heralded the ascendancy of the digital paradigm into the forces of production, Herzl's innovations suggested that the real engineering challenge in building a successful NC product lay in developing better analog technology.

The Gyro Group's management had given the mandate for numerical control to Sperry Canada, but another group within the Sperry Rand empire retained ambitions in the same area. Harry F. Vickers had founded the Vickers Company in 1920 to produce specialized hydraulic machinery.[37] The Sperry Corporation had acquired Vickers in 1937. By the time of the merger, the Vickers Group produced all manner of hydraulic equipment for industrial, mobile, marine, ordnance, and aircraft applications. Unlike the Gyro Group, the Vickers Group had strong interests in the civilian market. It had devoted "a large share of its program" to "the development of [hydraulic] servo-controlled machines."[38] Since hydraulic servo controls were key components in an important class of numerically controlled machine tools, the Vickers Group's desire to commercialize this technology became a natural extension of its long history in the hydraulics business.

In a 1959 examination of all the product development being undertaken by the Gyro, Vickers, and Ford Instrument Groups, the Sperry Rand Committee on Fields of Endeavor noted that Sperry Canada and Vickers overlapped in machine tool controls. But the committee made no recommendation to encourage or discourage this overlap in product jurisdictions. The Vickers Group had greater financial resources and more corporate clout than Sperry Canada, but any ambitions of the Vickers Group to lay claim to NC had to face Sperry Canada's clear technical lead. While the Vickers Group was still struggling to develop an appropriate electro-hydraulic valve, Sperry Canada had already designed and built an entire NC system. In 1959, Sperry Canada demonstrated a commercial NC system that impressed observers with its ability to move an 8000-pound machine tool bed quickly while still maintaining a repeat accuracy of 0.00005 inch.[39] To what extent the Vickers Group continued to spend money on its own NC ambitions is not known, but 10 years later its interest in the NC business resurfaced and it became embroiled with the Canadian government in a struggle to ship technology developed by Sperry Canada back to its plant in Detroit.

Having won the right to develop NC, Sperry Canada now faced the daunting task of finding the money to do it. The absence of parent-firm funds to directly support Sperry Canada's development of NC was symptomatic of a corporate culture shaped by the funding from the US military. In 1959, for example, the US Department of Defense paid for 99 percent of all the R&D undertaken by the Sperry Corporation.[40] Military R&D entailed few financial risks. With the military the dominant source of R&D revenues, gambling its own money to create new civilian markets did not appear to be a part of Sperry's corporate ethos. But this was exactly what Sperry Canada hoped to do with numerical control.

Even if the other divisions had been open to underwriting portions of Sperry Canada's NC program, Sperry Rand's deteriorating profits during the late 1950s made this possibility even more remote. Sperry Rand's percent of return on capital investment fell from 9 percent for the 1955–56 fiscal year to 3.5 percent for 1958–59.[41] The Remington Rand portion of Sperry Rand saw its returns on investment tumble from 7.32 percent in 1955 to 0.53 percent in 1958. The Sperry half exhibited similar declines. Sperry Gyroscope's treasurer, E. M. Brown, called his group's long-term decline in profits "alarming." From 21.91 percent in 1954, the Gyro Group's return on capital invested steadily declined to 6.9 percent in 1959.[42]

An uncertain Canadian defense market did not help matters. Sperry Canada's capacity to finance NC development depended entirely on the its ability to siphon money off any surpluses generated by its defense contracts. Sperry Canada's Industrial Relations Manager lamented the "horribly" competitive military equipment market in Canada, in which there were "too many companies competing for the very few Defence contracts available."[43] The uncertainty of military procurement manifested itself in the company's financial planning. In 1958, Sperry Canada's board of directors rescinded $683,000 of previously approved capital appropriations. In 1961, another $109,000 was rescinded. In these circumstances, Sperry Canada's ability to find "approximately $460,000 in the development of numerically controlled equipment and techniques" by the end of 1961 attests to the company's determination to gamble on its technological vision and its market ambitions.[44]

Though the Sperry Rand corporate federation did not contribute directly to R&D on numerical control in Canada, it did play an important part in offsetting some of the instability in the Canadian defense market that Sperry Canada faced. Throughout the late 1950s and the early 1960s, the various divisions in the Gyro Group subcontracted work to their Canadian partner. Sperry Rand used the US-Canada Defence Production Sharing Agreement, whose objective was to give Canadian companies access to US military procurement, to bring more US military spending into the overall corporate coffers. In 1961, 75 percent of Sperry Canada's $7 million in sales was from exports, an unusually high percentage for the Canadian manufacturing sector.[45] Most of these sales had originated from intra-firm purchases by the other divisions of Sperry Rand.[46] The subcontracting work entailed nothing more than routine parts manufacturing and assembly. Nevertheless, it was an important source of revenue for the Canadian subsidiary that generated the internal funding for NC development.[47]

Despite the early success of Sperry Canada's first experimental prototype (1957) and its first commercial system (1958), several years passed before Sperry Canada's investment in numerical control bore fruit. In 1961, Sperry Canada sold 60 NC systems.[48] For the 1960–61 fiscal year, the value of Sperry Canada's NC systems added up to $1 million, about one-eighth of the company's total revenues. For the 1961–62 fiscal year, sales climbed to $2 million.

After several years of work and $460,000[49] of investment from its limited surpluses, the acceleration of sales engendered great enthusiasm and optimism. One Sperry Canada official concluded that the 1962–63 sales would double again to $4 million. To enhance its competitive position, Sperry Canada maintained a constant rhythm of NC development engineering. Three years after the appearance of Mark I, Sperry Canada had improved its technology three times, with Mark II, Mark III, and Mark IV. With most of Sperry Canada's NC sales in the United States, Germany, and Switzerland, the company's president boasted that his company was proving that made-in-Canada technology could succeed in global markets.[50] In the spring of 1963, to penetrate the European market, Sperry Canada signed a licensing agreement with an Italian firm, Nuova San Giorgio S.P.A. of Genoa-Sestri, to manufacture the Canadian NC technology in Europe.[51] As one trade journal noted, the alliance with the Italian company was undertaken "to sidestep the rising scale of duties and competitive manufacture in European [Common Market] which could eventually shut out the Canadian built equipment."[52] There was even talk of Sperry Canada's licensing its technology to a Japanese firm.[53]

Two contracts, one from the Societé Genvoise d'Instrument de Physique (SIP) and one from the US aerospace company Rocketdyne, symbolized Sperry Canada's worldwide competitiveness in NC technology. SIP, an internationally renowned Swiss maker of machine tools, was synonymous with ultra-high-precision engineering. Sperry Canada had initially used SIP's small jig borer to prepare the first accurate measuring scales for its NC systems. SIP played an active part in helping Sperry Canada adapt its machine tools to manufacture these scales. In the process, Herzl built up a close working relationship with SIP's chief engineer. SIP already had an NC technology for its very large and expensive jig borers. In 1961, Herzl convinced SIP's technical people that Sperry Canada could produce a simpler and cheaper NC system to accompany SIP's smaller line of jig borers.[54] The machining tasks of the jig borer fit perfectly into Sperry Canada's market expertise of point-to-point NC technology.[55]

Both the president of the Gyro Group in the United States and Wence King, the president of Sperry Canada, considered the SIP deal to be *the* prestige contract and the pivotal opportunity Sperry Canada needed to gain industry-wide recognition. After all, if SIP decided to attach Sperry Canada's digital controllers to its machine tools, how could lesser lights refuse to follow suit? But if Sperry Canada failed to meet SIP's high standards of precision, the marketing consequences would be disastrous. Bob Cox, Manager of Engineering Development at Sperry Canada, later recalled how everyone felt that the SIP contract could make or break Sperry Canada's NC diversification strategy.[56] When it was completed, the Sperry Canada NC system guaranteed a remarkable overall accuracy of 80 millionths of an inch in the positioning of the jig borer's table and spindle head. Sperry Canada controls became a standard feature of SIP's smaller 3KCN and 5ECN jig borers.[57] The first such system sold by SIP went to Aviation Electric, a subsidiary of the Bendix Corporation (itself a prominent manufacturer of NC equipment).[58]

Developing digitally controlled inspection systems offered another outlet for Sperry Canada's point-to-point NC technology. In 1961, the Rocketdyne Company[59] of Los Angeles received $400,000 from the US Air Force to build two prototype numerically controlled inspection machines.[60] In the aerospace industry, inspection was a critical step in the production process. Discussions of NC usually focus on the speed and precision needed to machine very complex one-piece metal structures. An equally important aspect of the early development of NC was the development of digitally controlled automatic inspection machines. Depending on the complexity of the machining, the task of ensuring that the final metal part corresponded within very precise tolerances to the specifications given in the engineering diagram could be far more labor intensive and time consuming than the actual production of the piece. Rocketdyne restricted bidding on the prototype of the numerically controlled inspection systems to four companies, Sperry Canada being one of them. One can assume that Rocketdyne's selection reflected its satisfaction with an NC system that it had already bought from the Canadian company.

To win the Rocketdyne contract, Sperry Canada turned for the first time to the Canadian government for financial subsidy. Sperry asked the Department of Defence Production (DDP) to cover any losses arising from a below-cost bid. Of the $400,000, it was estimated that only $160,000–$200,000 would be available for the subcontract to design and build the digital controller for the inspection unit.[61] Sperry Canada argued that if it were to forgo any profit the design and construction of the two prototypes would cost about $236,000. To come in at a bid of $160,000, Sperry Canada asked DDP to cover the difference of $76,000. After assuring himself that $400,000 was all the US Air Force was going to pay Rocketdyne, the Deputy Minister of DDP, D. A. Golden, sent a long letter to the Treasury Board supporting Sperry Canada's request. Golden argued that Sperry Canada, having already financed much of its NC program entirely from company funds, was an exemplary Canadian corporate citizen that deserved assistance. Even more important, Golden noted, the development of these two prototype NC inspection machines could, in the short to medium term, be the basis for $1 million in additional export sales. DDP provided the subsidy, without which Sperry Canada would never have won the Rocketdyne contract.

The Rocketdyne inspection machine contract reveals that Sperry Canada's efforts to diversify into the "civilian" sector nevertheless depended greatly on the US military-industrial complex. In the area of digitally based controls for machine tools, the defense-oriented companies were the largest consumer group of numerical control technology in the United States. By the end of 1962, the US Air Force had supported the purchase of $60 million worth of NC technology. In 1963, the aerospace industry alone accounted for 36 percent of all NC installations in the United States. One survey estimated that by the end of 1964 this figure would climb to $100 million.[62]

During the early 1960s, a sizable portion of Sperry Canada's NC business arose, directly or indirectly, from the US defense program. After Sperry Canada completed the two prototype NC inspection machines for Rocketdyne, the US Air Force paid the Canadian company for two more such systems, one to be delivered to the Boeing Aircraft Company in Seattle and the other to the McDonnell Aircraft Company in St. Louis. Pleased with Sperry Canada's abilities in NC technology, the US Air Force also granted the Canadian company a $400,000 industrial R&D contract.[63] Rarely before had the US Air Force granted an NC development contract to a company outside the United States. In addition to systems paid for by the Air Force, Sperry Canada shipped NC technology to fill the machining needs of the US Navy and Army and those of programs related to the US Atomic Energy Commission.

Despite the preponderantly American defense-based demand for NC systems, Sperry Canada had believed from the outset that the Canadian market could be an important source of revenue. The company's desperate effort to market its technology as a "natural" for Canada tried to play off the findings of the 1957 Royal Commission on Canada's

Economic Prospects, which pointed to a small domestic market as one of the principal obstacles to the growth of a competitive manufacturing sector.[64] Ninety-four percent of Canada's secondary manufacturing output was for the domestic market. This reliance on an internal market, first nurtured by the National Policy, constrained economic opportunities. If Canada had had a huge domestic market, this reliance would have been more an asset than a liability. After all, the United States' successful economic development stemmed from its large domestic market—a market that had created a supportive environment for the development of the American System of manufacture, which later grew into the mass-production paradigm. The expanding domestic market also supported the rise of the multi-divisional firm and the rise of the American system of management.[65] In Canada, dependence on a small market prevented secondary manufacturers from "obtaining the maximum economies from mass production and specialized operations."[66]

Another structural weakness of Canadian secondary manufacturing, which the report of the Royal Commission also tied to the small size of the domestic market, was the proliferation of shorter and more varied production runs: "In comparing manufacturing plants in Canada and the United States, one is struck not so much by the relatively greater size of the United States production units—although United States plants are usually somewhat larger—but by the fact that the Canadian plant in practically all cases produces a much greater variety of products for its size."[67] To illustrate the problem, the report cited the testimony of one manufacturer: "There is potentially as great a variety of tastes and quantity requirements in a market of 15 million as there is in a market of ten times that size; but only in the larger market will the minority requirements be large enough to warrant economical production in specialized plants. Where an American plant manufactures a single type of brick, we have to produce five, six or more different colors, grades and finishes."[68]

Sperry Canada's Marketing Manager, Hugh Hauck, understood that the dominance of Fordism, with its specialized tools and high setup costs amortized over large production runs of one product, had turned Canadian small-batch production into a competitive handicap. "Our market is too small to support mass production with specially designed automated machine tools, and our labor costs are too high to make us competitive with our high-labor content methods of manufacturing," Hauck wrote in 1962.[69] In mass production, the costs of retooling are relatively small because they can be amortized over very large production runs. But with production runs of 20–100 units—characteristic of most

Canadian machine shops—the costs of retooling accounted for a large portion of the total costs.

A political solution to Canada's manufacturing dilemma was to seek privileged access to large foreign markets through bilateral or multilateral arrangements that eliminated tariffs, as the Auto Pact was to do in 1963. Another approach was to develop appropriate manufacturing technology.[70] Hauck preached numerical control as the ideal answer to the competitive challenges facing Canadian machine shops. NC technology, he argued, "allows standard machine tools to be automated for economical short or medium-run production runs. By simply changing a tape you change your product."[71] The 1957 study by the Royal Commission on Canada's Economic Prospects had underscored the inability of Canadian manufacturers to justify specialization on economic grounds as an important impediment to competitiveness. Hauck argued that the flexibility of numerical control would at last level the playing field for Canadian manufacturers.

When Sperry Canada first set about developing its technology and business strategy, in the 1950s, the company was convinced that Canada would be a major market for numerical control. To Sperry Canada's surprise, Canadian machine shops did not beat a path to this new digital technology, which promised more flexibility. From 1954 to the end of 1962, only nine NC systems found their way onto Canadian shop floors, and only two of the nine were made by Sperry Canada. In the United States, more than 2300 NC systems were in place by 1962.[72] By any measure, NC technology was extremely underrepresented in Canada's industrial infrastructure. Though the relative distribution of machine tools between the United States and Canada closely matched the ratio of their manufacturing-based gross domestic products, the ratio of NC systems was about 230 to 1.[73]

Sperry Canada's efforts to market its NC products domestically were frustrated by the Canadian metalworking industry's perception that NC was only a mass-production tool and hence was not appropriate for Canadian machine shops. Why Canadian industry had this perception is not known, but it lingered beyond the 1960s.[74] Sperry Canada's Manager of Engineering Development, Bob Cox, encountered this misconception when he went to work for RCA in 1965. Cox, who brought the first NC systems into RCA Montreal, recalls that "when [he] tried to sell NC to Jack Sutherland, who was then VP and General Manager of RCA's Government and Commercial Systems Division, Sutherland would say we were not in mass production."[75]

The high initial equipment cost of NC also reinforced the wait-and-see attitude of traditionally conservative machine shops. In the United States, the startup costs associated with NC had been systematically defrayed by defense contracts.[76] With no such equipment-support programs in Canada, the machining industry may have sensed a flaw in the economics underlying the advantages of NC for small production runs. One study found that the average machine tool in the United States was in operation only 25 percent of the time.[77] But an NC machine, according to the same study, had to be in operation at least 75 or 80 percent of the time before it could pay for itself. In the context of small production runs, to keep a machine tool operating 80 percent of the time would require production of a steady stream of items. Such production probably was not widespread in the Canadian metalworking industry.

Sperry Canada's numerical-control-based diversification strategy failed to create the hoped-for 50-50 split between military procurement and civilian markets. In 1963, 9 years after Sperry Canada decided to gamble on the then-nascent technology of numerical control, the civilian market accounted for only 20 percent of the company's sales.[78] Sperry Canada's overoptimistic forecasts of rapidly expanding sales became the victims of what Schumpeter had called "swarming": a rapid proliferation of imitators who follow the pioneering innovator-entrepreneurs. Swarming reduces the high profit margins that early innovators enjoy as a result of their monopoly over the technology.

In the mid 1950s, when Sperry Canada first got into the business of numerical control, only a few companies were designing and building NC systems. By the early 1960s, there were well over 100 companies in North America competing for NC business. The effect of this competition was felt most at the lower end of the NC market, where Sperry had concentrated its innovation strategy. Sperry Canada was a Schumpeterian firm whose once-exclusive North American technological and market niche was overrun by imitators, mostly American. Faced with this competition, Sperry Canada decided once again to be a Schumpeterian firm and to define a new exclusive niche for itself through radical innovation. This time it would take aim at the upper end of the market.

In a bold move, Sperry Canada decided sometime in late 1962 to bring the programmable electronic digital computer directly into the hands of the machinist on the shop floor. Computer Numerical Control (CNC) was born. None of the component technologies of CNC were new. Computer technology was well established. Sperry Canada already had the digital measuring and actuating technology. However, integrating all the elements into a small system that could be put on the shop floor was a radical innovation. Computer hardware designed to run in clean, air-conditioned rooms had to be adapted to the harsh environment of the machine shop. Considerable innovation was needed to make the integration work.

The design of Sperry Canada's new generation of CNC equipment had to walk a fine line between the pursuit of sheer technical performance and the realities of the market. Performance sought at any cost, though acceptable for strategic military projects, would not work in the civilian market. Sperry Canada's strategy of using software to compensate for compromises in hardware proved problematic. In the end, the hardware engineers' lack of appreciation of the complexity of real-time programming and management's inability to subject software development to the same discipline found in hardware development led Sperry Canada into a technical and financial minefield.

UMAC-5: Performance and Market Flexibility through Software

Despite the claims of great flexibility, there was no universal numerical control system that could be simply plugged into any machine tool. Considerable re-engineering was needed to adapt a basic NC system to each new machine tool. "As the need for more sophisticated control systems is recognized by users, and as more machine tools are adapted to N/C," a prominent US trade journal summarized, "many control systems need extensive modification after initial assembly. Hardware must be adapted to suit specific drive and operational characteristics. Valuable shop and engineering time is consumed in these modifications."[79] The high costs of these hardware modifications increased the prices of NC systems. Custom-engineering NC to match the various new generation of multi-axis machine tools and the new "machining centre" became problematic.[80] The more complex the machine tool, the greater the modification costs.

Sperry Canada took solving the problem of adaptability as the basis of its future NC business strategy. In late 1961, the company decided to develop a new generation of NC systems that could be adapted easily, with a minimum of re-engineering, to any two-axis, three-axis, four-axis, or even five-axis machine tool. Such a radical innovation, the company reasoned, would lower its production costs significantly, speed delivery considerably, and expand the flexibility of its NC systems—in short, it would increase the company's competitiveness.

Sperry Canada's management and engineers were unsure of the technical framework for the innovation of this new generation of NC systems. Yet, determined to create this market niche, the company began hiring more Canadian development engineers. One of these men was Paul Caden, a graduate of the University of Toronto's Engineering Physics program. Caden's first job had been on the Defence Research Board's Velvet Glove project for the De Havilland Aircraft Company during the Korean War period.[81] Borrowed from De Havilland for a year, he went to the Canadian Armaments Research and Development Establishment (CARDE) in Valcartier, Quebec, to study the technical feasibility of certain aspects of the proposed Arrow project. When he returned to De Havilland, the Velvet Glove project had been abandoned in favor of the United States' Sparrow missile project. Caden then went to work on the Sparrow project, again at De Havilland. When Canada's part in the development of the Sparrow was canceled, Caden moved on to A. V. Roe to work on the Arrow. There he did tactical simulation studies. With the demise of the Arrow, he moved on to Canadian Aviation Electronics Ltd. (CAE) in Montreal, where he designed flight simulators for the Argus, a long-range patrol aircraft.

In late 1961, Bob Cox, now head of Sperry Canada's Engineering Development, called ask Caden if he wanted to work on a new generation of numerical control systems. The idea of getting involved in sophisticated digital control systems for machine tools struck a chord in Caden, and he accepted Cox's offer. Sperry Canada's management had not even considered the use of a general-purpose computer when it hired Caden. Instead the company felt that the answer to greater adaptability somehow lay in more sophisticated and clever hard-wired designs. Caden was hired because of the electronics experience he could bring to bear on the problem. But, as a result of his work at CAE, Caden had become increasingly fascinated by the possible industrial control applications of electronic digital technology. [82]

After considerable study, Paul Caden concluded that the adaptability sought by Sperry Canada could best be implemented by embedding a programmable computer in the heart of the numerical control system. In this way, Caden reasoned, one could adapt an NC system to any machine tool, no matter how complex, by simply reprogramming it.

Sperry Canada's management greeted Caden's idea with caution. Skeptics objected that general-purpose computers tended to be too large, too expensive, or too fragile to be placed next to a machine tool on the shop floor. With no experience in designing, testing, and producing computer systems, senior management worried whether Sperry Canada could undertake to build a computer specifically for the machine tool environment. The company's military systems group had already built and installed a hybrid digital-analog tactical simulator for the Canadian Navy (in 1961),[83] but an off-the-shelf Packard Bell PB-250 had been used. Developing the software had not been difficult, but the experience with the simulator still served as a reminder that programming was not to be taken lightly. Caden's proposal entailed far more than tinkering with off-the-shelf technology. One government expert noted: "While the incorporation of a small digital computer into a multi-axis numerical control may appear at first hand to be a straightforward task, in fact, it becomes a special purpose digital system development. Not only is the equipment itself special, but the type of computer programming and the approach that must be taken thereto in order that it can be widely used also necessitates special development. Specifically, programming suitable for this special computer must be written to marry the control to every different machine tool."[84] Despite the software challenge, Bob Cox supported Paul Caden's proposal. In late 1962, after many studies and presentations, management finally accepted the programmable computer approach.

In the summer of 1963, Sperry Canada boasted that it would soon have a commercial system ready to go. Manager of Industrial Contracts Hugh Hauck told a reporter from the *Montreal Gazette* that the company had developed a new generation of numerical control that represented "a major breakthrough in the industry."[85] Though no system had yet been delivered, the Senior Editor of *American Machinist/Metalworking Manufacturing* announced that, with UMAC-5's general-purpose computer, Sperry Canada had solved the versatility problem that had hitherto plagued numerical control systems.[86]

Sperry Canada sought to develop a single NC system that could be interfaced to any complex machine tool without the need to redesign the circuits. The UMAC-5's hardware was standardized and was fairly independent of the machine tool it was to control. When a customer ordered a UMAC-5, all the detailed information about the configuration and operation of the customer's machine tool would be loaded as a program into the computer before shipment from the Sperry Canada plant. If for any reason the customer wished to change the configuration of the machine tool, UMAC-5 had only to be reprogrammed.

By November 1963, Sperry Canada found itself having to push back the first deliveries of the UMAC-5 into the new year.[87] By October 1964, 32 orders had been received[88] and pressure to deliver the first complete working system was mounting. But UMAC-5's development was mired in software problems. Once Sperry Canada had locked onto the idea of flexibility through programming, there was a tendency to want to design UMAC-5 to be all things to all people. The company wanted to pre-package more complex multi-step machining operations into subroutines which would simplify the job of the parts programmer.[89] In its aggressive marketing promises, the company had grossly underestimated the time and money that would be needed to develop all this software. The mistake lay in the notion that after the hardware had been designed and built everything else would be simply a matter of programming—"SMOP," as the overconfident UMAC-5 engineers called it. After 2 years, in which time the software overruns had exhausted the company's financial resources, "SMOLOM" (simply a matter of a lot of money) would have been a better acronym.

The difficulties in UMAC-5's software development arose from the inexperience of the design team, the craft-like nature of programming, and the cost-driven compromises made in the computer's hardware. The programmers had little understanding of metalworking and little knowledge of the intricacies of gearboxes, servomechanisms, and other aspects of constructing and operating machine tools. Gila Bauer, who headed the programming group, was a mathematician.[90]

It is revealing that the development of the software, which was the essence of Sperry Canada's new approach to numerical control, was given to inexperienced computer programmers. This decision seems to be evidence of a naive belief that the programming would be a relatively simple and routine matter. In time the programmers gained the requisite experience, but it was a long and costly process. The consequences of programming mistakes could be far more dramatic than a simple error message. "It was all too easy for someone who was a pure programmer," Bob Cox later observed, "to say 'well I'll just engage this clutch and that clutch' and have the spindle gear box going forward and reverse at the same time. The end result was broken gears all over the floor."[91] On one occasion, a programmer failed to realize that the control software for a large horizontal boring machine had to include a limit switch. During a demonstration of UMAC-5, a software bug caused the table to keep moving in one direction. In the absence of a limit switch, the 6-ton table went off the lead screw and through a brick wall.[92] After trying to get programmers to learn about machine tools, Sperry Canada discovered rather belatedly that faster results could be had by teaching machine tool technicians how to program.

"Despite 50 years of progress," W. Wayt Gibbs observed in a 1994 article in *Scientific American*, "the software industry remains years—perhaps decades—short of the mature engineering discipline needed to meet the demands of an information-age society."[93] The gist of Gibbs's argument is that software development has long been based on "craft" and has remained outside the ethos of modern engineering (which values the systematic, the disciplined, and the quantifiable). To this day, software is "hand crafted" by artisans using techniques they cannot measure and cannot repeat consistently. One computer scientist likens current software development to the craft production that preceded the rise of interchangeable parts manufacturing.[94] During the 1960s, when UMAC-5 was being developed, the craft-like nature of software development was even more pronounced.

With little experience in software development, Sperry Canada's management found it even more difficult to subject the craft of programming to strict engineering control. "The software people tended to be a bunch of mavericks and gurus," Cox later recalled. "They did not follow any discipline in documenting or modularizing their programs. They made changes in an uncontrolled fashion. There was no configuration management on software. People would make changes in the middle of the night and not tell the next guy."[95] One Sperry Canada document reported that "it was virtually impossible to keep accurate records of what individual machine programmes consisted of."[96] Sperry Canada's usual design estimation and control procedures, which had proved reliable in hardware engineering, failed in the case of software development.

The marketing idea behind UMAC-5 was that cost and time could be saved by replacing hard-wired re-engineering with "soft-wired" reprogramming. The fact that it took 11 weeks to load and debug a program for the prototype did not bode well for profitable sales.[97] In addition to the factors cited above, the absence of a high-level programming language lengthened the software development cycle. All the programming was done in a very un-intuitive and cumbersome machine language. "As we had no means of automatically assembling master programmes," one company progress report noted, "each of the prototype machines had to be programmed, optimized and loaded command by command. This in turn caused further problems. It took a great deal of time and money to program the controls. A great number of errors were made, some of which were of such an insidious nature that we still have not removed them from the Avey turret drill we own."[98] For Sperry Canada's programmable computer strategy to succeed, the company had to develop techniques that could generate and document the software in a quick, predictable, routine manner. To do this, a high-level language had to be developed for UMAC-5; the conventional compilers found on mainframes were unsuitable.[99] The absence of any efficient program-debugging tools also limited the productivity of UMAC-5's programmers.

Sperry Canada's design strategy created a conflicting set of design objectives whose final resolution compounded the difficulties of software development. The computer had to be fast enough to monitor and control machining processes that demanded split-second response. At the same time, the computer had to be small and resilient enough to be placed next to the machine tool in a harsh shop-floor environment. The production costs for the computer had to be low enough to allow the company to market it as an economic alternative to existing hard-wired NC technology. Finally, the design had to be simple enough to ensure that the company could finance development and bring a product to market in a reasonable time.

Although Sperry Canada's approach to computer numerical control was driven by technology, the design of the UMAC-5 computer was pragmatic and was driven by cost. After some analysis, Sperry Canada decided that a ceiling of $5000 had to be placed on the cost of the computer's hardware inputs if it was to be priced competitively.[100] To keep costs down and simplify hardware design, a serial rather than a parallel architecture was chosen. This compromise in performance did not pose any serious problems. However, the decision to go with an all-drum memory was a different matter. Magnetic cores offered superior performance, but with this memory running about $0.25 per bit this option would push costs far beyond the $5000 limit. The choice of drum memory created as many problems as it solved. Drum memory complicated software development because of the need for a very efficient programming code to overcome the relatively slow access provided by the drum's non-random-access memory. The technical question was how information was to be stored on and read off the drum.

The memory on the drum was divided into read-only and read/write tracks. The machine tool operator's program went on the read/write track. The "firmware," which took up most of the drum's memory, was put on the read-only track at the Sperry Canada plant. This firmware contained all the information and the subroutines needed to compile the part program and then operate the various parts of the machine tool precisely and in the proper sequence. In a revolving-drum system, any piece of data appears under the read head once per revolution, and in this case a revolution took 17 milliseconds.¹⁰¹ For some machine tool operations, a maximum access time of 17 milliseconds was acceptable; for others, it was not. Servo commands were very time sensitive; controlling spindle speeds and turning the coolant on and off were not.

The UMAC-5 prototypes turned out to be too slow owing to "the inefficient distribution of data on the drum and the absence of 'fast tracks.'" When UMAC-5 prototypes were demonstrated at Sperry Canada, representatives of the machine tool industry were quick to spot the smallest hesitation in the computer's control of the movement of the cutting tool or the work piece. Paul Caden's efforts to time optimize the software in order to overcome the limitations of drum memory met with partial success. The pursuit of better real-time performance from a drum memory also underscored the need to reduce the size and complexity of all the programs that had been developed, and to optimize their integration.

As software problems mounted, Sperry Canada turned for assistance to UNIVAC, another division of Sperry Rand. A leading manufacturer of computer technology, UNIVAC had shown little interest in applying its expertise to the machine tool market. It had more pressing concerns: losing a great deal money, it was in a desperate fight against IBM for survival in the mainframe computer business. Nevertheless, as a division of Sperry Rand, UNIVAC was quite open to intra-firm transfer of its know-how. It provided Sperry Canada with the printed circuit boards for UMAC-5's construction. In return, Sperry Canada provided the NC technology for UNIVAC's machine shops¹⁰² and for the high-speed automated wiring of UNIVAC's mainframe computers.¹⁰³ In late 1963 or early 1964, a software team from UNIVAC came to Montreal to help accelerate the pace of software development.¹⁰⁴ Whatever its contribution, the UNIVAC team did not cure the software ills of the UMAC-5 project.

In 1965, after 2 years of work and $475,000 of investment, UMAC-5's development was still incomplete.¹⁰⁵ Sperry Canada's two prototype installations still exhibited "several important design deficiencies in hardware and software."¹⁰⁶

To make matters worse, the Liberal government's cancellation of a defense contract for frigates underscored how Sperry Canada's financial capacity to underwrite the completion of UMAC-5's development still rested precariously on military procurement. The cancellation of the frigate contract reflected the change in Canada's perspective since the country had came into its own in the North Atlantic Triangle alliance as result of its role in the battle for the North Atlantic.

After World War II, Canada's military was determined never to be caught unprepared again. The Royal Canadian Navy, which saw its future strategic role as continuation of its wartime escort role, made plans to build a sophisticated escort fleet. Ferranti-Canada's DATAR technology was one component of this strategy. Another component was the construction of modern frigates. These ships were to be the modern descendants of the Canadian Corvettes that had defeated the German submarines in the North Atlantic. However, not until 1962 was the construction of eight frigates, all to be equipped with the latest weapons systems, approved (by a Conservative government).

In 1963, the new Liberal government questioned the frigate program on the ground of its high cost. In addition, the RCN was now wondering about the operational limitations of frigates in modern anti-submarine warfare. Nuclear submarines could circumnavigate the globe without ever coming up for air, they could disappear easily under the Arctic ice, they were faster than frigates, and they could launch missiles from the ocean's depths. Frigates were at risk of obsolescence. There was even talk that the RCN wanted to replace the eight new frigates with three submarines or even with some hydrofoils. On 24 October 1963, Minister of Defence Paul Hellyer stood up in the House of Commons and announced the cancellation of the $452 million frigate contract, one of the largest defense contracts since the Korean War.[107] Members of the Conservative opposition howled; Gordon Churchill called Hellyer's actions a "crippling setback."[108]

The cancellation of the frigate program left Sperry Canada with no money to complete UMAC-5. Manufacture of the ships' navigation systems was projected to provide the company with 25–35 percent of its total 1964 and 1965 revenues. The funds to support the company's diversification into NC and then CNC had been expected to come from its defense revenues. Sperry Canada's cash flow difficulties were compounded by the fact that its parent firm was no longer in a position to make up the shortfall through intra-firm subcontracting.[109]

In its efforts to survive in the civilian market, Sperry Canada again turned to the government for help. On 6 November 1964, the president of Sperry Canada, Wence King, asked the Liberal government to share the $125,000 cost of completing UMAC-5's development. He argued that this support was critical if Sperry Canada was ever going to gain a firm foothold in a $30 million export market for high-performance point-to-point CNC.[110] Once again, King used the export argument to win government support. Management was convinced that, with its computer strategy, Sperry Canada could win as much as 17 percent of this global market by 1968.[111]

Sperry Canada's approach to high-performance numerical control involved more than point-to-point applications. In point-to-point machining, the cutting tool moves only in straight lines relative to the work piece. Contour machining entails machining along arbitrary and often complex curves. Sperry Canada's management believed that a programmable computer also had distinct applications to contour machining.[112] However, considerable investment was needed to perfect this line of innovation. In addition to servomechanical innovations, new hardware and software would have to be designed to meet the faster and more intense computational demands of controlling contour machining. Unable to spend any money in a new round of CNC development, Sperry Canada tried to get the Department of Industry, Trade, and Commerce (DITC) to underwrite this technological ambition. It confidently forecast that a new UMAC-5 for contour applications would capture 10 percent of an estimated $18 million global market.[113] By 1968, Sperry Canada predicted, 40 percent of its CNC sales would come from contour applications.

About $600,000, according to company estimates, would be required to develop and commercialize a contour version of UMAC-5. Most of this money was to cover the costs of designing, developing, testing, and commercializing new computer hardware and software. Preliminary design plans for the new UMAC-5 computer called for the latest micro-circuit technology, parallel architecture with core memory, a new assembler, Fortran compilers, and a library of real-time software subroutines to handle the complex task of machining over arbitrary curves. Drawing from military systems development, Sperry Canada also proposed the creation of software to test the system in simulation. Sperry Canada estimated that it would need 20 months to design the prototype computer systems. An additional 16 months would be needed to combine the computer and the new servo-control technology in a sophisticated "machining centre" and then test the entire system.[114]

Sperry Canada's 50-50 cost sharing proposal was a poorly disguised attempt to get the Canadian government to carry all the costs of developing the advanced version of UMAC-5 for contour applications. The cost of completing the point-to-point version of UMAC-5 and building a new contour version was $725,000. The company, however, proposed a 50-50 split of $1.2 million. Rather than asking for 50 percent of future costs, Sperry Canada wanted the government to share the costs of UMAC-5 retroactively to the very beginning of the program in 1962–63. In effect,

Sperry Canada was offering to pay only $125,000 for the completion of the current UMAC-5 version. The government's $600,000 share conveniently matched the total estimated costs of developing and commercializing a new continuous-path UMAC-5 system.

When DITC balked at this ploy, Sperry Canada revised its proposal to a 60-40 split.[115] When DITC questioned why the company's wealthy parent firm did not contribute to UMAC-5's development, Sperry Canada underscored the divisional and decentralized corporate structure of Sperry Rand: "Each division of the corporation, including Sperry Canada," Director of Marketing R. H. Little explained, "is treated as a discrete cost entity. Each division must provide its own funding for its own operation."[116] The other divisions had no reason to divert their own precious R&D funds to Canada.

Within the Sperry Rand empire, Sperry Canada had won the world product mandate for "numerical and industrial controls." But without government aid, there was little likelihood that Sperry Canada could turn its ambitions of global technological leadership into profits. DITC, though receptive to Sperry Canada's export-market argument, also had reservations.

In the end, it was concern over the fate of Canada's metalworking industries that shaped the government's response to Sperry Canada's call for development cost sharing. As early as 1958, *Canadian Machinery and Manufacturing News* had warned of the impending obsolescence of Canada's manufacturing infrastructure: "The machine building industries—machinery, machine shop, machine tools, iron casting, office and household machinery, electrical apparatus—might be expected to act as bellwethers of modern production practice. There are more machine tools from eleven to twenty years old than there are under eleven years. Despite the tremendous expansion since the war almost 60 percent of today's machine tools in these industries date back to war's end or earlier."[117] This warning, which appeared shortly after the report of the Royal Commission on Canada' s Economic Prospects, added to the government's concern over the future competitiveness of the Canadian manufacturing sector. In 1964, the Economic Council of Canada, in its first report, again raised the issue of increasing the productivity of the nation's industrial infrastructure: "The productivity of a country's economy . . . is at the heart of its economic welfare and prosperity of its people. Productivity gains are also the essence of economic growth and are the real source of improvements in average living standards. Moreover, without adequate productivity growth, a nation's competitive position . . .

may be subjected to disturbing pressures and strains."[118] The Economic Council suggested that, in order to reverse the decline in the nation's productivity registered from 1956 to 1963, new emphasis be placed on increasing the technological preparedness of Canada's industries.

With the Economic Council Report only a few months old, Sperry Canada's UMAC-5 funding proposal resonated strongly within DITC. The remarkably slow diffusion of numerical control technology into Canada's forces of production was one symptom of the productivity problem.[119] *Canadian Metalworking/Machine Production* warned that Canada's metalworking industry would "crumble" if the diffusion of NC installations did not accelerate dramatically.[120] In 1965, in an effort to interest Canadian industry in numerical control, DITC participated in seminars on productivity sponsored by the Economic Council of Canada at which NC was the specific theme.[121] DITC saw the capacity to produce this technology domestically as crucial to furthering the spread of NC in Canada. When the government asked what DITC was doing to "promote the use of numerical control in Canada," the Deputy Minister replied: "Action by the department has taken two forms: first, the support of the development of numerical control in Canada where applicable, and the second, the encouragement of the acquisition of numerical controls by industry."[122]

Fearful that Canada's aging stock of machine tools seriously impaired the ability of the civilian sector to respond to military procurement needs, the government created the Defence Industry Modernization (DIM) program. Under the DIM program, Deputy Minister Simon Reisman said, "companies are encouraged to enter the defence export market which demands that their products be competitive in quality, delivery and price with the best American." "For relatively short run production," Reisman continued, "this means usually numerically controlled machines. . . . DIM is used with discrimination to favor such modernisation of facilities rather than simply to expand Canadian production facilities to meet particular American requirements."[123]

Sperry Canada's ambitions to perfect computer numerical control offered the only opportunity for DITC to promote a domestic capacity to design and sell this technology. But was Sperry Canada's CNC design agenda the optimal answer to the inadequate spread of NC technology in Canada? The poor experience to date with NC made it doubtful that many Canadian machine tool shops would be interested in the even more expensive CNC.

The premise of Sperry Canada's entry into computer design had been to escape from the swarm of competitors producing cheaper and less sophisticated NC systems. "It is the opinion of our company," Sperry Canada had announced in the first and second proposals, "that the low-performance point-to-point market, due to the extremely active competition and relatively unsophisticated nature of the equipment involved, makes it an unsuitable product for a company of our technological background."[124] The company's desire to create a new CNC product for contour machining was a further indication of its strategy of using computer technology as a means of pursuing the high-performance market.

Sperry Canada's CNC approach made sense only for the portion of the market that used large and complex machine tool setups, of which machining centers were a perfect example. Spending $30,000–$70,000 for a UMAC-5 CNC system could be justified only if it were used with $100,000–$500,000 machine tools. The contour-machining version of UMAC-5 promised to be even more expensive. Not only were "machining centers" rare in Canada; the bulk of Canadian purchases of metalworking machinery were of less complex equipment.[125] DITC's first detailed analysis of machine tool imports suggests that smaller, lower-priced machine tools dominated Canada's secondary manufacturing sector.[126] Another internal DITC memorandum notes that the Canadian market for metalworking machinery was about $75 million but that the Canadian market for more sophisticated numerically controlled machines was forecast to be only about $10 million in 1967.[127]

Although Sperry Canada expected to sell some UMAC-5s in Canada, its marketing strategy for computer numerical control was global. DITC, however, felt that there was a greater need for cheaper, high-performance NC technology for the smaller machine tools that made up most of Canada's metalworking industry. To spend $125,000 for the completion of the point-to-point CNC system seemed a reasonable expenditure of Defence Industry Modernization funds. After all, Sperry Canada had already spent a considerable amount of its own money on UMAC-5. DITC also recognized that UMAC-5 was of great potential value to those Canadian metalworking companies that produced very short runs of sophisticated parts on complex machine tools. On the other hand, DITC was not going to spend an additional $600,000 on the development of an even more advanced CNC system when the vast majority of domestic machine shops needed a less costly alternative.

Two months after asking DITC to share the $600,000 cost of new CNC development, Sperry Canada made an almost complete turnaround in its technological and marketing strategy. In a revised cost-sharing proposal dated 7 April 1965, Sperry Canada reaffirmed its desire to complete the software development for the point-to-point UMAC-5 system but completely dropped its plan to design a contour version of UMAC-5. Instead, embracing a new technological strategy to reflect the priorities of the government, Sperry Canada proposed that, with DITC funding, it would develop a new family of low-cost NC systems.[128] To get the needed assistance from DITC, Sperry Canada had to dance to DITC's tune. Sperry Canada explained this as follows in its revised proposal: "The smaller manufacturer, who cannot afford the substantial investments represented by the major numerically controlled machines, nevertheless, is beginning to recognize the advantages to be gained from flexible short run automation, and is investing in the machines within his financial reach. These more economical machine tools demand two and three axis numerical control systems in the $10,000 to $16,000 price range."[129]

The 50-50 cost sharing arrangement was approved on 30 June 1965, 8 months after Sperry Canada first asked for support for the completion of software development for UMAC-5.[130] During that period, Sperry Canada, confident that it would get some form of government assistance, continued work on the UMAC-5 computer.[131] By the time the Treasury Board actually approved the development contract, Sperry Canada had completed much of the software work.[132]

When the UMAC-5 was finally available, Sperry Canada found that its CNC innovation had priced itself out of the market. As Sperry Canada's President Wence King, recalled some 20 years later, "UMAC-5 was just not saleable." Only the most demanding point-to-point applications needing very complex machine tools could justify the high cost of UMAC-5.[133] This market turned out to be far smaller than Sperry Canada had estimated. About 50 units were sold in the United States, Europe, and Canada. In the United States, UMAC-5 was sold mainly to defense contractors, such as Boeing, Lucas Aerospace, Rocketdyne, and McDonnell Douglas. The demands of automated inspection machines, often for control in ten axes, offered another market for the expensive UMAC-5, particularly when the US Air Force was underwriting the purchase of the costly equipment. In Canada, only a few machine shops bought UMAC-5, and these were very large operations (such as Rolls Royce, Orenda, and Westinghouse).

Sperry Canada's strategy of designing a computer-based numerical control system that would be "adaptable to any machine . . . any machine requirement" was technologically important, but as a commercial product CNC failed to earn Sperry Canada the civilian revenue it needed to diversify its business. The failure to anticipate fully how much software development UMAC-5 required, the initial inexperience of Sperry

Canada's software team, and the mismatch between the design goals and the limitations of the available technologies all contributed to this failure. By 1967, the burden of initial technical tradeoffs in UMAC-5's design became onerous in the face of advances in basic computer hardware. Soon, Sperry Canada ceased selling UMAC-5s.

A Retreat from CNC Back to NC: The Final Chapter of Sperry Canada

The Department of Industry's desire to fit technological change to what it perceived to be the needs of Canada secondary manufacturing helped push Sperry Canada out of computer numerical control and back into numerical control. The new NC system was called UMAC-6. From the viewpoint of computer history, UMAC-6 was a regression from a programmable solution back to a hard-wired one. But from a business perspective, UMAC-6 seemed to have more market promise. Once again, Sperry Canada's capacity to carry out its plans for engineering development and commercialization balanced precariously on the fate of the company's defense contracts.

Work on the design of UMAC-6 started in March 1966. The prototype was ready in 2 years.[134] Designed to work with a Brown & Sharpe machine tool, UMAC-6 was publicly demonstrated in May 1968 at a machine tool trade show in the United States. The new low-cost, high-performance, point-to-point UMAC-6, with the latest integrated circuit technology, elicited considerable interest from machine tool manufacturers. Soon after the prototype was completed, Sperry Canada successfully negotiated with four prominent manufacturers of machine tools to build preproduction versions for extensive testing.[135] If the trials proved successful, these machine tool manufacturers agreed to buy at least 40 UMAC-6s by the end of 1968.[136] Optimism once again ran high. "The future of this program looks most promising," wrote Sperry Canada's director of planning.[137] "With an unquestionably successful product technically, and in price," he added, "we are anticipating export sales approaching $10 million annually by Fiscal 1973." But serious financial difficulties lay just below the surface of this optimism.

Market forecasts had been overoptimistic, and the Canadian defense market was as unreliable a source of revenue as ever. Over the years, Sperry Canada had exhausted all its surplus revenues—first in the pursuit of NC, then in the costly pioneering of CNC, and finally in the development of UMAC-6. In 1968, the company's military equipment business had sunk so low that the plant was running at less than 35 percent of its capacity.[138] Once again, Sperry Canada was plagued with the same cruel irony. To break away from its vulnerability to the uncertain Canadian military equipment market, it needed stable revenues from this very same market. With no profits, it was unable to finance the development and construction of the four pre-production models it promised to make available to machine tool manufacturers for testing. Even if the testing proved successful, the company did not have the resources to sustain the initial marketing and support of UMAC-6.

Desperate for military subsistence, Sperry Canada officials were "shocked" when the government failed to consider their company for the contract to build the Primary Display Units in the ASW/DS Systems, which were to be used as part of the Canadian Navy's SQS-5-5 sonar system.[139] Having spent considerable money to develop this unit, Sperry Canada was angry that it was denied the production contract. The revenues from the manufacture of the Primary Display Units were crucial to the future commercialization of the UMAC-6. Playing on DITC's financial involvement in UMAC-6 and on DITC's commitment to accelerate the diffusion of NC in Canada, Sperry Canada threatened to drop out of UMAC-6 unless it was given the Primary Display Unit contract or offered other military production work to replace it.[140] But Sperry Canada got neither.

By the summer of 1968, Sperry Canada's financial situation was desperate. Plant loading (i.e., utilization) was down to 25 percent.[141] Vainly searching for ways to use the plant's capacity, Wence King tried to convince UNIVAC's management to transfer some computer production to Canada. No doubt King argued that UMAC-5 had demonstrated that Sperry Canada could handle this production responsibility. DITC advised King that it "would be very willing to assist him in whatever way feasible to gain more manufacturing and R&D responsibility from Univac in the United States."[142] How far King got in his discussions with Univac management is not known. In the end, however, nothing came of it. In view of Sperry Rand's divisional corporate ethos and of a Canadian tariff structure that made it cheaper for US firms to export entire computer systems to Canada than to build them there, King was grasping at straws.[143]

The extant records show quite clearly that Sperry Canada, in its bid to raise development capital, never sought help from Canadian banks. The fact that Ferranti-Canada too never approached Canadian banks strongly underscores the remarkable absence of Canada's banking institutions in meeting the capital needs of leading-edge industrial innovators throughout the period 1945–1970. With no defense contracts forthcoming from the Canadian government, and with Canadian commercial banks not an option, Wence King turned to the parent firm and negotiated a loan to finance the four pre-production versions of UMAC-6. In such uncertain economic circumstances, borrowing money from the parent firm was a big gamble: Sperry Canada's autonomy was at risk if it defaulted on the loan.

Only a few months after the loan was approved, Sperry Rand announced that it was closing down all its Canadian manufacturing operations and transferring UMAC-6 to its Vickers Division in Troy, Michigan.[144] Wence King was angered by the news. Having spent 18 years of his life trying to build Sperry Canada into a truly Canadian operation, he resigned. Paul Caden was sent in to shut down the operation and oversee the transfer of UMAC-6 to Vickers.[145]

Sperry Rand stated that $3 million would be needed to turn UMAC-6 into a profitable product. The parent firm argued that the Canadian operation did not have the cash flow to carry this out. Furthermore, the small Canadian market could not support the development of UMAC-6. Sperry Rand also contended that locating UMAC-6 operations closer to the market was essential for a successful sales effort. Since the United States was the world's largest market, the Vickers Division was the natural choice.[146] There was one hitch: The terms of government funding, under which UMAC-5 and UMAC-6 were developed, forbade the transfer of design rights to foreign firms without governmental consent. Sperry Rand pressured the Canadian government to assign the design rights of UMAC-6 to Vickers quickly. If the company delayed too long in getting the UMAC-6 test versions to builders of machine tools, a window of commercial opportunity would close.

Sperry Rand's efforts to move the UMAC technology out of Canada ran up against DITC's new-found determination to nurture the growth of a Canadian computer industry. Before 1968, DITC's willingness to help Sperry Canada develop UMAC-5 and UMAC-6 had little to do with any broad policy framework of digital technology because no such framework existed. DITC had always wanted to promote domestic R&D and to help firms develop new products with good export potential. But the DITC's support for NC and CNC technology had one objective: modernization of Canada's metalworking industry.

During latter part of 1967, DITC's myopic view of NC and CNC technology began to broaden. For the governments of Japan and Western Europe, the existence of a strong domestic source of computer technology had become the crucial measure of a nation's capacity to ensure its own future material well-being. In view of the growing dominance of the

US computer industry, a sense of urgency gripped Japanese and Western European policy makers, and they rushed to create new policies to counter their countries' excessive dependence on US manufacturers. DITC policy makers eventually got caught up in this international scramble.

As was noted in the preceding chapter, Canada's Minister of Industry, Trade, and Commerce, Charles Drury, presented his department's first policy on computer technology in the form of a memorandum to the cabinet in January 1968. For DITC, Sperry Canada's pursuit of a market niche in numerical control was an excellent example of the kind of industrial activity in computer technology that DITC hoped to promote with its new policy. Here was a foreign-owned subsidiary that had found a global market niche in digital technology. Sperry Canada had a world product mandate. This new mindset came too late to save Sperry Canada's CNC ambitions, but DITC wanted to make UMAC-6 one of the first causes-célèbres of its new policy initiative.

It was a ready-made success story. When Drury announced to the press in March 1968 that his department was determined to expand Canada's industrial base in computer technology, he mentioned NC technology as one promising area of specialization that Canadians companies should pursue. Naturally, the unexpected news that Sperry Rand wanted to move the first potential showpiece of Drury's new computer policy to the United States did not sit well with DITC.

DITC officials were incensed that Sperry Rand wanted to move technology developed in Canada and paid for by public funds to the United States. DITC's support of Sperry Canada's CNC and NC development program had been a cornerstone of Canada's efforts to modernize its industrial infrastructure. What was at issue was not the loss of the public funds spent on Sperry Canada; it was the disappearance of a domestic technological capacity that had taken more than 10 years to build up. If UMAC-6 were allowed to go to Vickers, the entire 18-person design team at Sperry Canada would be transferred to Troy, Michigan. The Canadian government was faced with two options: to continue to subsidize the UMAC-6 project in order to keep it in Canada and to find another Canadian company to take over UMAC-6.[147]

As DITC tried to explore each option, it ran into a very uncooperative Sperry Rand, which was determined to force an action in its favor. Sperry Rand had originally told DITC officials that it had seriously considered how to keep the UMAC-6 in Canada but had found no realistic way to do it. However, after pressing the parent firm for details, DITC officials discovered that Sperry Rand had prevaricated. One senior DITC official concluded that Sperry Rand had barely given the option any thought.[148]

In an effort to keep the technology in Canada, DITC proposed that Sperry Rand sell UMAC-6 to Honeywell Canada, which had expressed interest in the technology. The Apparatus Controls Division of Honeywell, headquartered in Minneapolis, had been negotiating for more than a year with the Nippon Electric Company for the right to market NEC's NC technology in the United States. But from Honeywell's perspective, NEC was demanding too much. As a fallback position, Honeywell contemplated developing its own NC expertise and products. Despite the superiority of NEC's system, Honeywell felt that, if NEC refused to lower its price then the Sperry Canada team and UMAC-6 technology could serve as the basis of its own product. Honeywell was willing to leave the NC development and production operation in Canada and to supply the US market through its Apparatus Controls Division in Minneapolis. Before making any decision, however, Honeywell wanted to study the engineering data for UMAC-6.[149]

When DITC tried to broker an arrangement between Sperry Rand and Honeywell, the former went to great lengths to discourage the arrangement. On 3 September 1968, a meeting of DITC, Sperry Rand, and Honeywell representatives was held. While Honeywell was serious in its desire to explore the possibilities of negotiating a cooperative takeover of Sperry Canada, Paul Caden of Sperry Rand "threw up obstacles at every turn."[150] The meeting convinced DITC officials that Sperry Rand would have to be forced to keep UMAC-6 in Canada.

In a long letter to Sperry Rand, DITC argued that Sperry Rand's case for moving UMAC-6 to the United States was flawed and was based more on "corporate convenience than on economic necessity."[151] Under the terms of the cost-sharing arrangement for the development of UMAC-5 and UMAC-6, design rights could not be transferred outside Canada without the Canadian government's approval. DITC served notice that there was no way it would release these rights. The choice that Sperry Rand faced was clear: either negotiate in good faith to sell UMAC-6 technology to another company or stay in Canada. If Sperry Rand stayed in Canada, there were various financial assistance programs, DITC suggested, to support further work on the development of numerical control products.[152]

In the end, Sperry Rand decided that the place of manufacture could be profitably separated from the market. The rights to UMAC-6 were

DITC's ace in the hole, and Sperry Rand gave in to DITC's pressure and agreed to keep UMAC-6 in Canada. As a part of the arrangement, the Sperry Gyroscope Company of Canada, after 18 years of operation, was closed down. All the UMAC technology was transferred to a new Vickers UMAC Division that was to be set up in Pointe Claire, Quebec.[153] Vickers planned to develop NC marketing, sales, and service facilities through its existing offices in Cleveland, Detroit, Cincinnati, New York, Philadelphia, Milwaukee, Chicago, Seattle, and Europe.

Manufacture of the first production units of UMAC-6 began on 1 December 1968.[154] In 1971, Sperry Canada introduced UMAC-7. Claiming to have borrowed from minicomputer technology and to have "pushed the limits of integrated circuit technology to the limits of integration," Sperry Canada had packed more performance into a more compact computer.[155] By this time, Sperry Canada was functioning within the Vickers Group as an entirely separate entity with a unique corporate mandate: to design and build digital controls for machine tools. DITC had succeeded in maintaining a world product mandate for digital controls in Canada.

The Importance of the Sperry Canada Story

In a special 1982 commemorative issue marking the fiftieth anniversary of the American Society of Manufacturing Engineers, *Manufacturing Engineering* celebrated how the United States had "built an industrial juggernaut such as the world had never seen."[156] This issue portrayed the United States as the stage upon which all the important postwar technological dramas in manufacturing were played out.[157] No doubt most Canadians concurred with this American self-assessment. However, when *Manufacturing Engineering* proclaimed that computer numerical control had been born in 1967 in the United States, J. Edward ("Ted") Crozier felt it was time to pull out his Canadian flag.[158]

Crozier, then vice-president and general manager of the Canadian Institute of Metalworking, informed *Manufacturing Engineering* that CNC's parentage was Canadian, not American. Sperry Canada, Crozier wrote, had designed and commercialized this pivotal technology in the early 1960s. Editor-in-Chief Daniel Dallas was skeptical.[159] Dallas conceded that Sperry Canada had been, in 1963, among the first to conceptualize CNC. "But," he added, "it is also my understanding that their system did not work, and that modern CNC came into being via a different evolutionary route."[160]

Crozier's response was unequivocal: "The development work took place between 1961 and 1963 and resulted in the Sperry UMAC-5 computer numerical control system. I know this to be a fact because I spent three long years in the project as a field engineer and initially on the computer design interface team."[161] Crozier's assertion that 50 UMAC-5 systems had been sold by 1968 stood in blatant contradiction to Hartwig's contention that CNC first "began to move from the . . . laboratory to the production floor" in 1968.[162] After some further study, Dallas apologized that "the Golden Anniversary issue did not correctly identify the originators of this technology."[163] Dallas tempered this retraction by adding that he had not been able to "find any documentation better than [Crozier's] letter concerning CNC's true origin."[164]

Though important in fostering Canadian pride, the historical importance of the UMAC-5 story does not lie in who really developed CNC first. Rather, the Sperry Canada narrative offers yet another much-needed case study of how Canada tried to jump onto the postwar wave of innovation in digital electronics and computers.

Sperry Canada's autonomy first arose from Canada's flirtation with military self-reliance. Canada's efforts to promote local sourcing of military equipment led to Sperry Canada's manufacturing mandate. Any thought, on the part of Canadian management, of the necessity of an in-house R&D program lay dormant under the large volume of Korean War defense procurement of technology from the parent firm. Sperry Canada's pursuit of a design mandate was provoked by the need to diversify after the collapse of Canadian military self-reliance. For Sperry Canada, building a business on NC and CNC represented an opportunity to overcome vulnerability to an unpredictable defense market. However, Sperry Canada could succeed in this quest only because of the policy of corporate decentralization and divisionalization engendered by the merger of the Sperry Corporation with Remington Rand. Product-based divisionalization was supported by decentralized R&D. Sperry Canada's autonomy lay in staking out a product line not already claimed by one of the other divisions. The ensuing autonomy offered an outlet for the corporate and technological ambitions of Sperry Canada's management and engineers.

By the 1960s, Sperry Canada had gone from being the stereotypical branch-plant manufacturer to being a center of innovation of global stature. In the process, the Canadian subsidiary had created its own R&D agenda and had aggressively tried to shape its own global product mandate. In asserting its own technological and corporate raison d'être within the empire of its American parent firm, Sperry Canada ventured into the uncharted waters of CNC technology. The result was UMAC-5, described by Sperry Canada as "the most powerful production tool ever developed for point-to-point automation." Sperry Canada's efforts to embed a real-time programmable computer in machine tools was a harbinger of the profound effect the computer would have on the industrialized world's forces of production.

A coordinated effort to promote a domestic industrial capacity in computer design and manufacturing was not the driving force behind DITC's interest in Sperry Canada's CNC efforts in the years 1965–1968. Rather, DITC's focus was on the modernization of Canada's metalworking industries. As a result, DITC's funding priorities pushed Sperry Canada to abandon the development of advanced and costly CNC systems and, instead, to design less expensive NC products. Had DITC articulated a computer technology policy in 1965, it surely would have supported Sperry Canada's pursuit of both NC and CNC. It remains an open question whether Sperry Canada, with government assistance, could have developed a viable business from a more intense exploitation of the programmable computer in machine tool automation.

Did the low-cost NC option offer a more viable business strategy for Sperry Canada? Underlying DITC's strategy were two assumptions: that Canadian shops would rush to a made-in-Canada NC solution and hence facilitate the diffusion of this technology, and that producing a low-cost NC product to meet the needs of the Canadian market would generate healthy export revenues. Whether Sperry Canada believed that dropping CNC for NC was the best commercial strategy is not known. Desperate for money, Sperry Canada had little choice but to bend to DITC's funding priorities. However, in 1968, filled with a sense of urgency to promote a domestic computer industry, DITC was eager to implement its new computer policy. Even though UMAC-6 represented a retreat from computers, saving it became a test of DITC's new policy.

How successful was this NC strategy? The numbers of NC installations sold in 1966, 1968, 1970, and 1974 were 20, 288, 400, and 1200, respectively.[165] This growth had little to do with the presence of any domestic source for NC technology. Rather, it stemmed from DITC's aggressive program of subsidies to offset the costs of installing advanced machine tools in the early 1970s. One industry estimate claimed that DITC had provided subsidies for as much as 75 percent of the NC installations in Canada.[166] Despite this growth, the diffusion of numerical control in proportion to the total number of machine tools was far less pronounced in

Canada than in the United States. In 1977, an editorial in *Canadian Machinery and Metalworking* complained that inadequate diffusion of this technology continued to impede the competitiveness of Canada's small production runs.[167] Once the funds allocated to assist in the modernization of machine shops ended, the rate of diffusion slowed. This problem continued into the 1980s.[168] In 1982, Canada's CAD/CAM Council lamented how Canadian management had failed to see that other nations were using digital machine tool automation to gain market share, whereas "Canadian industry tends to view the new manufacturing technologies in a more tactical sense, mainly as a way to shed direct labor even though labor costs in manufacturing typically accounts for only 10 per cent of total costs."[169]

No doubt DITC's funding of NC procurement helped to spur Sperry Canada's domestic sales of UMAC 6 and (especially) UMAC-7s.[170] But it helped foreign suppliers even more. Sperry Canada's share of installed systems declined steadily. In 1966 Sperry Canada's products accounted for 8.3 percent of the NC systems installed in Canada. In 1968, the company's share was 6.6 percent; in 1974 it was 3.5 percent; in 1977 it was 2.9 percent.[171]

In the late 1960s, competition in the NC business intensified as new entrants from the United States and Japan rushed in. Having retreated from its strategy of competing at the high end of the market through "radical" product innovations, Sperry Canada could not survive in the price-sensitive middle market. By the early 1980s, as a shakeout hit the industry and few large suppliers emerged, Sperry Canada's (i.e., Vickers UMAC's) NC business faded into obscurity. The irony is that, as Sperry Canada left the scene, advances in microprocessors led to an explosive growth of CNC.

Support of Sperry Canada's NC efforts may have been DITC's first initiative under the computer policy framework enunciated in the 1968 in a memorandum to the cabinet. Faced with what appeared to be insurmountable entry costs, DITC believed that Canada could not aspire to have a domestic mainframe computer industry. Sperry Canada's development of numerical control, in contrast, was felt to exemplify the kind of specialization that Canadian companies should go after. That is how it appeared to DITC officials in 1968. But as soon as DITC renounced the mainframe as an element in its computer strategy, serendipity and the prestige of the mainframe seduced DITC into reversing its position. In late 1969, quite unexpectedly, the US-based Control Data Corporation approached DITC with a proposal to design and manufacture at its

Canadian subsidiary, on a 50-50 cost-sharing basis, the company's next generation of mainframe computers. Without any "moral suasion," a leading US computer multinational was offering to move a pillar of its product-design expertise to Canada. Designing a new mainframe product line was risky and expensive, but Control Data's international reputation and its willingness to cover 50 percent of the costs seemed to reassure DITC of the project's success.

The "joint venture" with Control Data was DITC's first big initiative to expand Canada's technological competence and managerial competence in computer technology. Whereas DITC's assistance to Sperry Canada had been quite modest, DITC pulled out all the stops for Control Data. The prospect of creating a world-class mainframe design and manufacturing center in Canada was too enticing for DITC to refuse. But why did Control Data, which to that point had centralized all its design and manufacturing activities in the Minneapolis-St. Paul area, decide to decentralize such a crucial part of its technological competence abroad?

7

The Dilemma of "Buying" Mandated Subsidiaries: The Case of the Control Data Corporation

In 1970, the Department of Industry, Trade, and Commerce pulled out all the stops to help the Control Data Corporation, headquartered in Minneapolis, set up the first computer design and manufacturing facility in Canada. Without a coherent policy in place to promote the Canadian computer industry, DITC had not been able to respond to Ferranti-Canada's aspirations to carve out a niche in the commercial computer business or to Sperry Canada's ambitions to convert its pioneering efforts in computer numerical control into a lucrative business. Soon after articulating its first industrial policy on the computer, DITC did help Sperry Canada salvage a numerical control business from the ruins of its CNC enterprise; however, the assistance had been very modest, and NC was still a very small segment of the computer industry. Nearly 2 years had gone by since DITC had enunciated its computer technology policy, and still Canada did not have any important design and manufacturing capacity in computer technology. DITC needed a success story, and the Control Data project became its first big industrial support program for computer technology.

Before 1970, Control Data's Canadian subsidiary had no experience in either R&D or manufacturing. And yet, for a price, the parent firm wanted to catapult the operation overnight into the forefront of computer design and manufacturing. DITC felt comfortable with the idea that a foreign parent firm was going to provide a turnkey solution to Canada's search for greater technological and managerial competence to design, adapt, test, manufacture, and market the products of the post-industrial revolution. DITC's strategy for promoting the growth of the computer industry thus revolved around trying to persuade parent firms, through a strategy of "moral suasion" and public subsidies, to hand out mandates to their Canadian subsidiaries. Like the economic and management scholars of the day, DITC officials believed in a one-way relationship in which parent firms gave mandates to subsidiary firms in a top-down manner.

This story of the journey that led the Control Data Corporation to establish an important design and manufacturing capacity in Canada, and about the fate of the enterprise, commences with the parent firm's efforts to diversify its mainframe business and with the scramble, during the late 1960s, to define a new product line.[1]

Faced with IBM's capability to finance product development, CDC looked to strategic alliances as a way to multiply the returns on its relatively modest R&D budget. The growth of technological and managerial competence at Control Data Canada (henceforth referred to as CDC-Canada) is thus also a story about the parent firm's experimentation with different organizational forms of international R&D collaboration. Perhaps more than any other computer manufacturer of the day, CDC tried to craft collaborative inter-firm R&D arrangements. Some of these were of a pre-competitive nature; others dealt with reinforcing technical complementarities. The same imperative that pushed CDC to seek out strategic alliances with US and foreign computer firms also prompted CDC to explore the idea of moving some of its product development activities abroad in the hope of tapping into local technological and industrial resources and support programs.

With considerable help from the Canadian government, CDC set up its first computer R&D and manufacturing facility outside the United States. The new Canadian Development Division then came to play a pivotal role in the design and manufacture of CDC's most successful computer line, the Cyber 170. The Canadian subsidiary then struggled to sustain a strong R&D presence within the overall corporation. The reasons for these difficulties included intra-organizational constraints, a rapidly changing technological context, and the shifting business strategies and fortunes of the parent firm.

The Rise of the Control Data Corporation

In a little more than 10 years after its creation in June 1957, the Minneapolis-based Control Data Corporation became the second most profitable computer manufacturer in the world. Its revenues soared from $3 million in 1958 to $578 million in 1969.[2] Unlike Sperry Rand, RCA, Burroughs, Honeywell, NCR, and General Electric, which struggled to make a go of their computer businesses, CDC consistently showed growing profits—$39,000 in 1958, $18.4 million in 1969. Although IBM continued to stand alone as the dominant power in the market, CDC had become a force to be reckoned with. Convinced that CDC was poised for even greater growth, investors rushed to it. In 1967–68, the value of CDC's stock rose by nearly sixfold.

For founder and CEO William Norris, the company's success was no doubt a great vindication of his initial strategy. From the outset, Norris had concluded that it was foolhardy to compete toe to toe with IBM for the business electronic data processing market. To flourish, CDC had to identify new markets in which it could exploit weaknesses in IBM's sales and technical know-how. Furthermore, Norris was firmly convinced that conglomerates had little chance of succeeding in the computer business. "To run a computer company," Norris once explained to business journalist, "it's necessary to have top executives who understand computers. People are afraid of what they don't understand, and losing money makes a man doubly afraid. Top management in the conglomerates doesn't understand computers, and when a division is losing money, they won't take a risk. They won't make the necessary budget allocations because they don't understand."[3]

The basic lines of Norris's strategy did not come to him in a sudden flash of insight. They were the hard-won lessons of more than 10 years' experience before he even created CDC, first as the co-founder of Engineering Research Associates and then as a vice-president in Sperry Rand and general manager of its UNIVAC division.

Not only had the experience with Engineering Research Associates convinced Norris of the market opportunities in specializing in high-end machines for large research organizations; it was also a reminder of what could be achieved if a company's technical and managerial energies and financial resources were exclusively focused on the computer. The Sperry Rand experience had taught Norris the perils of head-on competition with IBM and the pitfalls of multi-market companies' efforts to set up profitable computer businesses. A multi-market company (such as Sperry Rand) was not, in Norris's view, well suited to run a successful computer business, and events in the period 1950–1970 seemed to prove Norris's point. Most of IBM's revenue was generated through its computer business. On the other hand, Sperry Rand, Honeywell, Burroughs, RCA, NCR, and GE derived much smaller portions of their revenues from the computer business.[4] All these multi-market companies, however, were losing money in their computer businesses. In Norris's mind, success required total corporate commitment to the computer business. However, being dedicated exclusively to the computer business was not enough.

CDC chose a point of entry into the computer market where it felt IBM was most vulnerable: the use of computers for large-scale, computation-intensive scientific applications. Universities and large government and industrial research laboratories made up this market. Though it was nowhere near as large as the business market, the scientific market offered considerable room for growth. IBM's hold on scientific users was far looser than its grip on the business data processing market. IBM had become the undisputed master of selling standardized computers to the technically unsophisticated business market. CDC developed a keen understanding of selling powerful, customized computers to scientists and engineers.[5] Preferring to buy rather than lease, the scientific market also mitigated the cash-flow problems of trying to compete with IBM's leasing strategy.

CDC marketed its first commercially available computer, the 1604, in 1960. Although the 1604 did not outperform the competing IBM product, the 7090, it was considerably cheaper.[6] The 7090 sold for $2.4 million; the CDC machine offered comparable performance for only $1 million.[7] CDC wisely decided to make the 1604 compatible with IBM's peripheral devices. Producing about two 1604s a month, CDC managed to sell about 50. The success of the 1604 gave CDC's revenues a healthy boost. High margins ensured healthy profits. But, equally important, the 1604 established CDC as credible producer of large scientific computers

Within 10 years, CDC had emerged as the only company to have found a way to coexist with IBM and still show high annual growth rates. Although its revenues and its profits were still far behind IBM's, Norris's company led all its other competitors. CDC was still a distant second in total market share, but its mainframes dominated the market for high-performance machines.

By the end of the 1960s, the CDC name had become synonymous with "supercomputers," a term that had emerged in the trade literature with the appearance, in 1965, of the CDC 6600. Conceived to tackle the complex sets of partial differential equations encountered in the modeling of weather or in the design of nuclear weapons, the 6600 had been a big technical and financial gamble for CDC.[8]

Seymour Cray had been the technical mind behind the 1604 and the 6600. After completing his Master's degree in electrical engineering in 1951, Cray went to work for Engineering Research Associates. Six years later, when Norris left Sperry Rand's UNIVAC Division, Cray joined him as a founding member of CDC. After the 1604's success, Cray devoted himself, with an uncompromising obsession, to designing the fastest machine that circuit technology could support.[9]

With the 1604, CDC had wedged its foot in the door of the scientific market. With the 6600, CDC it burst through. The rapid rise in the 6600's popularity was an important factor in the company's stellar financial performance during the late 1960s. Earnings per share more than tripled in the period 1967–1969, and the value of the company's stock soared.

Moving from the Scientific to the Business Market

Despite CDC's phenomenal rise, William Norris believed that if the scale of operations did not rise to much higher levels the company's existence would remain precarious. "We are now living at IBM's sufferance," Norris admitted to *Fortune*:[10] Pointing to the US automotive industry, Norris noted that General Motors dominated the industry but could not put Ford out of business. The message was clear: CDC had to become the Ford of the computer industry. But could CDC's almost exclusive focus on the "supercomputers" and on the scientific market sustain the kind of growth needed to achieve this goal? Getting much bigger meant diversification, not into multiple markets, but into new niches within the computer market.

For the next 25 years, managing diversification through rapid growth and in maturity remained CDC's most difficult challenge. In a path-dependent process of experimentation, core technical competencies had to be expanded and shifted, innovative approaches to collaborative R&D had to be found to counter IBM's tremendous resources, new marketing perspectives had to be developed, optimal reallocation of resources had to be redefined continually, appropriate organizational structures had to be discovered, and suitable instruments to finance expansion had to be devised.

Control Data had begun in the early 1960s to experiment with venturing into new facets of the computer business. It developed a peripheral-equipment business, which over the long haul provided a reliable source of revenue.[11] It also ventured into data services, a move that many years later would profoundly change the company. But in terms of selling computers, the business electronic data processing market provided the largest opportunity for diversification. Heady with success in the scientific market, Control Data began to contemplate a serious encroachment on IBM's territory. In 1968, Norris declared the business market "our next area for expansion."[12] Again, Norris tried to identify the weakest point in the flank of IBM's business EDP market—the point where CDC's core competencies could be best exploited.

During the 1960s, the use of computers in business revolved around the automation of personnel files, payroll and accounting records, and inventory control. Norris was convinced that large-scale information systems would be an important growth area within the business market. Seen as important tools for corporate decision making, these systems would allow management to use the computer—in a real-time, interactive, manner—in the analysis of vast amounts of information. Along similar lines, Norris's approach to the business market also targeted real-time, interactive, transaction-based applications in banking and stock markets. Central to CDC's strategy for the business market was the development of time-sharing technology, which allowed a large number of staff members to have simultaneous access to a computer. For such systems to work, Norris reasoned, the computer would have to be very powerful and very fast—a perfect market for CDC's supercomputer technology.[13]

But there were many uncertainties associated with Norris's strategy. Could a computer explicitly designed for high-speed scientific computation offer a competitive price-performance ratio for business applications? Developing a good time-sharing operating system would be a very challenging undertaking. Many companies struggled with this, even those, like General Electric, whose business strategy centered on time sharing. Did CDC have the resources and marketing ability to move business users from the security blanket of IBM systems to the unknown of supercomputers? Could CDC's engineering culture produce competitive machines that specifically targeted the business user? How did CDC intend to counter IBM's ability to offer, in the 360 series, a complete and integrated product line of computers that allowed users to upgrade without losing their considerable investment in applications software? Ironically, although it was the success of 6600 that gave Norris the confidence to talk about expanding into the business market, Cray's design group had little to offer in the way of answers to these questions. Cray had only one interest; building an even better supercomputer.

The challenge of developing a broader range of high-end computers for the business market fell to a second design tradition within CDC, one that had evolved quite separately from Cray's group. Between Cray's 1604, in the early 1960s, and the successful 6600, in the late 1960s, CDC developed a series of machines known as the "3000 computers." The revenues and profits from the 3000 computers played an important role in CDC's financial success through much of the 1960s. Collectively called the "3000 group," these engineers had developed a broader market perspective on computer design than Seymour Cray's group had.

The development of the first of the 3000 computers, the 3600, was initiated at the same time that Cray started work on the 6600.[14] With the risks of the 6600 so high, CDC had no way of knowing when and if the successor to the 1604 would appear. CDC could not afford to let the marketing momentum of the 1604 stop while waiting for the Cray supercomputer. The 3600 offered reasonable continuity for those who had bought the 1604. Cray wanted to use more radical circuit technology in the design of the 6600. The 3000 group chose a more conservative circuit that Cray had already rejected as too slow.[15] While CDC positioned the 6600 as a dramatic technological advance, the 3600 was to be a follow-on to the 1604 that offered architectural improvements and a better price/performance ratio.[16] The product strategy behind the 3600 thus adapted a more conventional approach to competing against IBM. Being a more conservative design, the 3600 appeared on the market several years before the 6600. Although CDC's senior executives saw the 3600 simply as an interim product until the 6600 became available, the designers of the 3600 thought they were also building a machine for a broader market in which there was some overlap between scientific computation and business data management.[17] For example, the development of COBOL (a business-oriented program language) was as much a part of the 3600 project as was the development of a 3600 version of Fortran (a language used in scientific and engineering applications).

From 1963 to 1967, the 3600 and one of its variant, the 3400, had generated $150 million in sales with good margins.[18] The impact of the 3600 on CDC went deeper than just good profits. Work on the 3600 fostered the growth of CDC's software division in Palo Alto. But, more significant, the success of the 3600 secured the existence of a development capability that was independent of Seymour Cray. The 3000 tradition within CDC, however, included more than just the 3600 and 3400. Straddling scientific and data-management applications, the 3600 was still a large computer with a relatively high price point.[19] As the 3600 came onto the market, another team of engineers began work on a mid-range machine called the 3200. The 3200 evolved from earlier attempts to market small, relatively inexpensive computers called the 160/160A.

In the early 1960s, CDC sold two computers that were located at opposite ends of the product spectrum: the very large 1604 for $1.5 million and the very small 160/160A for $60,000. Why Norris adopted this strategy is not clear. Did he feel that IBM was vulnerable at two extremes, or was the 160/160A meant to generate revenue quickly? In any case, although these very small computers found uses as off-line processors, their commercial success was marginal. In 1963, with the acquisition of Bendix's computer division and the successful introduction of the 3600 in sight, CDC revisited the idea of developing a followup to the 160/160A. The Bendix acquisition brought a new customer base, along with the sales and service manpower to support it. Thus, a followup to the 160/160A could not only offer a replacement to those CDC customers who had bought the 160/160A; it could also find a market among the former Bendix customers looking to improve their older systems. In 1963, using the same circuit technology as the 3600, CDC began work on the 3200 computer.

The 3200 was far more than a followup to the 160/160A.[20] Rather than targeting a low price point, the specifications for the 3200 expanded into a mid-range computer that responded to the needs of a fairly broad customer base, which included engineering applications, business data processing, and real-time control. The success of the 3200 led to other improved models: the 3100, the 3300, the 3170, and the 3500. Collectively, these computers became known within the company as the "Lower-3000" (3000L); the 3400, the 3600, and the 3800 were referred to as the "Upper-3000" (3000U). Revenues and profits from the 3000L computers were considerable and were remarkably long-lasting. Together, the 3000L and 3000U computers provided CDC with higher profits than the computers produced by Cray's group.[21]

Collectively, the people behind the 3000L and 3000U computers—the "3000 group"—offered an alternative design and marketing philosophy to Cray's group. The 3000 group had to confront the design compromises inherent in any system that targets diverse market segments. The diversity of the 3000 computers models compelled the 3000 group to be more sensitive to compatibility and to the obstacle of migrating customers up the product line. Although the 3000L and 3000U computers had similar hardware design and used common peripherals and controllers, there was no software compatibility between the two.[22] As designers, Cray's group had a far more straightforward market imperative: build the fastest computer possible. The market was relatively uniform: scientific users with deep pockets looking for the fastest computer. When Norris boasted in 1968 that his company's next big challenge was to expand into the high end of the business data processing market, the 3000 group was poised to lead the charge.

In moving into the high-end business EDP market, CDC faced IBM's formidable 360 series. Where CDC had three incompatible lines of computers (the 3000L, the 3000U, and the 6000), IBM had one line (the 360) that spanned all price points. In 1965, a group of senior managers within CDC's Computer Division met to discuss the company's future computer development strategy in relation to IBM's 360 strategy.[23] It was quite evident to this group that "the impact of IBM's new product line on Control Data was substantial," and that "instead of facing a series of non-related equipment like the 1401, 7070, 7090, 1620, etc. Control Data is confronted with an integrated line, compatible from top to bottom. [The 360 series] was purported to satisfy all users, scientific and business, small and large, no matter what their particular application. . . . In the face of this competitive situation it has become a primary concern of Control Data to evolve a product line strategy that can be competitive in the market place in terms of price and in terms of market acceptance."[24]

CDC, however, had neither the financial resources nor the market share to emulate IBM's 360 strategy. The R&D costs of the 360 series dwarfed CDC's R&D budget. IBM also devoted considerable time and money to migrating customers from the older computers to the 360 machines. In addition to retraining, IBM also offered software for the 360 that emulated older models in order to make the migration less painful. Launching the 360 was indeed, as one business historian put it, comparable to "massing the troops and equipment for the D-Day invasion of France during World War II." CDC could never mount such a frontal assault on the market. Constrained by relatively limited financial resources and by the legacy of earlier design decisions, the senior managers who analyzed the impact of the IBM 360 strategy had little choice but to propose blending IBM's compatibility approach with CDC's existing strategy of building machines for specific market segments. The 6000 series of computers would be devoted to large-scale computation; the 3000L series would be medium-price machines for a broader data processing market.[25] However, within each of these two series, CDC would copy the IBM strategy and ensure that there was total software compatibility. Although a pragmatic decision, the rejection of any software compatibility between the new 3000L line of computers and the 6000 line also mirrored the deep division in product-design philosophy that had grown between the 3000 group and Seymour Cray.

The weak point in this hybrid strategy was the 3000L series itself. The 6000 series was still new; the 3000L would soon become obsolete. Although it was recognized in 1965 that a new generation of computers would soon be needed to replace the 3000L and the 3000U, consensus on an effective technical strategy proved elusive. However, in 1968, when the issue of obsolescence could no longer be avoided, a team of individuals who had worked on the 3000L and the 3000U was formally put together to start thinking about the next product cycle. For the next year, this group grappled with defining the system specification for the next series of compatible computers to replace the 3000L, the 3000U, and possibly the 6600.[26] In 1999, Derrel Slais, who was central to this effort and who later became the company's Vice-President of Computer Development, recalled the growing realization back in the late 1960s that CDC had a series of incompatible computers and no way to move its customers from the 3000L to the 3000U and then to the 6000. At this time, Cray was also starting to work on the successor to the 6600, the 7600. "There was no way of getting Seymour interested in a wide product line," Slais recalled. "He was out to get the fastest scientific processor that he could with the architecture and circuit logic capability that was available at the time."[27]

By early 1969, the 3000 group had at last articulated a framework for the new generation of computers: the MPL ("medium priced line"). In addition to protecting CDC's 3000L customer base, the marketing objectives of the MPL were also "to lessen dependence on Federal government procurement by penetrating the large-scale commercial market, to meet the needs of the 70s by providing a balanced performance on commercial data processing, communications, data management, enquiry retrieval, scientific computing and real-time applications, and to provide an 8-bit replacement system for the non-CDC, non-IBM scientific market represented by the Burroughs 5500, the G.E.625 and the Univac 1108."[28] Although these objectives mirrored Norris's new computer strategy, senior executives were nevertheless reluctant to release funds to start development of the MPL series. There were concerns about the high costs needed to migrate the existing 3000L and 3000U customers over to the MPL. There were also doubts about Control Data's financial capacity to support three very different product development efforts: Cray and the 7600, the 3000 group and the MPL, and a third group that had just been created to build a special supercomputer for the Lawrence Radiation Laboratory.

The Lawrence Laboratory had been looking for a new high-speed computer system that could tackle a range of scientific computations involving very large arrays of floating-point numbers. Originally, Livermore had asked CDC to bid on the design and production of a computer system based on the "Solomon" concept, which entailed a highly parallel machine. Declining to bid, CDC came back with its own concept to solve the laboratory's specific computational needs: vector streaming.[29] The lab accepted that concept, and by July 1967 CDC had started the formal design process. By the end of 1969, some of the major components had been completed and testing had begun.[30] James Thornton, who had worked with Seymour Cray on the 6600, was put in charge of the new Livermore project.[31]

The new supercomputer was to be of a radically new design, very different from Cray's approach to the 7600. The central feature of the new design was string arrays, from which the computer derived its name: STAR 100.[32] The strength of the STAR 100 lay in its ability to perform memory-to-memory operations on very long vectors of operands.[33] For a class of large-scale scientific problems, such as designing nuclear weapons or predicting weather, vector processing was supposed to rip through the computations faster than any machine in existence.[34] With its very-high-bandwidth memory, the STAR was also to be the first mainframe ever built to run 100 million instructions per second.[35]

By mid 1969, senior management's ambivalence toward the simultaneous development of the STAR and MPL architectures had passed and the MPL specifications were abandoned. CDC, however, had not given up on its ambitions to expand into the high end of the business market. The replacement to the 3000 series would be designed around the STAR architecture. The details of how this decision was reached remain unknown. Those who had worked on MPL were not enthusiastic about STAR. However, it seems clear that James Thornton was instrumental in selling senior management on a wider role for the STAR architecture within CDC's product development. After all, the STAR specifications did appear to offer wide applicability. Endowed with a rich instruction set—which included scalar processing, vector processing, business data processing, and bit string handling instructions—the STAR 100's basic architecture seemed versatile enough to be adaptable to business applications. Thornton argued that the STAR 100 was a dramatic advance not only in scientific computation but also in commercial data processing.[36]

The STAR-based strategy was seductive. At long last, CDC would be able to respond to IBM's 360 strategy with its own line of compatible computers. But CDC had no intention of getting into a suicidal competitive battle with IBM over the entire range of the 360. The new STAR-like line of computers included the same market objectives as the MPL: from the high end of business applications to the low end of scientific applications. But now the top end of the new line would extend into supercomputing and be anchored by the STAR 100. The new series was called NPL ("new product line"). "PL-50" was the internal name assigned to the first computer of the NPL series that would be built.[37]

The marketing advantage of the 360 was that it could draw in customers who wanted to start small. Unfortunately for CDC, when these customers wanted to move up to larger systems they would stay with IBM for reasons of software compatibility. To increase the attractiveness of its high-end product line, CDC actively sought an ally that could design and market products that were equivalent to the lower half of the IBM 360 series but also compatible with the PL-50. In this way, CDC and its partner could be free to develop their strategies within their respective markets, yet they could gang up on IBM with a fairly unified alternative to the 360 line. To this end, CDC aggressively sought cooperative and collaborative arrangements with other companies; either to promote compatibility through common standards or to share R&D costs for technologies of common interest.[38]

Although the new STAR-based strategy rationalized three independent computer architectures into two, it never reduced the financial and technical risks of the MPL. CDC was still faced with the market, financial, and technical risks associated with trying to migrate the 3000L, and 3000U and 6000 users over to PL-50, and now there was an additional risk. No doubt the MPL project would have faced the challenge of trying to engineer the product to be competitive, on a performance basis, at a specific price point. But that would have entailed far fewer technical uncertainties than the more ambitious PL-50. The PL-50 design was unexplored territory. Loaded with a lot of sophisticated hardware to optimize very long string-array or vector processing, the STAR 100 was to be a very expensive machine. The PL-50 team would have to transform a highly specialized, highly customized, and very costly supercomputer design into a competitive, mid-range, standardized machine for both business and scientific EDP markets. This difficulty was compounded greatly by the fact that the STAR 100 itself had yet to be built and fully tested. Unable to observe the STAR 100's performance over a range of applications, CDC had no way to assess all the technical ramifications of this radically new design. The big question facing the 3000 group was whether one could shrink the STAR 100 down to a machine that could sell profitably. Although these difficulties were recognized and discussed, they were never subjected to meticulous scrutiny. The impetus to forge ahead was strong. The company needed a new product line, and the STAR strategy had seduced upper management

With Seymour Cray's work on the 7600 still in progress, and with a contractual commitment to finish the STAR 100, financing the entire PL-50 project would put a severe strain on CDC's finances, particularly if the new product was to get out in a timely manner. CDC needed to find ways to leverage its constrained R&D budget. One option was to expand its international organizational capacity and leverage foreign funds for R&D. This imperative would eventually lead CDC to Canada.

CDC's Experiments with International R&D Ventures

In 1961, with the 1604 a firmly established success in the US market, CDC set its sights on foreign markets. Western Europe's recovery from World War II had begun to bring new demands from universities and research organization for powerful computers. Being much cheaper than the IBM 7090, the 1604 had considerable potential in Europe. Here was a clear opportunity for CDC to expand its attack on IBM's flank to the international arena. William Norris pondered how to organize this global campaign. CDC was still a relatively small organization with no infrastructure for international marketing, sales, and service. Rather than embark on the daunting and costly task of setting up subsidiaries, Norris had initially favored a joint venture with a large multinational corporation. Philips Industries, which had not yet decided to enter directly into the computer business, was eager for a partnership.[39]

CDC and Philips seemed like a good fit. Philips, one of the world's foremost electronics manufacturers, had a vast international network of subsidiaries. After extensive negotiations, Philips and CDC had reached a tentative agreement: CDC would handle the US market, and Philips would market, sell, and service CDC computers in Europe. But at the last moment the deal fell apart. Multinationals have evolved diverse organizational structures with varied networks of intra-firm relationships.[40] Philips was a highly decentralized multinational with the relatively autonomous subsidiaries. Having gotten wind of CDC's pending deal with the Netherlands-based parent firm, Philips USA demanded that it become part of the deal. Philips USA insisted that a new joint venture with CDC be set up to handle the US market. Norris flatly rejected the offer. Sharing the US market did not sit well with him.[41] Philips USA, on the other hand, argued that, although CDC would lose a portion of the US market, CDC would gain far more through Philips's international sales. In addition to sharing markets, Norris and his executives had concerns about being suffocated in the embrace of the much larger Philips USA. Norris knew that many of his senior executives would resign because of this fear.[42] Despite the efforts of the parent firm to find an acceptable compromise, Norris refused to budge. The collapse of the deal convinced Norris that the time had come for CDC to develop its own global organization capacity.

As many other US companies had done since the late nineteenth century, Control Data chose Canada for its first international foray.[43] Geographic proximity, language and customs, good transportation links, common business practices, and a very good market for computers made Canada a safe and easy place in which to experiment with setting up subsidiaries. CDC-Canada was set up in Toronto on 11 June 1962.[44] Run by a "country manager," CDC-Canada had rather tightly circumscribed responsibilities: to act as regional sales and marketing divisions for the company.

The challenge of funding the PL-50 project eventually pushed CDC executives to reconsider the exclusive marketing and sales character of the company's multinational structure. Should the company move some of its product development activities abroad in the hope of tapping into local technological and industrial resources and support programs? In September 1969, CDC chose Canada for its first experiment in moving R&D abroad.[45] Even if it meant winning R&D subsidies, no other US computer manufacturer would have proposed such a daring partnership—especially one that involved transferring a core competency to another country. Although provoked by the financial needs of the PL-50, CDC's search for international partnerships had become central to its corporate philosophy. In a market dominated by IBM, strategic alliances offered CDC opportunities to leverage smaller R&D budgets. More than any other computer manufacturer of the day, CDC actively sought out collaborative R&D ventures with other firms in the computer industry.[46]

In considering where to locate its PL-50 project, CDC was ready to exploit international concern over IBM's stifling of competitors. But most of the Western European nations were desperately seeking to accelerate their domestic capacities to design, manufacture, and market mainframes. The existence of a domestic computer industry had become a high-profile political issue wrapped in strong nationalist sentiments. The governments of the advanced European nations were too preoccupied with ensuring the success of their own "national champions" to be interested in helping another US company. In the political climate of the late 1960s, building local ownership was inseparable from nurturing domestic technological competence. Although Canada got swept up in the same policy currents, its political economy could not support a single-minded crusade led by a Canadian-owned "national champion." Having to balance the call for Canadian control with the historic importance of direct-foreign investment in industrial development, Canada was more receptive than other nations to CDC's search for a government partner.

A propitious political climate was not the only thing that led CDC to consider Canada as place to move its PL-50 program. The risks of the PL-50 project were compounded by the fact that CDC was, for the first time, experimenting with moving product development abroad. Having the project take place in "close proximity" to the parent firm no doubt mitigated some of these uncertainties.[47] In 1968, the proximity advantage was further reinforced by CDC's purchase of complete control of Computing Devices of Canada (ComDev), one of Canada's leading military electronics firms.[48]

In 1956, ComDev turned to Bendix for equity financing when it was unable to find enough private venture capital in Canada. By the mid 1960s, Bendix's initial 40 percent ownership in ComDev had grown to a controlling interest of 66 percent.[49] In addition to marketing Bendix computers, ComDev manufactured defense-related Bendix products under license. In return, Bendix marketed ComDev's sophisticated defense products in Europe, Asia, and South America. In 1963, when CDC bought Bendix's computer business, it also acquired controlling interest in ComDev. The president of Bendix had spoken highly of ComDev to William Norris.[50] Norris soon came to agree. In 1968, Control Data acquired all ComDev's outstanding shares.

Through its ComDev operation, CDC learned of the Canadian government's determination to foster the growth of a domestic computer industry and of DITC's policy strategy of encouraging multinationals to give their Canadian subsidiaries broader mandates.[51] Around September 1969, well aware that the Canada desperately wanted to grow a computer industry, CDC invited C. D. Quarterman, the senior DITC official responsible for the entire electronics sector, to come to Minneapolis to discuss the possibility of a more ambitious mandate for CDC-Canada. Senior executives explained to Quarterman that CDC had reached a point in its growth where it was time to consider expanding the company's manufacturing and R&D presence beyond the borders of the United States.[52] In particular, CDC raised the idea of shifting the PL-50 project to Canada if the government was willing to share in its costs. On 13 February 1970, the proposal was fleshed out in a full-day meeting in Minneapolis by senior executives of CDC and ComDev and senior officials of DITC.[53] James Thornton took the Canadian delegation on a tour of CDC's research facilities in Arden Hills and outlined the STAR-100 project. Director of Advanced Product Planning Ron Manning then presented

CDC's vision of the PL-50 and the company's belief that this project could be successfully carried out in Canada. The CDC, ComDev, and DITC participants then developed an action plan to guide the preparation of a detailed feasibility study. In the weeks that followed the meeting, committees were formed to examine the hardware, software, manufacturing, manpower, financial, legal, and marketing aspects of the proposed project. The work of these committees was to be submitted to William Norris.

Over the next 4 months, through numerous meetings and communications, CDC and DITC tried to hammer out a mutually acceptable agreement. DITC officials were convinced that funding the PL-50 project would create an important R&D capability, but they were more worried about the Canadian subsidiary's freedom to exploit the worldwide commercial opportunities of the PL-50 product. DITC pushed for guarantees that all production of the PL-50 would be done in Canada; CDC wanted the flexibility to use its worldwide production facilities efficiently, which meant not letting its existing US production facilities be underutilized.[54] Of course CDC was eventually going to design and manufacture most of the subsequent models in the PL series, but until then CDC wanted some of the PL-50 production shifted to its US plants. Ownership of intellectual property was another issue. Fearful that valuable technology developed in Canada (with the help of public money) would be "repatriated" to the parent firm, DITC set the standard terms of its assistance program (called the Program for the Advancement of Industrial Technology) so as to prohibit the parent firm from freely using technical know-how developed in the Canadian subsidiary. CDC argued that, because the success of the PL-50 would depend on close collaboration among the parent firm, its software division in Palo Alto, and its Canadian subsidiary, restrictions on the flow of technical know-how would be counterproductive. CDC also reminded DITC that a great deal of technical know-how was being freely transferred from the parent firm to the subsidiary.[55]

The extra-territorial jurisdiction of the US government over US firms' foreign subsidiaries was a particularly thorny issue for DITC. The Canadian government did not want U.S legislation to shape the disposition of innovations developed in Canada through public programs. The PL-50 was to be a state-of-the-art machine. In the name of the national interest, the US government forbade US companies from exporting sophisticated technology to Eastern Bloc nations and to any other nations that it felt endangered US security. The ban also applied to the foreign subsidiaries of US-based companies. Not only would this ban constrain the freedom of CDC's Canadian subsidiary to export the PL-50; it also compromised Canadian sovereignty. This policy effectively imposed US foreign policy and US law on Canadian territory. CDC was concerned over the Canadian government's practice of insisting that Canadian subsidiaries be free to pursue all international markets for any product developed in Canada with the assistance of public funds.[56] In 1965, the US government had prohibited CDC from selling the 3600 computer to France's nuclear research program.[57] Since CDC specialized in high-performance scientific computers, its marketing efforts were more likely than not to run up against US security concerns. Aside from security issues, the US government was fearful that exporting high-tech know-how—even to Canada, the United States' largest trading partner and closest ally—would diminish the competitive advantage of the United States.[58] Although much of the world views US-based multinationals as promoting US hegemony, US political leaders once worried that these multinationals were conduits moving technology and jobs out of the US.[59] CDC executives were quite sensitive and sympathetic to the Canadian position, but they knew that their hands were tied.

In view of the scope and scale of the proposed project and the complexity of the issues, it is remarkable that CDC and DITC hammered out a detailed project proposal in only 4 months. Both parties were desperate for an agreement. By the spring of 1970, neither side could afford to let this initiative fail. For the second time in CDC's history, the company's computer operations were heading for a loss. Unlike the first loss, this one was going to be big. A month before the Canadian government agreed to the proposal, CDC knew its extent. From a profit of $18.4 million in the 1968–69 fiscal year, the financial performance of CDC's computer business plummeted to a $41.4 million loss in the next year. The price of a share of the company's common stock swung from a high of $159 to a low of $28; it may have fallen even further had CDC not also owned Commercial Credit, one of the largest US credit companies.[60] Control Data was forced to institute extensive cost cutting. For the first time in the company's history there would be extensive layoffs. The number of employees in Control Data's worldwide operations had grown every year, approaching 48,500 in 1969. Within a year, CDC would eliminate more than 10,000 positions.[61]

This financial tumble was provoked primarily by an inventory problem with the 3000L computers.[62] Demand fell rapidly, and the company got stuck with an oversupply. This underscored the growing obsolescence of the 3000L computers and the need to develop the PL-50 as a replacement line. Yet the financial downturn would force the PL-50 program to come to a grinding halt if there were no deal with Canada. Without Canadian funding, CDC faced the distasteful prospect of laying off talented engineers and, in the process, losing valuable expertise.

Bringing the PL-50 project to Canada was equally important to DITC. More than a year earlier, the Minister of Industry, Trade and Commerce had first underscored, in a memorandum to his cabinet colleagues, the urgent need to foster the growth of a domestic computer industry. DITC also doggedly pursued its "moral suasion" in the hope of getting IBM to give its Canadian subsidiary some sort of R&D and manufacturing role. These efforts, however, had yet to produce any tangible results. Furthermore, by early 1970 there was new pressure on DITC's bureaucrats and minister to be more aggressive. The newly formed Department of Communications and its new minister, Eric Kierans, had succeeded in raising the issue of Canada's technological competence in computers to national prominence. Kierans was determined in his efforts to claim the leadership role on this high-profile portfolio. Eclipsed by the publicity given to Kierans and his department, DITC and its new minister, Jean-Luc Pepin, needed success stories. The PL-50 was the kind of success story that DITC yearned for.

In June 1970, CDC made a formal application to the Canadian government for financial assistance, under DITC's Program for the Advancement of Industrial Technology, to develop and manufacture the PL-50 in Canada. Owing to CDC's and DITC's efforts to build a consensus through extensive informal discussions, approval of the formal application came quickly (in July).

The PL-50 and Canada

In a press conference in Toronto, Jean-Luc Pepin, William Norris, and senior executives of CDC spoke in glowing terms of the win-win situation that the new R&D facility had created for both CDC and Canada. Still, DITC, recalling Ferranti-Canada's demise and DITC's inability to act decisively, was not going to let this opportunity to create Canada's only center of technical and managerial competence in mainframe computers slip away. The *Montreal Gazette* quoted one government source as saying that the federal aid for this project was "the largest amount of government assistance ever given to such a manufacturer."[63] "Control Data's decision to establish research, development and manufacturing facilities in Canada," Pepin explained, would "give Canada an opportunity to participate in a world market for medium to large mainframe computers which my officials expect will represent a significant share of the overall computer market."[64] Norris reminded the audience of reporters that the new R&D facility in Toronto would be "engaged in advanced computer technology leading to the design, development and manufacture of powerful, general-purpose digital computer systems having a wide range of applications in industry, business, education and government."[65] He also underscored the software-oriented mission of this new R&D center.

The estimated cost of bringing the PL-50 from an idea to full-scale production was about $53 million by 1975, of which $39 million was directly attributable to R&D.[66] Insofar as CDC had spent a total of $91 million in R&D on all its products and services in the previous 5 years, the PL-50 represented a considerable commitment.[67] The Canadian government agreed to contribute $23.1 million, spread over 5 years, toward the costs of developing the hardware and the software and obtaining the capital equipment needed to start manufacturing. To produce the PL-50, CDC planned to create a world-class R&D facility in Canada with a staff of more than 400. According to CDC's marketing estimates, 716 systems would be sold worldwide over the lifetime of the PL-50. CDC committed to manufacture 407 of these Canada.[68] In terms of the net value added that would accrue to Canada, the PL-50 project held out the promise of $350 million.[69] The R&D and the initial manufacturing would be done in a facility to be built in Mississauga, near Toronto. Most of the PL-50s, however, would be manufactured in a new 135,000-square-foot plant in Quebec City.

A large contingent from the 3000 group was to be the vanguard of CDC's new Canadian R&D facility. Seventy of this group's software developers, hardware engineers, and technical support people, along with their families, were to move to Canada. Working side by side with these people, the new Canadian recruits to the R&D facility were to gain experience in the many technical and managerial facets of computer design, testing, and production. After 2 years, most the Americans were to return to Minneapolis, leaving behind a cadre of managers, engineers, and technicians who would finish the development and bring the PL-50 to market. When PL-50s were coming off the Mississauga production line, the bulk of the production would shift to the facility in Quebec City.

It could be argued that CDC's management, in setting up the PL-50 project in the Toronto area, wanted to integrate mainframe design with its existing marketing and sales subsidiary in Toronto. But the reporting relationships that linked the "Development Division" to the Canadian subsidiary and to the parent firm do not support such a motive. Although the Canadian Development Division was legally a part of CDC-Canada, in reality it operated as an independent organization that reported directly to the parent firm on matters of R&D strategy and budget. The president of CDC-Canada, charged with overseeing marketing and sales of CDC products in Canada, played no role in either short- or long-term management of the Development Division. Operating strictly as a cost center for the parent firm, the Canadian Development Division could never be mobilized by the president of CDC-Canada as a part of any profit strategy at the subsidiary level.

Since CDC had already purchased complete control of ComDev, Ottawa (where ComDev was located), rather than Toronto, would have been the logical place for CDC to locate its new Canadian "Development Division." In addition to being one Canada's most important centers of military R&D, ComDev had been (along with Ferranti-Canada) one of Canada's earliest industrial centers of R&D in digital technology. Since ComDev had a considerable pool of design and production expertise in electronics technology, having the new mainframe design group in close proximity to ComDev could have promoted a mutually beneficial two-way flow of technical and marketing know-how between the two. Although the development and sale of computers for commercial EDP and for specialized military applications were organized into different business units within the parent firm, the boundary between the two units had always been very permeable.[70] For example, CDC had quickly applied the expertise it had accumulated in producing the 1604 to execute a series of important contracts to develop and advanced geo-ballistic computer for the Polaris submarine's fire control system.[71] CDC then used the know-how it had used to build 24-bit Polaris computer to develop the 904, a commercial computer for large real-time applications.[72] In the early 1970s, CDC was developing specialized computers for US missile systems. With military products producing about $45 million in revenue, with healthy cost-plus margins, CDC's upper management strove to maximize the interchange of technical know-how between the company's civilian and defense computer development efforts.[73] CDC, however, did not transfer this organizational strategy to Canada. Instead of promoting strong interaction between its two Canadian R&D organizations, CDC mistakenly isolated the mainframe design group from the defense-oriented digital design expertise at ComDev. The Canadian Development Division was means to an end: the PL-50. It appears that CDC's senior executives, focusing on the new product line, gave little thought to the long-term sustainability of a separate R&D organization in Canada.

Why did CDC, having decided that the Canadian R&D for the PL-50 would be done in Toronto, then decide to have most of the production done in Quebec City? From the outset, the PL-50 project was caught up in the politics of regional development and in the policy ambitions of a new department. Inter-province disparities have always been a contentious issue in the politics of Canadian federalism. With the rise of the Quebec separatist movement in the late 1960s, regional disparities and linguistic divisions produced an explosive political climate in Canada. Pierre Elliott Trudeau, the leader of the Liberal Party, having campaigned vigorously on a vision of national unity during the 1968 federal elections, argued that regional economic disparities were as destructive to national unity as the French-English confrontation. Only a strong, interventionist federal government, Trudeau proclaimed, could prevent the Balkanization of Canada.

Soon after becoming Trudeau's Minister of Forestry and Rural Expansion, Jean Marchand set about creating an institutional framework to make regional policy an instrument of national unity. Marchand was one of Trudeau's most trusted political allies. Before entering politics at Trudeau's request, Marchand had established a highly respected reputation in the labor movement. To his cabinet colleagues, Marchand explained: "We are not achieving anything like a just society in a united Canada if opportunities are so disparate that there is heavy and persistent unemployment and under-employment in some regions even when the economy is buoyant. . . . Our unity is not secure if people in some extensive regions have to put up with opportunities and standards well below those of other Canadians. . . ."[74]

To better address the regional disparities, Marchand reoriented the regional programs from all manner of assistance spread out across an entire region to incentives aimed at promoting the growth of a small number of designated "industrial centers."[75] Giving incentives to industry became one of the pillars of helping economically depressed regions.[76] "Growth poles" became the principal policy instrument for addressing regional disparities.[77] This top-down, "centrist" approach, championed by the French economist François Perroux, viewed economic growth and modernization as originating in a major city and then diffusing to the rest of the region. Regional development first required urban economic development. To better reflect the new mission, the Department of Forestry and Rural Development was renamed the Department of Regional Economic Expansion.

The government had decided that DREE should focus all its energies on the eastern portion of Quebec and the four Atlantic provinces.

For its first major incentive, the newly formed Department of Regional Economic Expansion offered, through its Industrial Research and Development Incentives Act (IRDIA), to cover 40 percent of the capital costs if CDC would build its main manufacturing facility in Quebec City, which was to be a growth pole for the eastern portion of the province of Quebec.[78] In effect, DREE promised to throw a large portion of its budget—$10 million—into the PL-50 project.[79] Like DITC, DREE was looking to the PL-50 as an all-important first success story. Although Quebec City had high unemployment, there were cities in the Atlantic provinces that were more in need of measures to spur economic development. Had DREE's minister been a Maritimer, he or she would have surely insisted that the PL-50 plant be located in Halifax or some other city in Atlantic Canada. But Jean Marchand's riding was in Quebec City. Broader political concerns may also have been at work. Marchand was sensitive to the linguistic tensions in Quebec. With separatists proclaiming that Quebec's economic and industrial aspirations could not be met within the confederation, there were obvious political advantages to the federal government's being seen as bringing Canada's most important computer manufacturing facility to the heart of Quebec: Quebec City.[80]

How did CDC's American executives feel about the Canadian government's intense lobby to get them to set up in Quebec City? It is hard to imagine the tough-minded William Norris accepting any terms he did not like. Low-cost labor would have surely been attractive.[81] The proposed Quebec City plant was going to employ about 560, but nearly 400 would be women doing semi-skilled assembly work.[82] DREE's promise to cover 40 percent of the construction and fixed-capital costs would have been even more attractive to Norris, particularly since his company's Computer Systems Division[83] had suffered considerable financial losses. However, in addition to the obvious financial advantages, Norris would have also been sympathetic to the idea of using business investment to give opportunities to economically disadvantaged regions. He was a firm believer in the corporation's social responsibility to address the poverty that existed in America's inner cities.[84]

The agreement between CDC and the Canadian government set out four objectives: to create a permanent mainframe R&D facility in Canada, to design the PL-50, to create the infrastructure and transfer the know-how needed to manufacture the PL-50, and to foster the emergence of Canadian sourcing for the various components used in manufacturing mainframes. The fact that CDC sent some of its best men to Canada indicates that it was committed to these objectives. But CDC had no other option. It urgently needed a new product like the PL-50.

The project's general manager, Ray Nienburg, had joined CDC in 1968 as General Manager of the Advanced Software Development group. Before coming to CDC, Nienburg had been Manager of the IBM Time-Sharing Group, which had worked on the 360/67. At CDC, he had risen to General Manager of the entire Advanced Products Development Division. Michael Sherck, brought in by Nienburg in 1969 as Director of CDC's Systems Integration Group, was assigned to head up the systems integration work in Toronto. Sherck had previously worked for IBM on the systems integration of the 360/67. Derrel Slais (in charge of the PL-50's hardware design) and Gerald Schumacher (in charge of software development) had been important figures in the development of the 3000 lines of computers and in the MPL system definition. Bob Olson, General Manager of CDC's US manufacturing operations, was chosen to set up the manufacturing operation for the PL-50. The nearly 70 individuals who accompanied these men represented the heart of the 3000 group's know-how. One cannot imagine any other US computer manufacturer embarking on such an unprecedented international reallocation of manpower.

Set at a price point that corresponded to IBM 360/65, the 8-bit-oriented, 64-bit-word PL-50 was to have "four times the performance of the CDC 6400 for scientific data processing and four times that of the IBM 360/65 for business data processing."[85] The Canadian team had the responsibility for developing the central processing unit, the memory storage unit, the storage access control, and the software (which included the time-shared operating system, Fortran and COBOL compilers, and emulation software that would help entice 3000L and 360/65 users to migrate to the PL-50).[86] Initially, the design for the PL-50 called for a 1.2-microsecond core memory, then considered state of the art. But, in parallel with the Canadian team, a group in Minneapolis was developing a new solid-state MOS (metal oxide semiconductor) 400-nanosecond memory that, when incorporated in the second version of the PL-50, would give that computer 16 times the performance of the IBM 360/65.

By the end of 1971, Nienburg, Slais, Schumacher, Sherck, and Olson had put a strong R&D group together at CDC-Canada, and all was progressing well.[87] The prototypes of the central processing unit had been built and were undergoing intensive testing.[88] The first 1.2-microsecond core memory unit and the input/output control unit had both undergone initial checkout. The software group had made important advances in developing emulation techniques that would allow them to do the initial development work on the PL-50 on several CDC 6400s that acted as dedicated emulators. They also were working on PL-50 simulation techniques that would allow prediction of the system's performance. To house both the new "Development Division" and the first manufacturing plant, CDC-Canada constructed a new 165,000-square-foot facility in Mississauga.

Finding a sufficient number of skilled Canadians to fill the needs of the crash R&D program was problematic at first, and recruiting was done in the United States and in Europe. By 1972, however, 281 of the 397 staff members of the Development Division were Canadian. The Canadians tended to occupy entry-level and intermediate engineering positions, but they nevertheless included some Canada's brightest young computer engineers. "At the time," recalls one of the US engineers who came to work on the PL-50, "Control Data was the only company to build a mainframe in Canada. When we announced that we were building operating systems and compilers the kids from [the University of] Waterloo would come to work for free. I mean it was the greatest game in town. So we got the cream of the crop out of Waterloo."[89] Finding qualified senior engineers who could assume project-management responsibilities, or who could be senior software engineers, remained an ongoing problem.

Fostering the growth of an upstream semiconductor components industry had been an integral part of DITC's strategy for the PL-50 program. Back in the early 1950s, Canadian military enterprise had viewed the existence of a domestic semiconductor as essential to national self-reliance. But as the cost of military electronics had climbed and defense budgets had tightened, Canada's armed forces had lost their desire to subsidize a costly domestic capacity to produce semiconductor components. Instead, as was discussed in chapter 3, Canadian military enterprise shifted its focus on self-reliance from a capacity to produce transistors to a capacity to do leading-edge circuit design. But there were still some within the Defence Research Board and the Department of Defence Production who believed that a domestic capacity to produce semiconductor components was essential to Canada's development as an advanced industrial nation. By the mid 1960s, many of these men had shifted over to the newly formed Department of Industry, which had been added to the existing Department of Trade and Commerce to form

DITC. The dream of a domestic semiconductor industry took on new strength within DITC's Electrical & Electronics Branch. Realizing how high the entry costs were, DITC had hoped to convince an established US producer to set up a plant in Canada. Despite the possibility of subsidies, US semiconductor companies had no interest in coming to Canada. DITC understood all too well that the Canadian market was too small to justify the startup costs of component manufacturing. US companies, on the other hand, did not see any advantage in DITC's suggestion that they build a plant in Canada to export semiconductor components when their plants in the US could export just as well.

In the fall of 1968, DITC decided to push for the creation of a home-grown microelectronics facility. Surveying Canada's industrial landscape, DITC concluded that Northern Electric, the manufacturing arm of Bell Canada, had the need, the resources, and the expertise to be an active partner in the creation of a separate microelectronics business.[90] Until 1962, Northern Electric had done little to advance its competence in the transistor revolution. That changed in 1962, when Canada's Defence Research Board and National Research Council asked Northern Electric to join in an collaborative R&D program in semiconductor devices.[91] From this joint program, Northern Electric's Advanced Devices Centre emerged. In 1968, DITC wanted the interdepartmental Advisory Electronics Board to support the idea of establishing a Canadian components industry. Claiming that "most of the companies with the required capabilities were American [and] none appeared interested in establishing in Canada a facility based largely on serving export markets," DITC recommended Northern Electric as the best hope.[92] By the time DITC approached Northern Electric with its plan, the Advanced Devices Centre had nearly 600 highly skilled employees and was producing more than 260 different semiconductor devices.[93]

The DITC proposal called for the creation of a new company to develop the know-how to design and manufacture "hybrid integrated circuits and monolithic devices" for telecommunications.[94] This new company was to build an export business around Northern Electric's internal competence. According to the plan, Northern Electric was to sell its Advanced Devices Centre to the new company in exchange for sizable equity in the new company.[95] Of the $60 million needed to finance the new company, the Canadian government would provide $36 million in R&D subsidies and $12 million in the form of an interest-free loan for capital equipment.[96] On 11 March 1969, DITC's plan led to the creation of the Ottawa-based firm Microsystems International Limited (MIL).

Initially, DITC hoped that the existence of MIL would lead to an innovation network that would link the nascent microelectronics company to the telephone monopolies.

From the earliest elaboration of the PL-50 proposal, DITC had hoped to integrate MIL into the project. The appearance of CDC had created the opportunity for a larger and more powerful network of innovation in microelectronics, communications, and computers. With more than $100 million in private and public investments tied up in just the MIL and PL-50 initiatives, DITC wanted MIL and CDC to work as closely together as possible. MIL's capacity to produce the advanced hybrid and memory circuitry for the PL-50 seemed more promising when, at about the start of the PL-50 project, Intel's board of directors, over the objections of the company's chief of operations, Andy Grove, gave MIL the license to second source its new 1K DRAM memory chips.[97] And Canadian Marconi expressed an interest to produce the PL-50s multi-layer, printed circuit boards. Canadian Marconi was eager to get the contract, but its first samples proved substandard and too costly. Working closely with CDC, Canadian Marconi began to improve its manufacturing capability. Two other Canadian firms also participated: Elco Connectors agreed to produce the 64-pin connectors, and Canron was to develop and manufacture the PL-50's 125-kVA motor generators.

Buoyed by the initial progress in the PL-50 project, CDC and its Canadian subsidiary tried to win a privileged role in the Canadian government's efforts to develop a domestic computer industry. CDC wanted to extend its cooperation with Canada beyond the PL-50. In January 1972, CDC expressed its ambitions in a 17-page submission to Minister of Industry, Trade and Commerce Jean-Luc Pepin. Astutely couching its own corporate interests in Canadian nationalist political rhetoric, the company argued:

This nation has suffered for the past 100 years from industrial and thus economic domination, first by the United Kingdom and more recently by the United States. As most Canadians know too well, there are no easy solutions nor any single solution to this dilemma. There are however, some conclusions which can be drawn . . . : our ability to grow industrially is heavily dependent on our ability to keep up with the advanced nations in the automation field. . . . One of the primary automation tools is the availability of digital technology and its resulting computer capacity.

Control Data is committed to the view that Canada will seriously jeopardize her ability to solve her problems . . . if she lags rather than accelerates in industrial development. We further believe that an indigenous computer industry is a vital prerequisite of industrial expansion.[98]

Underscoring the existing DITC-CDC partnership, the submission concluded:

We believe that the [PL-50] program can be made the focus and vehicle to develop a Canadian capability including the necessary support industries.[99]

This document's implicit critique of America's dominance did not stop the company from wrapping the US-owned CDC-Canada in the Canadian flag. With the PL-50 project underway, CDC-Canada believed that it was best suited to champion the emergence of strong domestic computer industry. However, the federal government would have to lend a hand with its power of procurement. Although CDC-Canada was going to produce about half of the PL-50s sold globally, it was not responsible for the international marketing of the PL-50. The subsidiary's marketing responsibility was confined to Canada.

As the only facility for designing and manufacturing mainframes in Canada, CDC-Canada naturally felt that the government's procurement practices should better reflect this presence. IBM did not design or build any computers in Canada, and yet it completely dominated the federal government's computer purchases. For many government departments faced with a costly computer purchase, the IBM mystique offered security. To break this hold, CDC-Canada asked DITC to push for the centralization of the federal government's computer procurement decisions. Furthermore, CDC-Canada's submission to Pepin recommended that the federal government adopt a standardized, open-systems approach to procurement. (The parent firm had long pushed for common standards at the international level as way of breaking IBM's dominance.) Although blatantly self-serving, the submission to Pepin underscores the company's clear understanding that international cooperation required sensitivity to the national interests of the country in question, at least as articulated by government policy.[100]

The CDC submission grabbed Jean-Luc Pepin's attention. He immediately asked Minister of Finance Edgar Benson to comment on it. Department of Finance officials no doubt scrutinized the document carefully before advising the minister. In the interim, John Turner had replaced Benson as minister of finance. Turner recommended that Pepin pursue this matter further with the company's executives in Minneapolis. "Control Data Corporation, in its past international operations," Turner concluded, "has shown that it is willing to cooperate with host governments to further their mutual interests. I support the position that we should demonstrate to the company our willingness to seriously consider a viable proposal."[101] Turner's endorsement, however, was cautious, reflecting the growing economic nationalism in Canada, particularly on the question of foreign ownership. Turner advised Pepin of the need to press CDC to give its Canadian subsidiary more effective autonomy and to accept up to 35 percent Canadian equity ownership in the subsidiary.[102] Whether William Norris would have accepted such conditions as a basis for more extensive cooperation remains an open question; in any case, there is no evidence that Pepin pushed the matter further.

The Collapse of the PL-50 Project

At the close of 1971, all aspects of the PL-50 project seemed to be advancing well. Nienburg, CDC's manager of the project, went so far as to tell the DITC officer assigned to it: "The check out is proceeding so smoothly that it scares me."[103] But by the spring of 1972 Nienburg's earlier optimism had given way to growing apprehension. Nagging doubts about the PL-50's feasibility had begun to undermine the Canadian Development Division. Though there was never a question that the PL-50 could be built, the hardware designers began to fear that it could never be built at a competitive price. Two fundamental questions had always surrounded CDC's decision to develop the PL-50 computer: Could the costly STAR architecture, whose expected superior computational power was inextricably linked to a very specialized and elaborate architectural configuration, be scaled down to the price point of the IBM 360/65 and yet exceed the performance of the IBM 360/65 for business data processing? Could this be done at a price consistent with CDC's traditional high profit margins? Since the STAR 100 had yet to be built and used, the performance implications of its architecture were not known. Though only very limited theoretical analyses and simulations had been done, the need for a new product line turned everyone into optimists.[104] But only after a year of trying to work out the logic design implications of downsizing the STAR and of doing very extensive simulations, it dawned on the Canadian team that everyone had grossly underestimated the effect of this architecture's "overhead costs" on the PL-50's performance-to-price ratio and its profitability.[105]

The strength of the STAR approach lay in the use of memory-to-memory operations on very long vectors of operands. There was an overhead cost, a loss of time, each time the computer went to memory to prepare the vectors for the next set of computations. Although register-to-register operations were faster than memory-to memory operations, registers could not accommodate very large arrays. If the vectors remained long, the setup-time penalty was negligible when put up against the enormous speed advantages of moving extremely large arrays of calculations through the CPU. As long as the STAR 100 could spend most of its time processing very long vectors, it performance was superior to any other supercomputer. In other words, the STAR worked best when, in relation to the size of vector computations, it did not have to do too many random calls to memory.[106] To handle the higher calls to memory found in conventional scalar-based processing or in computations using short vectors, the STAR 100 was packed with special circuitry to make it match the performance of register-to-register machines.

The Canadian team's mission, however, was to design a stripped-down version of the STAR architecture. If the PL-50 was going be competitive at the IBM 360/50 price point (which was equivalent to the $30,000–$60,000 monthly lease category), it could not avail itself of all the STAR 100's very sophisticated but extremely costly hardware features. Very quickly, the team became painfully aware that attempting to implement the STAR architecture on a smaller machine was no way to match the scalar performance and the cost targets that had been originally set for the PL-50. The team used a CDC 6400 to do preliminary emulations and performance analyses to compare the PL-50's performance to that of its competitors.[107] The PL-50 always came up lacking

As it tried to increase PL-50's performance for scalar processing, the Canadian hardware team was inexorably moving to more complex and costly circuit designs. For example, the team introduced much more extensive pipelining features and pre-fetching. Performance then went up, but so did cost. The Canadian Development Division searched frantically for a technical cure to the overhead problem that plagued the PL-50 project, but every fix seemed to bring another problem. With CDC now desperate for a new product, the Canadian Development Division could not admit defeat. Yet there seemed no way to reconcile the costly hardware demands of the STAR architecture with the expectations of performance and profit that had been set for the PL-50. In early 1972, unable to design a PL-50 with high profit margins, CDC's Canadian team convinced DITC to change the goal of the program from the PL-50 to a higher-performance, higher-price PL-65.[108] Even this change could not resolve the fundamental problems.

The strategy for the new STAR-based product line, of which the PL-50 would be the first of several models, still lacked a suitable approach to migration. The PL-50 project emerged from the need to replace the 3000 series. Developing migration aids to move 3000L customers to the PL-50 had been part of the mandate for the Canadian Development Division. Initially, the PL-50 strategy, like the MPL concept that preceded it, had little if any connection to the 6000 and 7000 lines. This separation had been drawn in 1965, when the senior management of CDC's Computer Division in Minneapolis had advocated a hybrid response to IBM's 360 strategy. The wedge between a need for a followup to the 3000 line and the 6000 and 7000 lines was reinforced by the emergence of two supercomputer design philosophies: Seymour Cray's and James Thornton's. Uncertain as to which was more promising, CDC ran with both despite the enormous burden this ambivalence put on the company's R&D resources. While Cray was planning to work on the 8600 (his next-generation supercomputer after the 7600), the PL-50 was the first step in a migration path to a radically different supercomputer: Thornton's STAR 100. Since the technological trajectories of the 6000/7000/8000 and the STAR 100 were completely independent, migrating users to the STAR was never on the table. There was no reason to worry about moving people from Cray's machines to the PL-50. However, in 1972, with work on the 8600 well underway, Seymour Cray walked away from CDC to form his own supercomputer company.

Without Cray, the likelihood of a successful followup to the 6000 and the 7000 diminished considerably. CDC could not afford to abandon users of those machines. They were the large institutional clients that had been, and continued to be, the principle source of CDC's computer business. With the popularity of the Cyber 70 series, the number of 6000- and 7000-based customers continued to grow. Suddenly, it became imperative to give serious thought to the necessity of migrating these users over the new product line being pioneered in Canada. Transforming the PL-50 design into a machine that could automatically accept 6000 and 7000 software seemed impossible. On the other hand, providing conversion aids and software support to help customers migrate their applications software raised the specter of soaring costs. Neither did CDC have the market dominance to force clients to accept this incompatible line, as IBM had done with the 360. Furthermore, CDC could not easily use emulation, as IBM had. Having a 6000 emulation running on a PL-50 would have slowed performance considerably and negated the power of the PL-50. Although IBM's business customers could accept this degradation in performance, CDC's customers would not easily do so. CDC's market niche demanded high performance.

The unexpected success of the Cyber 70, however, provided the Canadian team with an escape from the overwhelming technical problems of the PL-50. In late 1970, just as work on the PL-50 was starting, CDC also decided to try to squeeze more life out of the 6000 line. This was clearly intended as an interim measure to generate revenue until the PL-50 and its successors were on the market. But the older technology proved more resilient than anyone had expected. The 6000 computers used discrete transistor components and core memory. Since the first appearance of the 6600, the price of these components had dropped considerably and the performance had increased. Although integrated circuits and microelectronics were clearly the emerging component technology for mainframes in 1970, significant improvements in the manufacture of discrete transistors and core memory modules allowed CDC to squeeze new life out of its older technology.

Using the cheaper and better transistors, CDC brought back the 6000 line and marketed it as the Cyber 70 line.[109] Software was the biggest cost in developing a new computer. The Cyber 70 allowed the company to recycle software previously developed for the 6000 line. Offered at a competitive price, the Cyber 70 computer surpassed CDC's sales expectations.[110] So high was the initial demand that CDC had to increase its production schedules twice in 1972.[111] Canada was also called in to manufacture the Cyber 70. With the Cyber 70 series selling well, the sense of urgency that drove the company's pursuit of the PL-50 had been temporarily removed. In view of the Canadian Development Division's inability to solve the overhead problem, the Cyber 70 bought CDC desperately needed time to reevaluate the technical and market feasibility of a new line of computers based on the STAR architecture.

Perhaps, if the PL-50 could have been designed to meet the original performance and cost targets, CDC might have been willing to accept the technical and financial risks that migration raised. But in the fall of 1972, faced with overhead and migration problems and with the unexpected success of the Cyber 70, CDC had little choice but to abandoned the PL-50 project.

Though toward the end one might well have seen it coming, the cancellation of the PL-50 was a terrible blow to the morale of the Canadian team. For Derrel Slais, whose hardware team was directly burdened by the overhead issue, there was also a sense of relief. Slais, like his colleagues, had devoted 2 years to the project. Admitting defeat was painful. At the same time, with the cancellation, Slais knew that Canadian team would no longer be shackled to the impossible objective of scaling down the STAR architecture to a high-end line of multipurpose computers.

From the Ashes of Failure Rises a Success

The Cyber 70 had bought CDC some time. But if a new line of improved computers did not replace the 3000, the 6000, and soon even the 7000 models, CDC's computer business would be in desperate straits. Furthermore, CDC had committed itself to designing and manufacturing a new computer in Canada. With Derrel Slais now the general manager, the Canadian Development Division scrambled to reinvent its mission. Being located relatively far from the parent firm and the constraints of entrenched practices, that division looked at the problem of product definition with new eyes. Any new product line, however, could not stray too far from the software investments of CDC's existing customer base. The Canadian Development Division was all too painfully aware of this object lesson.

Starting with a clean slate, the Canadian Development Division had the freedom to experiment with new the generation of digital circuitry of monolithic integration. As late as the early 1970s, the use of discrete components continued to dominate CDC's circuit-design and circuit-production practices. Even the Cyber 70s, still being sold in 1975, used discrete components. The higher performance and lower prices of the Cyber 70 series were due to the availability of cheaper and better discrete components. The introduction of the Cyber 70 did not entail any new circuit-design work, nor did it alter the way CDC manufactured its computers. The Canadian team had already started up the microelectronics learning curve.

Around the time the Canadian Development Division was grappling with the PL-50's overhead problem, Motorola announced a new approach to the manufacture of large-scale integrated emitter-coupled logic (ECL). Hoping to transform the original PL-50 concept into a more competitive design, the Canadian design team experimented extensively with Motorola's new ECL-10k integrated circuits. Although the PL-50 had been dropped, the Canadian team had not abandoned its belief in the price and performance advantages to be gained by switching all design and production practices to large-scale integrated circuits.

After assessing the strengths and weaknesses of all the architectures used in CDC computers, the Canadian team proposed a plan to put out a modified version of the 6000 architecture with logic circuits totally redesigned around Motorola's ECL 10K chips. Rather than cores, the computer's memory would use metal oxide silicon (MOS) technology. Larry Jodsaas, who at the time was a senior executive and who later became president of CDC's Computer Systems and Services Group, recalls the Canadian Cyber 173 initiative as follows: "ECL 10K was not an approved logic that we would use within Control Data at the time. TCL was what we were supposed to be using for all computer designs. [Derrel Slais's] group started screwing around with the ECL 10K stuff up in Toronto and put together enough data that said we should do this thing. The seeds came out of what I call 'skunk works,' which are often times the best kinds of seeds one is ever going to get."[112]

The Canadian group argued that its proposal offered the quickest route to a new and profitable line of computers. Furthermore, since the proposed line of computers did not force users of the 6000, the 7000, and the Cyber 70 to sacrifice their investment in software applications. The Canadian Development Division's idea also bought the company considerable time to come up with another product that would offer an acceptable solution to the migration issue. The Canadian group had done a great deal of preliminary analysis to justify its proposal. Approval was swift. The company desperately needed a new product, and the "only game in town" was taking place in the Canadian Development Division. "I do not think," Derrel Slais later recalled, "that a PIA [Project Initiation & Approval] had ever been walked through Control Data as fast!"[113] In 1973, the Canadian Development Division began work on what would be the Cyber 173. Slais had returned to the United States, where he resumed his responsibilities as general manager of CDC's advanced computer development program.

DITC looked for ways to salvage Canada's first effort to implant a mainframe R&D and manufacturing capacity in Canada. When shown the proposal to develop a totally new version of the 6000 architecture using state-of-the-art integrated circuit technology, DITC agreed to reassign money that had been earmarked for the PL-50 to the new project.

The success of the Cyber 173, which was on the market in late 1974, was not due to any radical advances in computer design, as had been the case with the 6600.[114] It was due to careful design that considerably lowered production costs without sacrificing performance gains. The first saving came from minimizing the costs of software development. Better than any group in CDC, the Canadian development team understood the lesson of the PL-50: do not develop a radically new product line unless one has solved the migration problem. By starting with the 6000 architecture, the Canadian group avoided the costly process of developing new software and sidestepped the migration problem. In this regard the Cyber 173 development repeated the approach taken by the Cyber 70, but that is where the similarity ends.

Whereas the Cyber 70 involved nothing more than replacing the 6000's old discrete components with new ones, the Cyber 173's circuits had to be redesigned from the ground up. About 40 engineers worked on logic design, circuit design, mechanical design, and diagnostic and test design. The early stages of the Cyber 173's conception were truly a shared effort between the Canadians and the Americans in the Canadian Development Division. Americans took the lead in logic design, Canadians in circuit and mechanical design and in the logic design of the new peripheral processor subsystem designs that were to be used throughout the entire 170 line.[115] By early 1973, with the exceptions of General Manager Ron Manning and Engineering Manager Duane LaFortune, most of the members of the US contingent on the Cyber 173 had returned to Minneapolis, leaving the program largely in Canadian hands.[116]

The development of the Cyber 173 introduced a much tighter relationship between design and production engineering. Away from the inertia of the parent firm's engineering practices, the group in Canada had the freedom to rethink the logic and the economics of the entrenched design and production processes that dominated the company. The Canadian group decided to bring production engineers into the design process. Never before in CDC's history had these two groups worked so closely together on the design of a computer. This close collaboration ensured that, before any design was frozen, its spatial configuration and packaging implications would optimize the manufacturing process on cost, speed, and quality control. In using ECL 10K, the Canadian group pushed CDC out of its discrete-component mindset. Designers back in Minneapolis would have eventually embraced large-scale-integration (LSI) technology, but Cyber 173 accelerated the process considerably.

Ultimate performance was the hallmark of CDC's supercomputers. On the question of speed, the bulkier discrete components still offered superior performance. But LSI technology was evolving quickly. With the ECL 10K, the Canadian group was convinced that the speed advantage of discrete had vanished. The Canadian Development Division's use of ECL 10K marked the first time CDC used an industry-standard circuit. Before this switch, the company depended on its internal needs alone to justify the economics of developing and manufacturing its own circuit elements.[117]

From the outset, the use of standardized LSI components appealed to the Canadian team because of the automation it would engender.[118]

Before the Cyber 173, a significant portion of CDC computers were manufactured by hand. As a result, Minneapolis sent circuit boards to Mexico to be assembled; according to Bob Olson (who had been general manager of CDC's computer manufacturing operations before going to Canada to work on the PL-50 project), the cheap labor more than compensated for the inconvenience of having to put up with the bureaucratic quagmire and graft of Mexican customs.[119] The circuit boards were not designed to make the best use of standardized automation equipment. "[CDC] computers used to have discrete components placed in every direction, some of them not even square to the world," Olson recalled. "That's okay," Olson explained, "if you stick those in by hand, but get a machine to do it. That's an altogether different thing."[120] More than 25 years later, reflecting on the Cyber 173 and what the Canadian team had achieved, Olson observed:

We were leaders in automation. People back at the parent firm were amazed that we could come out with a very cost-effective package and that with automation we could cut throughput time and didn't have to send anything to Mexico to be hand assembled.... [Cyber 173] turned out to be the most reliable, most cost-effective, and profitable systems ever built in Control Data. Putting aside its higher performance for a minute and just looking at cost, the machine was well under half of what the [Cyber 70] cost to build and it could be produced in half the time. We had more standardization of modules so that the spare parts in the field were far less. We needed 25% less inventory to support this machine. And the machine was a lot more reliable.[121]

Olson, Slais, and Schumacher were instrumental in getting the parent firm to approve the Cyber 173 project. They fought hard to get senior people in Minneapolis to accept the new circuit approach and to get more than $10 million to cover the capital expenditures for the new automated production facility.[122] Although they had been in Canada for more than 2 years, all three of these men still occupied senior technical positions within the parent firm. Highly respected and with a close working relationship to the key decision makers in Minneapolis, they were able to fight effectively for the parent firm's support of the Cyber 173 proposal by means that were not available to Canadian management. For example, on one occasion, when Minneapolis management was reluctant to approve a particular circuit configuration proposed by the Canadian team, Olson told his boss (Dick Gottier, head of operations back in Minneapolis): "Well, I'm going to take a vacation. When you guys have decided what the hell you'll do, I'll come back to work." Believing that was a good idea, Olson's boss asked: "What are you going to do on your vacation?" Olson replied, "I'm going to build me a big tractor so if the decision comes out wrong, I'm going to leave and go into farming." There was dead silence in the room. A few days later, Gottier called Olson at home to tell him that upper management had decided that the Canadian group could go ahead with its circuit concept. No senior Canadian engineer within CDC-Canada could have been so bold and succeeded.

Norris proudly told CDC shareholders that the Cyber 170 line had been the "first well defined computer systems that [the company] could sell in standard configurations."[123] In addition to the circuit design and manufacturing innovations, the rationalization of operating systems was a factor in the profitability of the Cyber 170 line. As the first of Cyber 170 series, the Cyber 173 also marked an important first step in the technical and business rationalization of CDC's operating system strategy. Before work on the Cyber 173 began, CDC had committed itself to several operating systems. KRONOS and SCOPE 3.4, both of which had evolved out of improvements to the earlier Chippewa Operating System (COS), were CDC's main operating systems.[124] At the same time, CDC had embarked on a variety of other more customized operating systems.[125] CDC's dependence on a multiplicity of operating systems to further the sale of its computers had been extremely costly and had seriously undermined the profitability of the company's Computer Systems business. In an effort to control the costs of software development, Robert Price, president of CDC's Computer Group, decided that CDC had to refocus its efforts to develop operating systems around one standard. An improved version of KRONOS called NOS (for Network Operating System) became the operating system for then entire Cyber 170 series.[126] With the introduction of the 170 line, CDC recognized, Norris explained to his shareholders, "the growing importance of software by totally separating the price of the operating and applications software from the computer mainframe." He continued: "The costs to develop and maintain software are continuing to rise while hardware costs are not. Our separate pricing of software for the Cyber 170 recognizes that software is the most important and costly element in making further significant progress in the application of computer systems."[127]

In 1978, speaking at the annual meeting of shareholders, Price, proclaimed: "I have little doubt that the Cyber 170 will become the most successful computer line in our history."[128] Ray Allard, a senior engineer and manager at CDC, underscored that the Cyber 170 computers "ended a series of aborted attempts to develop whole new hardware and software product lines to replace [the] 6000 and then [the] Cyber 70." He continued: "Each of these plans either failed to meet requirements or projected costs too high to bear. They all shared inadequate solutions to the problem of converting software, both CDC and user generated, to a new incompatible product line."[129]

Saying that the Cyber 170 series saved CDC's computer systems business from near extinction does not reveal the whole picture. Another thing that helped bring the computer systems business back into profitability was the realignment of focus that John Titsworth introduced as the new head of the Computer Systems Group. CDC's meteoric rise during the late 1960s had given Norris enough confidence to attempt to encroach on IBM's business market, albeit at the high end. In the area of computers, this strategy did not produce any commercial success.[130] Titsworth's strategy brought CDC back to scientific and engineering applications. "In my opinion," Titsworth later recalled, "the reason [CDC's computer systems business] lost money was because its sales department had free reign to sell computers to anybody."[131] Titsworth elaborated:

> For example, take the Union Bank of Switzerland. Now imagine, I mean there's not a more classical business environment than a bank. At the same time I don't think bankers knew a hell of a lot about computers. Well, they knew about as much about computers as we knew about the banking business. We got into a humongous contract, tens of millions of dollars, to provide all the business computation and running the business using our large Cyber computers. When I took over, I told our marketing and sales people to stick to what they knew how to do. Just because our computer was faster than an IBM machine didn't mean that we knew anything about the commercial business. We didn't really understand the banking business. We didn't know how to develop the software, the application software, to go with the big computer. The customers certainly couldn't do it.
>
> In the past we worked with scientific customers, everybody from the weather bureau to the atomic energy commission. They could take our operating system and do their own applications and solve the problems, specific problems. That wasn't the case with the banking industry or several other IBM-type business applications that we got ourselves into. We probably didn't even estimate the cost of the applications software development. The result was it cost us many, many, many times more in millions of dollars than the computer contract itself. So my philosophy was get out of everything you don't understand. Go to your classic business. Go back to the scientific community which at the time was still a booming.[132]

With the help of the 170 line, Titsworth's strategy worked well. CDC had abandoned the strategy of aggressively expanding into the high end of the business EDP market. In going to the 170 line, which preserved the common architecture of the 6000, 7000, and Cyber 70 lines, CDC was prepared to sacrifice its 3000 client base. The Cyber 170 line also represented a retreat from the goal of developing a broader, integrated line of computers that would span the needs of users ranging from 3000 users to supercomputer users. But the retreat was only temporary. In the midst of the Cyber 170's success, CDC began work on a radically new line that, when completed, would cover the range from minicomputers to very large computers.

The PL-50 was a bold venture that failed. Thanks to an opportunity, a willingness, and a capability to explore the advantages of new microcircuit technology, and thanks to the continued assistance of DITC, the Canadian Development Division's hardware group survived the collapse of the PL-50 unscathed. But the software group emerged quite weakened and very vulnerable. The Canadian Development Division had assembled a large group of software engineers to tackle the PL-50. Very labor intensive, software development commanded a far larger portion of the PL-50 budget than did hardware.[133] Canada had the mandate to start from a clean slate and develop an entirely new operating system, a complete set of compilers for all the important programming languages, all diagnostic utilities, and migration aids to help move 3000 customers over to the new STAR-based product line. In retrospect, it was such an awesome mandate that one must wonder how Schumacher ever thought his team in Canada could have carried it off. The demise of the PL-50 orphaned the software group, and the Cyber 173 program did not adopt the group. Since 6000 and Cyber 70 software was going to recycled, Canada's Cyber 173 program was essentially a mandate for hardware and diagnostic software. Any fine tuning of the older software would naturally be carried out where it was first written: either in the headquarters in Minneapolis or in the company's other software group in Sunnyvale, California.

After the news of the PL-50's cancellation, the American software team returned to Minneapolis. Of the remaining number of software people who were hired in Canada, a few found positions in Minneapolis and Sunnyvale and many found jobs with other companies in Canada. But a sizable group stayed and scrambled to reinvent its mission within CDC. Obviously, the group had to find ways to contribute to the Cyber 170 line. Unable to attach itself directly to the Cyber 173 project, the software group wanted to position itself as a center of excellence over the entire line, but in well-defined specialties. What expertise could the group carry over from the PL-50 program that did not infringe on the entrenched mandates of CDC's two centers of software development (Sunnyvale and Minneapolis)? The Sunnyvale group, which was larger and more established than the Canadian group, had the responsibility for the more established programming languages, including COBOL and Fortran.

The engineer Amnon Zohar often had the task of selling the Canadian software group's talents to the parent firm.[134] Zohar had been a part of CDC's sales and marketing operations in Israel. He and with four other engineers (one from Denmark, one from France, one from Australia, and one from the UK) had come to Canada to participate in the PL-50 project with the goal of going back to their respective countries as the marketing specialists for this new product line. The opportunity to get in on the ground floor of such an ambitious and crucial project, and other associated career opportunities, had brought Zohar to Canada. The plan was to stay for 2 years, but he never went back to Israel. Today the CEO of a Canadian software company, Zohar looks back on those early post PL-50 years as an exciting time "because it was so entrepreneurial and innovative."[135] And yet the Canadian software group's horizons were seriously limited. Desperate to stay alive, the software group scouted the parent firm's needs in search of some small piece of new territory not yet claimed by Sunnyvale or Minneapolis.

The Canadian software team won the responsibility to develop the compilers for APL and BASIC.[136] Although they had far less market penetration than Fortran and COBOL, APL and Basic were becoming important languages for interacting with commercially available time-sharing services. Again drawing on the competence it had acquired during the PL-50 project, the Canadian software group became an important source of performance analysis for the rest of the corporation. This expertise entailed developing software tools that allowed the user to do comparative performance evaluations of compilers and operating systems, and that also allowed the user to optimize the use of a computer's resources. This performance expertise was not created from scratch in Canada. Working for the parent firm, Schumacher and his software team transferred some know-how to the Canadian operation. But the Canadians developed this performance expertise further. The Canadian software group had turned into an "outstanding small team" for doing performance analysis.[137] The Canadian software group also worked on some migration aids. In the PL-50 program, the challenge of migrating over the 3000 customers had a high priority and was a responsibility of the Canadian group. Although the needs of the 3000 users had low priority in the 170 program, they were not forgotten. The Canadian software group took on the task of developing some aids to help convert 3000 applications to the 6000 operating system.

The cooperative venture with the Canadian government, which began in 1970 and ended in 1974, proved to be an enormously beneficial experience for CDC. The PL-50 program, although it collapsed, was a valuable learning experience that prepared the Canadian team for the Cyber 173. The Canadian Development Division's approach to the Cyber 173 became a methodological template for the more advanced models of the Cyber 170 series that were subsequently designed and manufactured in Minneapolis.[138]

The failure of the PL-50 also taught CDC's computer development group a crucial lesson: One should not jump into the development of a completely new computer architecture until a sound migration strategy had been clearly articulated and validated. This lesson was crucial to an important 1973–1975 collaborative effort between CDC and National Cash Register.[139] A significant marketing advantage of IBM's integrated line of computers lay in the range of models it contained. Users could start small and could be confident that they could upgrade as their needs grew without fear of any software incompatibility. Since CDC was content to focus its design and marketing expertise on developing a line that would compete with the upper end of IBM's line, it was unable to capture those users who started small and wanted to upgrade to large systems. NCR, however, was interested in competing at the lower end of IBM's line. But NCR's marketing was at a disadvantage because it could not offer its potential customers the longer-term possibility of upgrading to larger systems. The CDC-NCR research effort set out to define some common architectural features around which each company could develop a computer line to meet the needs of its respective market and still ensure upward software compatibility between the two.[140] Together, then, NCR and CDC could out-specialize IBM in their respective ends of the market, and each still could offer a complete line of computers.

By 1976, a considerable amount of preliminary analysis had been done and it was time to move to the next phase: designing and building a prototype of the mainframe that would serve as the bridge between CDC's and NCR's future lines of computers. A high-level meeting was held to discuss this next step. Robert Price (president of CDC's computer systems and services group), Tom Kamp (president of CDC's peripheral products group), and John Lacey (a CDC corporate senior vice-president) met with three senior executives from NCR.[141] CDC came to the meeting armed with a detailed technical analysis of the migration issue. To ensure that Cyber 70 and Cyber 170 customers could migrate over to the new computer, CDC presented a list of essential hardware and software features that would have to be incorporated in the new design. From NCR's perspective, these features raised the R&D stakes significantly. CDC had given a lot of attention to the migration issue, but NCR had not. NCR wanted time to consider the matter further. In the months after the meeting, NCR examined the migration question from the perspective of its corporate strengths, its long-term product strategy, and its existing customer base. In the end, the company concluded that it should not be in the mainframe business—at least, not in the kind of ambitious and very costly project proposed by CDC. An intersection of strategic interests had brought CDC and NCR together in this collaboration, and a divergence of interests had ended it.[142]

CDC continued on alone. By the late 1970s, the company had embarked on a program to develop an entirely new line of computers that would become the Cyber 180 series. The new mainframes were to mark the first significant break from the 6000 architectural tradition. In 1984, 10 years after the appearance of the Cyber 173, CDC finally solved the migration issue with the introduction of the "Dual-State" Cyber 180 line.[143] The Cyber 180 ran three operating systems: the Cyber 170, NOS/BE, and the new NOS/VE (Network Operated System/Virtual Environment). But at any moment it could simultaneously run old Cyber 170 application software under NOS and newer application software under NOS/VE.[144] Because most of CDC's customers wanted high performance, the company could not resort to simple software emulation to facilitate migration of its 170 users to the 180. Emulation always produces degraded performance. The "dual-state" approach allowed NOS-based applications to run as if the machine were a 170.[145] When it finally emerged, the Cyber 180 line was the culmination of the 3000 group's 16-year quest to develop an integrated line of versatile high-performance computers. The important landmarks in this journey leading to the Cyber 180 were the MPL concept, the Canadian PL-50 project, the Canadian Cyber 173, and the NCR collaboration. Derrel Slais was unequivocal in his assessment that the accumulated experience of CDC's Canadian efforts was an integral and essential part of the path-dependent process that led to the Cyber 180.[146]

From Mainframe to Supermini

As the Cyber 170 development cycle wound down in the late 1970s, unsettling questions about the long-term viability of the Canadian Development Division loomed on the horizon. From a high of nearly 400 people engaged in R&D during the PL-50 days, the Canadian Development

Division had stabilized to 125 people in 1977. Despite the considerable drop, this number still represented a very large proportion of the Canadian subsidiary's work force of 430 employees, and the group was still Canada's only industrial center for the design of large computers. After the Cyber 173, CDC-Canada went on to put out the Cyber 172. Although still a large high-performance machine by general market standards, this computer occupied the low end of the 170 line. The 172 was a technical exercise in slowing down the performance of the 173 to a lower price point. Not much design was needed, although considerable testing and evaluation had to be done to make sure no new internal problems had been introduced by removing features.[147] In addition to the first releases of the 173 and the 172, the Canadian group maintained a program of ongoing incremental innovations.[148] As microelectronic technology advanced, the Canadian group readapted its designs and manufacturing process to new kinds of components. But by the end of the 1970s it was clear to all that this kind of design maintenance work could not sustain the Canadian group's size or its ambitions. The Canadian Development Division was naturally fearful that its chances to secure a new, meaningful mandate were rapidly diminishing as the parent firm embarked on 180 line in Minneapolis without any appreciable role assigned to Canada. Barry Robinson, who had become the first Canadian general manager of CDC's Canadian Development Division, pushed to get Canada involved in the Cyber 180 program. But what territory could Canada stake out that the development group in Minneapolis had not already claimed? Taking its cue from the legacy of the Cyber 170 program, Canada wanted the mandate to develop an entry-level model for the new Cyber 180 series.

The Canadian Development Division took CDC's product strategy in the direction of the minicomputer. The plan called for Canada to design and manufacture a minicomputer (code named S0) which would be completely software compatible with the Cyber 180's new NOS/VE operating system. From a marketing perspective, anchoring the Cyber 180 line at the low end to a minicomputer would introduce a much broader range of customers to CDC products. The idea marked a return to the marketing strategy that had prompted NCR and CDC to attempt joint development of a line ranging from small computers to very large systems. The proposal also addressed the need to respond to the growing inroads that the Digital Equipment Corporation's minicomputer strategy was making into the traditional mainframe markets. CDC's earlier efforts to collaborate with the German firm Nixdorf to develop a minicomputer had not gone very far. Since the goal was to bring in new customers, migration was not an issue in the design of the S0 as it had been for the 180 series. As a result, the S0 design would not have to be burdened with the costly "dual-state" features.

Also troubled about what to do with the Canadian Development Division in the post–Cyber 170 era, CDC was receptive to the Canadian initiative. By 1980, senior executives in CDC's Computer Group in Minneapolis were ambivalent about the future of the Canadian R&D group. On the one hand, CDC had invested a lot of effort and money in building up a first-class center of technical competence in Canada. Should such a valuable intellectual asset be discarded? On the other hand, could CDC still afford to sustain a large development center in Canada? CDC's supercomputer program and its Cyber 180 program put enormous demands on the company's R&D budgets. Run as a cost center, the Canadian Development Division drew its funding directly from the parent firm's computer systems R&D budget. Terry Kirsch, who had come to Canada to manage the Cyber 170 program, recalls all too well the ambivalence Vice-President of Computer Development Larry Jodsaas had to live with: ". . . from a business perspective, Jodsaas would have preferred not to have the Canadian operation. It was a headache for him to manage. He would admit this. It took resources away from his ability as VP to do other things back in Minneapolis. On the other hand, I will tell you, although he may have wanted to close the operation down, it was the only group from whom he ever got any new innovative products. He would admit that too. And yet the Canadian operation was a constant thorn for him. He had to go through all the exercise of justifying his R&D operation there."[149]

Clearly wanting to underscore importance of the Canadian effort to the survival of CDC's computer systems business during the 10 years that followed the appearance of the Cyber 173, Terry Kirsch surely overstated the case when he claimed that the Canadian group was "the only group from whom [Jodsaas] ever got any new innovative products." Nonetheless, the S0 proposal underscores that the Canadian Development Division had become a technically strong and innovative asset for CDC.

Although technically appealing, the ambitious S0 proposal faced a high financial hurdle: It required considerable funding. Having committed a great deal of money to the Cyber 180 series and to a new generation of STAR-based supercomputers, CDC could not afford to underwrite another development effort, particularly if it meant reallocating scarce resources to a subsidiary. Distance had removed the subsidiary from the parent firm's technical traditions and had given the Canadian team the flexibility to pursue the Cyber 173 concept. At the same time, distance also made it more difficult for the subsidiary, which operated as cost center, to pry more R&D funding from the parent firm.

With the help of Canada's DITC, a limited partnership involving CDC, the Canadian government, and private investment was put together to finance the S0. In all, approximately $28 million was raised for the S0 project.[150] DITC, through the Enterprise Development Board, provided an R&D grant of $11.5 million.[151] Under the conditions of the grant, all future products sold worldwide that had resulted from this R&D would have to be manufactured in Canada. Though CDC's contribution to the program cannot be estimated from the extant evidence, we must assume that it was at least equal to the Canadian government's contribution.[152] Although we do not know the number or the identities of the private investors, we can reasonable assume that the value of their contributions was between $3 million and $5 million.

Larry Jodsaas had left CDC's Computer Systems Division to head up its Peripheral Division just before the initial discussions between the parent and the subsidiary over the S0 idea took place. When he returned as president of CDC's Computer Systems and Services Group, Jodsaas found CDC committed to another new and ambitious computer development program. Detailed plans and financing for the S0 were all in place, but Jodsaas had doubts about the S0's specifications. With the geometric increase in microelectronic densities, the price/performance ratio of computation was dropping rapidly. Jodsaas was convinced that the originally planned price/performance ratio of the S0 no longer offered a competitive product in the minicomputer market.[153] CDC needed to push the S0's performance up by several factors without increasing the price. All through 1983, Jodsaas pushed a plan to restructure the S0 program. This plan meant that the appearance of the S0 would be delayed by at least a year. DITC accepted CDC's rationale for the changes. The Canadian investors, however, did not. They were incensed over CDC's decision to change the plan in midstream. In their eyes, delaying the S0's appearance meant losing the competitive advantages of early market entry. Perhaps they were right; however, CDC was convinced that the original specifications would not have been competitive, regardless of first-entry advantages.

With the S0 program on a new tack, the investors launched a lawsuit. The extant and accessible documents do not shed any light on the identity of these investors, and none of CDC's senior executives of the time can recall the details.[154] Though it would take several years to settle the lawsuit, the Canadian team set about designing and manufacturing the S0 according to the new price and performance requirements.

Under the revised plan, the S0 was scheduled to appear on the market sometime in late 1984 or early 1985 under the product name Cyber 180 Model 810. An upgraded version, the 830, was to appear a year later. Progress on the minicomputer had fallen behind schedule. That in itself is not unusual for such a big project, but in addition CDC had lost confidence in the Canadian senior manager's ability to stay on top of the entire S0 project. CDC decided to send an American to manage the S0 program. Since Terry Kirsch had been in Canada during the Cyber 170 program, he was sent again.

Despite these temporary setbacks, Larry Jodsaas retained considerable confidence in the Canadian subsidiary's design and production talents. Speaking before representatives of investment houses, he was quite optimistic about the minicomputer:

At the entry level [to the Cyber 180 series] we have the Model 810 which can be utilized for emerging applications in smaller departmental needs. We expect it to be used as part of a distributed system. It is operable in an office environment. ...We call it the CYBER Supermini 810, and it is being brought in to compete in the VAX environment. It is our clear intention to be able to compete with VAX at the department level and ensure that Control Data continues to meet the needs in distributed environments as well as the local user environments....The system, is available at a purchase price of $250,000 and is easily expandable up to a full-scale Model 830 system with two CPUs, two million words of memory and a 20 peripheral Processor Unit configuration.[155]

The Cyber 180 Models 810 and 830 were not to be middle-of-the-road minicomputers. As it had done with its mainframes, CDC targeted the high end of the product spectrum. Historically, CDC's competitive advantage lay in that niche, where margins on sales tended to be very high. The 810 and the 830 were, as Jodsaas pointed out, "super" minicomputers. In its initial marketing forecasts, CDC had estimated that over 10 years about 10,000 of them would be sold worldwide.[156] The terms of DITC's R&D subsidy had ensured that CDC-Canada would have the exclusive world product mandate for the 810 and the 830,[157] and a wonderful mandate it seemed. But the competitive structure of the computer industry had changed so rapidly during the 1980s that the mandate soon came to rest on a crumbling foundation.

The microelectronic revolution, as embodied in Moore's Law, wreaked havoc on the marketing assumptions underlying the Cyber 180/810 and

Cyber 180/830. As circuit densities and performance increased and prices dropped, the demarcation between minicomputers and superminicomputers and that between the low and middle ranges of mainframes began to blur. If in the early 1980s it thought that it could have exclusive ownership of the superminicomputer, by the 1980s CDC had found itself overwhelmed by a swarm of competitors. In late 1986 and early 1987, Digital introduced seven superminicomputers. In 1987, IBM launched its superminicomputer line, the 9370. To make matters worse, the established club of superminicomputer makers was challenged from below. Low-end producers began to push up performances with each advance in microelectronics. In 1985, superminicomputers were capable of 4–5 MIPS.[158] Just 2 years later, a number of companies introduced systems capable of 3 MIPS at a fraction of the cost of the earlier superminicomputers.[159]

The rapid technological change quickly robbed CDC-Canada's mandate of much of its currency. Hardware was becoming an increasingly cheap commodity. The number of competitors was exploding, and margins were getting ever smaller. The 810 and 830 superminicomputers did have the marketing advantage of being totally software compatible with CDC's high-performance mainframe line, the Cyber 180. But the competitive advantage of CDC's mainframe for scientific computation was under attack from the appearance of mini-supercomputers.[160] To sustain any meaningful mandate in the rapidly evolving computer hardware industry, CDC-Canada would have to innovate its way to high-performance workstations and compete against the likes of Sun Microsystems and Silicon Graphics. However, by 1986 the question of CDC-Canada's ability to compete in the new technological order and to sustain an important R&D and manufacturing mandate was irrelevant.

As CDC-Canada was shipping its first superminicomputers, the parent firm was hemorrhaging financially. In 1985, CDC lost $567 million on $3.7 billion of revenue.[161] Almost all aspects of its diversified operations were losing money. Under attack from the Japanese, CDC's supercomputer business consistently ran up losses. Even CDC's OEM peripherals business, a "cash cow" for more than 20 years, was bleeding profusely under intense price competition from an influx of domestic startups and Japanese companies. CDC's computer services business, which was to be the jewel of CDC's diversification strategy, was also performing poorly. To make matters worse, in 1985 CDC's bankers lost confidence and declared it in default of $383 million in loans.[162] As CDC fought to stay afloat, investors abandoned ship. The value of the company's stock fell by more than 50 percent. To survive, it sold its financial services and peripherals businesses and withdrew from the supercomputer business. Finally, CDC was forced to drop out of the computer design and manufacturing business. In 1992, it changed its name to Ceridian and spun off Control Data Systems. Ceridian specialized in information services and defense industry work. Control Data Systems became an integrator of large-scale information systems that used hardware made by other companies. CDC-Canada had no future in the computer business. The computer R&D and manufacturing competence that had been built up in the Canadian subsidiary had no place in either of the new companies. In 1992, the doors of one of Canada's most significant computer design and manufacturing facilities were shut.[163]

The engineers of CDC-Canada's Development Division had a 10-year accumulation of computer design and manufacturing know-how. The details of where this expertise went to after the Development Division was closed down are still not known. Anecdotal evidence, however, suggests that much of the technical expertise stayed in Canada and was snapped up by other companies. Some of the engineers went to ATI Technology, some to Mitel.[164] One went to Celestica.[165] Seven hardware engineers went to Edmonton to join the newly formed Myrias Research Corporation. Prompted by the scientific computational needs of the oil industry, Myrias designed, manufactured, and marketed a state-of-the-art, highly parallel supercomputer.

The Canada-CDC Joint Venture: Success or Failure?

By 1969, the inexorable expansion of the scale and scope of the R&D needed to compete with IBM, internal financial constraints, and the urgency of finding a new product line forced the Control Data Corporation to experiment with decentralizing its design and manufacturing competence abroad. At an organizational level, the Canadian experience with the PL-50 and the Cyber 173 encouraged CDC to broaden the international scope of its experiment in decentralization.

Control Data had always pursued collaborative firm-to-firm R&D arrangements.[166] The PL-50 project was its first cooperative venture with a foreign government. It used that venture to help close the gap between its modest internal capacity for R&D and IBM's huge resources. Canada hoped to use the venture to reduce the large gap between its inadequate capacity to design and manufacture computer and the comparable capacities of the other advanced industrialized nations, particularly the

United States. Pleased with the success of the Cyber 170 project, CDC hoped to reapply the Canadian experience to other national contexts. William Norris put it this way to the shareholders: "In different parts of the world local markets require specific strategies suited to their individual needs. Also, there is the problem everywhere in the world of availability of skilled, knowledgeable people. Cooperative ventures are one way of meeting these two problems head on."[167]

Norris focused his strategy of international cooperative ventures on the Socialist Bloc nations, which, like Canada, were desperate to expand the technological and managerial capacity of their domestic computer industries. For CDC, dealing with centrally planned economies held the promise of considerable state intervention that could ensure broad cooperative arrangements and (through the power of public procurement) privileged access to local markets. For reasons of US national security, the latest supercomputer technology was clearly off limits to any cooperative ventures. But there was still considerable latitude for cooperation in other aspects of computer technology. In 1973, while the US Congress was debating the transfer of technology to other nations (particularly communist ones), CDC signed a "ten year frame agreement with the USSR for broad scientific and technological cooperation."[168] That same year, in a joint venture with the Romanian government, CDC set up a company to manufacture peripherals. Norris was determined to push the cooperative notion further in order to gain bigger foothold in the other national markets of the Socialist Bloc. To underscore the merits of these far-flung ventures, Norris pointed to the Canadian experience: "Our cooperative program in Canada for development and manufacturing of computer systems is a beautiful demonstration how we can work in tandem with government to the mutual benefit of both parties. We believe that we can gain similar benefits form cooperative arrangements with Socialist countries."

Did the Canadian government share Norris's view of the benefits derived from the cooperative venture? One of the major goals of the Department of Industry, Trade and Commerce's policy to foster the growth of Canada's computer industry was to convince foreign multinationals, through the enticement of R&D and capital cost sharing, to give meaningful mandates to their Canadian subsidiaries. In the PL-50/Cyber 173 program, the Canadian government had invested about $15 million over 6 years in CDC-Canada. Though it is not much by today's standards, in the early 1970s this investment was considerable. Aside from Microsystems International Limited, the PL-50/Cyber 173 program accounted for the largest investment that DITC had made in a specific high-tech development program. Did this investment produce the results that DITC had hoped for? At the end of the program, Canada had its first and only industrial center of excellence in mainframe design, CDC-Canada was CDC's only foreign subsidiary doing mainframe design, and Canadian-made computers supplied a significant portion of CDC's global market for the 170 line.[169]

The original manufacturing scenario did not go exactly as the Canadian government had hoped. When the PL-50 project collapsed, so also did the grand plans to build a large manufacturing facility in Quebec City. The decision to locate in Quebec had been premised on the use of lower-paid female labor for low-skill assembly work. The innovations in standardization and automation introduced by the Cyber 170 series eliminated the need for a second plant. In giving up the Quebec City plant, CDC lost the $10 million subsidy from the Department of Regional Economic Expansion. Instead, CDC invested its own money in a more capital-intensive plant in Mississauga. The fate of the Quebec City proposal underscored the vulnerability of DREE's strategy. Automation or cheaper offshore labor was always ready to undermine subsidies intended to create low-skilled manufacturing jobs.

The Department of Industry, Trade, and Commerce had hoped that the emergence of a strong mainframe design and manufacturing group in Canada would reinforce DITC's microelectronics initiative. Efforts to foster the growth of a network of companies to supply the high-quality, competitively priced components that would be needed to manufacture the PL-50 produced mixed results. Finding suppliers for the less sophisticated components proved quite easy.[170] Finding suppliers capable of producing high-quality, price-competitive printed circuit boards and various circuit components proved far more difficult. Canadian Marconi was eager to get the contract to produce the PL-50's multi-layer printed circuit boards, but its first samples proved substandard and too costly. Working closely with CDC, Canadian Marconi did finally meet the standards. To reduce production costs, CDC even gave Canadian Marconi proprietary know-how from its US printed circuit board manufacturing facility.[171] Elco Connectors agreed to produce the 64-pin connectors for the PL-50, and Canron agreed to develop and manufacture the PL-50's 125-kVA motor generators. Despite the collapse of the PL-50, these companies were well placed to be suppliers to the Cyber 170 series.

But the real prize for Canadian sourcing was the PL-50's various semiconductor memory and hybrid circuits. Despite CDC's best efforts, no

Canadian company could provide these components.[172] Microsystems International Limited had been DITC's chosen instrument to create a national competence in advanced microelectronic design and manufacture, but after encountering initial difficulties in producing sample circuits MIL withdrew from the PL-50 project,[173] thus dashing DITC's hopes that synergy and cooperative learning would emerge between CDC-Canada and MIL. The MIL initiative incurred huge losses and proved disastrous for DITC.[174] When work on the Cyber 173 began, CDC could not find any Canadian sourcing for the sophisticated LSI circuit components. Bob Olson lobbied hard to convince Motorola to produce some of the components for the Cyber 173 in Canada, but to no avail. Texas Instruments showed some interest but then changed its mind.

In 1980, an internal DITC study expressed a deep sense of frustration with that department's inability to get foreign-owned multinationals operating in the computer and communications sector in such a way as to give their Canadian subsidiaries a meaningful product mandate.[175] This study decried the fact that too few multinationals gave their Canadian subsidiaries any significant manufacturing role, let alone a meaningful R&D role. After 10 years, the policy of "moral suasion" and the inducements of R&D subsidies had had little effect on any company other than IBM. By the 1980s, IBM Canada did manage to achieve a rough balance between the dollar value of its imports and that of its exports. But whereas IBM's plants in Europe produced 90 percent of the computers used there, IBM Canada did not manufacture a single computer. IBM Canada had, nevertheless, achieved global manufacturing mandates for its plants near Toronto, Ontario and Bromont, Quebec. These plants supplied IBM's worldwide operations with a few specific lower-tech components. Although the Toronto plant, for example, did not possess any meaningful capacity for R&D in product design, it grew into a remarkable center of excellence in manufacturing engineering. During the 1990s, the Toronto plant grew into one of the leading companies in the global electronics manufacturing services business.[176]

In the 1970s, with the PL-50/Cyber 173 program, Control Data proved to be an exception in the otherwise grim portrait of multinational behavior in Canada painted by DITC. "Where R&D is carried out under the auspices of the subsidiary in Canada," one DITC study noted, "it tends to consist of relatively unsophisticated items and a total computer systems capability has not developed in Canada with the exception of Control Data Canada Ltd."[177] In 1968, in its first policy memorandum to the cabinet, DITC argued that Canada should not try to match other countries by developing a mainframe industry. Canada could not support a domestically owned champion, but DITC hoped that IBM would give its Canadian subsidiary that role. Instead, contingencies, competitive imperatives, and a particular corporate culture brought CDC up to bat. On the issue of designing and manufacturing computer systems in Canada, CDC was the only multinational firm that was open to cooperation. DITC's support of the PL-50/Cyber 173 program created a top-notch center of technological competence that otherwise would not have been readily available to Canada. CDC had devoted considerable time, effort, know-how, and money to moving its Canadian subsidiary up the learning curve. The Canadian group did not work on bits and pieces; it designed and built entire systems. CDC-Canada now had CDC's most important computer development group and most important manufacturing facilities outside the United States. In 1980, DITC again concluded that Canada "[could not] aspire to make computer manufacturing a major Canadian-owned industry emulating the policies adopted by other governments."[178] Again DITC pointed to the need to cajole and/or force multinationals to do their share. And once again, when CDC proposed the superminicomputer, DITC rushed forward to cover half of the R&D costs.

The superminicomputer project, a very important step in the evolution of CDC-Canada, illustrates DITC's inability to free itself from a one-dimensional view of parent-subsidiary relationships. Like many management scholars of the day, DITC conceived of the multinational corporation only as a highly centralized structure, with decision making a unidirectional process flowing from corporate headquarters to the subsidiaries. It was this "central brain" paradigm of the multinational corporation that animated the Canadian government's policy on world product mandates. Public policy was thus aimed at persuading, cajoling, and arm-twisting multinational corporations to be good corporate citizens and hand over "world product mandates" to Canadian subsidiaries.

There is no doubt that the PL-50 program followed the classic central-brain model of the multinational corporation. The parent firm moved technological and managerial resources to the subsidiary to pursue the design of a product that had been completely defined in Minneapolis. But in the process it was nurturing an international expansion of its R&D capacity. This is how DITC expected world product mandates to work. But this top-down structure began to mutate. By the time the PL-50 program collapsed, a domestic pool of technical know-how had formed in the subsidiary that commanded respect and trust from the parent firm.

The Cyber 173 story reveals a nascent capacity to formulate and execute a product mandate. Although relatively short-lived, the software group showed considerable initiative in creating mandates within rather tight corporate-wide constraints.

Despite the emergence of a very strong technical cadre within CDC-Canada, management competence remained truncated. Functional line reporting effectively isolated the Canadian Development Division from the rest of CDC-Canada. The head of the Canadian Development Division reported directly to his superior in Minneapolis. All matters of R&D planning, technical or budgetary, were settled through discussions between the head of the Canadian Development Division (who was an American) and the vice-president of Computer Systems in Minneapolis. For example, the reporting structure did not permit the Canadian Development Division to work with the Canadian group responsible for computer services to exploit local opportunities. Furthermore, the subsidiary, operated as a cost center, could not accumulate surpluses to fund its own independent initiatives. Although CDC-Canada designed and manufactured a significant number of the Cyber 170s sold around the world, its management did not participate, and hence did not receive any training, in the international marketing of these computers. "We were not given any international marketing experience," a former president of CDC-Canada recalled.[179] In effect, the subsidiary could not propose, defend, or execute a coordinated plan of action.

Nevertheless, the superminicomputer mandate for CDC-Canada did show a clear transition toward stronger local management and toward a capacity to articulate a mandate strategy independently of the parent firm. CDC-Canada started the Cyber 180/810 and 180/830 initiatives. But, equally important, the subsidiary's ambitions pushed the company into a new product line: the superminicomputer. Could CDC-Canada have continued to expand its technical, marketing, and even managerial competencies? Could the subsidiary have successfully expanded the scope of its strategic initiatives? Had the mainframe paradigm continued to dominate and had the parent firm remained financially strong, CDC-Canada may have indeed turned out to be the kind of long-term success story dreamed of by Canadian industrial policy makers. Unfortunately, CDC-Canada was in a sector driven by rapid technological change and by intense Schumpeterian competition and characterized by a high mortality rate. Even the mighty fell. Look at RCA, Sperry Rand, and Digital; indeed, look at Control Data.

Conclusion

In the 10 years immediately after World War II, new technologies and industries emerged as powerful engines of global economic development. A century and a half after the Industrial Revolution, innovations in digital electronics were about to herald the Post-Industrial or Information Revolution. "Just as in the previous era," write Peter Hall and Paschal Preston, "automobiles and electrical machinery had determined the course of high technology economic growth and the relative fortunes of the major economic powers, so now the laurel would pass to the nation that proved the most effective in bringing electronic information systems to the market place."[1]

In 1950 it was not clear which nation, if any, would come to dominate the future technical development and commercialization of digital electronics. The United States, Great Britain, and even war-torn Germany all had the prerequisite scientific and technical skills to push ahead in digital-electronics-based information technologies. But by the 1970s, the United States had become the preeminent technical and industrial leader in all things digital. A large internal market, first driven by heavy defense spending and then reinforced by a prosperous civilian sector, contributed considerably to the United States' ability to win global dominance in digital electronic technology in the period 1945–1970.[2] While Europe's major industrialized nations struggled, in the wake of the American technological and economic juggernaut, to create some semblance of a domestic digital electronics industry, Japan loomed unexpectedly on the horizon as the principal rival to the technological and industrial hegemony of the United States. Though Japan did not have a military-industrial complex to support its computer industry, protectionist policies allowed its large internal market to reinforce the domestic, state-sponsored appropriation of technology transfer.

In the midst of the international scramble to jump on this rising global innovation wave, where was Canada? From a macroeconomic perspective, Canada was poised at the end of World War II to get a jump start in the race to profit from digital electronics. After all, while most of the industrialized economies lay in ruins, Canada's economy and infrastructure, along with the United States', remained unscathed. In fact, Canada came out of the war with a larger and more vigorous industrial economy than it had before the war. This initial advantage, however, was offset by the absence of any industrial tradition of scientific research and experimental development, a weakness that was particularly acute in the areas needed for innovation in digital electronics. The eight case studies presented in this book offer the first fine-grained view of Canada's participation in the fourth innovation wave. The salient features of Canada's ascent of the digital electronics learning curve were the unprecedented role of the military enterprise in the creation of a peacetime pool of scientific and technical skills in industry and government defense laboratories; the smaller, sometimes erratic, but nevertheless important role of civilian public enterprise in nurturing industrial learning after the military's retreat; and the surprising contribution of some plants in fostering domestic technical knowledge accumulation and skill acquisition at the frontiers of technical change. In many instances, this newly created pool of technical expertise, in its design decisions, refracted the global innovation wave through the Canadian context.

Military Enterprise Accelerates Canada's Jump to the Fourth Wave

Without a doubt, military enterprise was the primary force behind Canada's early participation in the digital electronics revolution. "Military enterprise," Merritt Roe Smith writes, "refers to a broad range of activities through which armed forces have promoted, co-ordinated, and directed technological change and have thereby, sometimes directly and sometimes indirectly, affected the course of modern industry."[3] Whereas American historiography, from the earliest days of the republic, is replete with studies of the role of the military in the process of peacetime technological and industrial development, Canadian historians have devoted little if any attention to this line of inquiry. Is this lacuna in Canada's national historiography simply an indication of the absence of anything of importance to write about? Has there ever been a Canadian peacetime "military enterprise" that matches Smith's sense of the term? This narrative has shown that military enterprise did indeed play a crucial and sizable role in shaping the direction and rhythm of technological change in Canada from 1945 to about 1958.

This book's narrative adds more substance to Keith Krauss's assertion that from 1945 to 1958 Canada "flirted" with military "self-reliance." In fact, one could say that it was a historically unprecedented "flirtation." However, there was a dearth of evidence to support Krauss's assertion. Except for studies of the AVRO Arrow, and to a lesser degree of the Mid-Canada and DEW Lines, the Canadian historical literature contains few details on the impact of the military's quest for self-reliance on the content and rhythm of Canadian peacetime technological development. This narrative has provided a richer portrait of this process. By linking national defense with national welfare and sovereignty, Canadian military enterprise greatly accelerated Canada's movement up the digital electronics learning curve. Through collaboration with academia, direct support of industrial experimental development, and the creation of in-house R&D facilities, military enterprise fostered the creation of a national capacity to design and manufacture products within the fourth innovation wave many years earlier than otherwise would have been the case.

The systematic exploitation of science for modern warfare, which began in World War II and grew in intensity after the war's end, called for an ever-increasing capacity to carry out high-speed, large-scale computations. By 1948, it had become clear to the more technically oriented elements of Canada's military that the absence of advanced computational facilities doomed any national effort to carve out some semblance of a self-reliant weapons development program within the overall North Atlantic Triangle alliance. First with UTEC and then with the purchase of FERUT, the military's desire to secure a domestic access to computational power was the driving force behind the creation of Canada's first computer science research facility and high-speed computation center, at the University of Toronto.[4] Although the purchase of FERUT precipitated the collapse of electronic computer design at the university, it nevertheless gave Canada a valuable head start in the growth of a national competence in high-speed computation techniques; even more important, it gave Canada an earlier opportunity to gain know-how in software design.[5] For nearly 5 years, FERUT served as a public training ground for academics, students, and people from industry and the military. And the military's sponsorship of academic computation centers did not end with the University of Toronto. The need to meet the computational needs of its geographically dispersed laboratories also led Canadian military enterprise during the 1950s to subsidize the creation of computation centers at other universities, including the University of British Columbia.

Through its early support of computation centers at the universities, the military again fostered the dissemination of digital electronic technology. During the 1950s, in its educational program, in its research, and in its interaction with industry, the University of Toronto helped forge a whole generation of expertise in computer science. Military assistance allowed the University of British Columbia (UBC) to develop into an oasis of software know-how in a region far removed from the eastern center of Canada's activity in digital electronics. In the years that followed, UBC's computer department nourished the growth of a high-tech sector in the Vancouver. During the late 1970s and the early 1980s, UBC's programming talent nourished the growth of software-based companies such as MacDonald Dettwiler and Associates, the Sydney Development Corporation, and the Basic Software Group.[6]

Despite its cancellation, the UTEC project contributed to moving Canada up the learning curve by fostering the early development of technical expertise. The hardware design team of Alfred Ratz, Josef Kates, Leonard Casciato, and Bob Johnston subsequently formed part of the nucleus of Canada's first generation of computer engineers. Although the latter three failed in their initial attempt to establish a computer manufacturing business, they nevertheless became an important source of know-how to the private and public sectors. Kates, Casciato, and Johnston, for example, advised Trans-Canada Air Lines on the engineering feasibility of a computerized reservation system. They interacted with manufacturing companies such as Ferranti-Canada and with users such as the federal government. Kates and Casciato later went on to form the computer engineering firm KCS Limited, which designed and implemented the world's first computerized road traffic control system (in Toronto).

The purchase of FERUT signaled the end of the Canadian military's interest in designing computers for scientific computations, but it did not diminish the strategic importance of designing modern weapons systems around digital electronics. The Royal Canadian Navy's pursuit of DATAR reflected a new vision of an automated "battlefield" dependent on high-speed information processing and communications. The RCN's very costly Naval Battle Simulator was further evidence of the importance the RCN placed on digital electronics. Not only could digital electronics be used to wage war; it could also be used to train Canadians for war.[7]

The dream of military self-reliance, through the mediation of technology, led the RCN to invest heavily in the development of a core of Canadian research, development, prototyping, and production engineering capacity. In seeking to elevate its role in the North Atlantic Triangle, the RCN sponsored Canada's first industrial ventures in the design and manufacture of digital electronics: Ferranti-Canada and Computing Devices of Canada. While the former built DATAR, the latter worked on the Simulator. While Ferranti-Canada tried to move its DATAR experience into the civilian sector, Computing Devices kept its focus on the military sector. As a result, though both companies started out as good candidates to create a domestic computer industry, by the end of the 1950s Ferranti-Canada represented Canada's only hope in this realm.

Success in the design of sophisticated computational systems also depended on the transistor. Without the capacity to miniaturize, the growing weight of electronic circuit complexity would have crushed any country's attempt at self-reliance in the design of advanced weapons systems. As a result, an ability to manufacture and to design with high-performance transistors became crucial for each member of the North Atlantic Triangle in the mid 1950s. In the case of Canada, the constraints of a small and open economy set limits on how much self-reliance the military could pursue in the area of miniaturization. Canadian military enterprise flirted with the idea of promoting a domestic source of transistors, but it could never ensure a sufficiently high level of procurement to stimulate any domestic R&D or production (branch plant or otherwise). Although the creation of domestic expertise in transistor manufacture proved impossible, Canadian military enterprise did consciously invest in expanding the nation's capacity to do innovative circuit design work within the semiconductor medium. The Defence Research Telecommunications Establishment's (DRTE) digital transistor work not only served as pole of innovation diffusion to the defense industry; it also contributed to the state's increased scientific and engineering capacity to be a direct agent of ground-breaking technological change, as in the case of Canada's space program.

In the early 1950s, Canada became, after the United States and the Soviet Union, the third nation in the world to design, build, and deploy an artificial satellite. Behind the guise of civilian government scientific research, the impetus to embark on a space program came from Canadian military enterprise.[8] Concern over secure and reliable communications across Canada's northern flank led to intensive study by DRTE of the behavior of the ionosphere in high latitudes.[9] The Alouette I satellite provided valuable scientific data. While the Alouette project involved the entire DRTE, it was the research program in silicon transistor circuits that helped create the in-house miniaturization know-how needed to build the satellite. Canada's subsequent satellite design programs (Alouette II, ISIS I and II, the Hermes program, Radarsat, and MSAT) drew on the human expertise first nurtured by military enterprise at DRTE in the 1950s. On the industrial front, the influence of DRTE extended through the 1980s. According to a study carried out by Denzil Doyle, the existence of DRTE influenced the creation of more than 32 high-tech electronics firms in Canada.[10]

The importance of military enterprise to the history of Canadian technology during the period 1945–1958 does not lie in the success or failure of particular projects such as UTEC, DATAR, and the DRTE computer. Rather, it lies in the military's role in channeling investment into high-risk R&D that otherwise would have had little chance of finding support within an economy dominated by the export of natural resources. The military accelerated the creation of a pool of skilled scientists and engineers who kept Canada within reach of rapidly advancing technological frontiers. Throughout the early stages of the digital electronics revolution, it was the "visible hand" of Canadian military enterprise that incited companies such as Ferranti-Canada, Computing Devices of Canada, Canadian Aviation Electronics (CAE), RCA Victor, and Canadian Marconi to increase their competence in digital electronics.

It is useful to consider the contrasting case of Australia. The initial political will to support the design and construction of Australia's first electronic digital computer came from the preoccupations of a staple economy, not from the computational imperatives of advanced weapons development.[11] Like Canada, Australia is a vast but sparsely populated land with a small domestic market in which the export of primary natural resources has dominated economic development. Since half of the Australian land mass is a hot desert and much of the other half is arid savannah, the availability of water has been of the utmost importance to Australia's agriculture-dominated export economy. In 1947, scientists at Australia's Council of Scientific and Industrial Research (CSIR) toyed with the idea of developing methods to induce and control rainfall. Radar, developed during the war, was a powerful new tool with which to probe the physics of cloud and rain formation. But in order to carry out this line of investigation effectively, considerable computational power was needed to model and analyze data on microwave propagation.

By the end of 1947, Trevor Pearcy, who accumulated considerable experience on large-scale computational issues while working on the wartime development of radar in the United Kingdom, convinced CSIR to fund the engineering development of an electronic digital computer of his own design.[12] By the end of 1949, Pearcy and his team of engineers at CSIR's Radiophysics Division had built the Mark I, Australia's first computer, and had performed crude programming tests on it. A year later, the Mark I was working somewhat reliably. Not animated by any military fascination with performance, the Mark I was of humbler design than University of Toronto's UTEC project. It was serial rather than parallel, simpler delay lines were used instead of electrostatic memory, and there were no attempts to pioneer vacuum tube miniaturization. The Mark I was a practical, cost-effective solution to an immediate computational need.

After its initial funding of the Mark I's development, CSIR's enthusiasm for computer engineering quickly waned, and in 1954 it pulled the plug on all further computer design work. The nascent field of digital electronics held little interest for an institution whose R&D ethos revolved around the needs of a nation's natural resource industries.[13] Computer development at CSIR's Division of Radiophysics thus became an isolated occurrence within the Australian technological landscape. There is no evidence that Australia's military ever seriously pursued a policy of national technological self-reliance during the 1950s.[14] As a result, there were no deep pockets in Australia to counter the investment logic of a staples export economy and to divert resources to creating of a national capacity to manufacture, adapt, and design along the fourth innovation wave.[15]

Clearly, military enterprise led to earlier and faster competence formation in digital electronics in Canada than in Australia. Whereas the University of Toronto (with FERUT) had become a important pole of technology diffusion in 1952, a computer did not appear at an Australian university until 1956. During the 1950s, Australia did not have military-sponsored industrial and governmental facilities (such as those at Ferranti-Canada, Computing Devices of Canada, Canadian Marconi, RCA Victor, CAE, Sperry Canada, and DRTE) pursuing sophisticated research, experimental development, prototyping, testing, and production engineering in a broad range of digital technologies. Did Canada's earlier and faster move up the digital innovation wave give Canada decided long-term benefits over Australia? An initial, cursory look at the evidence suggests that it did.[16] However, a far more detailed comparative study is needed to test this conclusion.

The evidence in this volume adds considerable weight to the sociologist Gordon Laxer's speculation that Canada was slow to develop a diversified industrial economy before 1945 because it lacked a strong peacetime military. The case studies presented above suggest that Canadian military enterprise, limited by Canada's small economy and close proximity to the United States, systematically tried to promote a national industrial competence in the constellation of technologies that would contribute to post-1945 economic success. The flirtation with military self-reliance, according to Krauss, ended in the late 1950s,[17] yet Canada retained a core of sophisticated, technology-driven engineering firms.

As the Canadian military enterprise's financial resources for experimental engineering development decreased, the Canadian firms mentioned in this volume exhibited remarkable ingenuity in trying to sustain knowledge, experience, and skills acquired under military patronage. They aggressively shifted the focus of their product development to the civilian sector. However, the civilian departments of the Canadian government failed to use their full power of procurement to reinforce the commercial viability of industrial R&D in digital electronics. Even so, on the occasions when public enterprise investment actually became available, it made an important contribution to the knowledge accumulation of the engineering firms. Meanwhile, the corporate commitment effort to nurture strong in-house technical expertise in leading-edge experimental engineering contradicts the stereotype of Canadian branch plants as passive or parasitic entities—a stereotype that was devised by nationalist political economists in the 1970s and the 1980s. The profile of the branch plant that emerges from a study of the postwar digital electronics industry in Canada is, as the final pages will argue, one of fierce local pride, independent initiative, daring product-innovation ambitions, and high managerial risk taking.

The Subsidiary and National Technological Competence

The role of foreign-owned subsidiaries in Canada's development remains a controversial theme in Canadian historiography and political-economic thought. In 1879, fearful that the US industrial miracle would flood Canada's small market with goods and thus block the growth of a domestic manufacturing sector, Sir John A. MacDonald's Conservative government enacted the "National Policy," under which tariff barriers were put up to encourage US industry to set up manufacturing facilities in Canada.[18] By 1929, "branch plants" had become a dominant feature of Canada's political economy—particularly in manufacturing activities associated with the third wave of innovation, which began with the twentieth century.[19] In the years that followed World War II, doubts about the capacity of foreign-owned subsidiaries to support Canada's long-term economic interests began to mount. The branch plant had clearly expanded manufacturing in Canada, but was it compatible with domestic political and economic aspirations? Did the subsidiary foster a greater local capacity to design, develop, and commercialize competitive products?

"Since the mid fifties," the economist A. Edward Safarian wrote in 1966,

> there has been considerable controversy over the role of the international firm and some new legislation regarding it. This controversy has focused particularly on the role of United States direct investment in Canada. . . .Critics state that there are, or can be, adverse consequences from [foreign] direct investment due to conflicts of interests, actual or potential, between the interests of the subsidiary and those of the foreign owners of the firm. Since the maximum feasible development of the subsidiary companies can be taken to be in the Canadian interest, at least so long as the resources involved are used efficiently, there is a conflict then between the interest of Canada and of the foreign owners of the subsidiaries. This situation arises because, it is stated [by the critics], the international firm seeks to maximize its global profit over time, an objective which is not necessarily in the best interest of the subsidiary.[20]

After extensive study, Safarian concluded that the performance of foreign-owned subsidiaries in Canada was so varied that "sweeping generalized criticisms simply [did] not stand up under even cursory examination of the available data." More fundamentally, where the behavior "[had] been defined as undesirable by public authorities does appear, it can often be related more closely to aspects of the economic environment of the subsidiary and only distantly, if at all, to the fact of foreign ownership."[21] At the heart of the performance issue is the subsidiary's capacity to innovate. Critics of direct foreign investment argued that the centralization of R&D facilities in the parent firm abroad inhibited the development of this expenditure in Canada, further restricted the sale potential of the subsidiary, and limited the development of technical and scientific personal. Safarian challenged the view that subsidiaries did less R&D than their domestically owned counterparts, and that what R&D they did was less sophisticated.[22] The rising tide of nationalism, however, drowned out Safarian's caution against sweeping generalizations.

By 1970, a new generation of Canadian nationalist intellectuals was arguing that Canada had failed to brake away from its excessive dependence on the exporting of raw materials. Inspired by the literature on

Third World underdevelopment, this "nationalist" discourse laid the blame of Canada's supposedly "truncated" or "arrested" techno-economic development on the presence of too many branch plants. Rather than acting as a conduit to transfer the parent firm's design capacity to Canada, "nationalist" political economists and historians argued, the branch plant accentuated Canada's technological dependence and blocked the emergence of a Schumpeterian class of entrepreneurs. During the 1970s, the Science Council of Canada—an arms-length, policy advisory body that answered directly to the cabinet—aggressively underscored the deleterious effects that subsidiaries had on Canada's overall innovative abilities. "To an increasing extent," one Science Council study of Canada's manufacturing sector argued, "technological capability is the basis of power in the advanced industrialized countries."[23] Here, for "technological capability," one must read "capacity to carry out sophisticated R&D." Foreign ownership, in the Science Council's view, had stifled Canada's industrial capacity to carry out advanced research and development. Viewing the ascendancy of the digital electronics paradigm as a far-reaching "transformative force" and as the nexus of the post-industrial revolution, the Science Council warned that a lack of capacity to design and manufacture within this technological revolution would endanger Canada's future economic, social, and political development.[24] Of course, one of the largest impediments to obtaining this capacity was the US-owned subsidiary.

In 1980 the Science Council, having grappled with the "branch plant" issue for 10 years, proclaimed its World Product Mandate strategy.[25] The idea of this strategy was not new, though the Science Council may have thought so. Since 1965, when the Auto Pact was put in place, DITC had doggedly tried to find new implementations of the WPM concept, although the term "world product mandate" had yet to be used. One can push the historical precedence for WPMs even further back in time. The Canadian military's strategy to secure an R&D and manufacturing role in weapons systems with its US and British allies was based on the promotion of specialization and international rationalization of production and of the needed transnational framework for standardization. Though the WPM concept had been around for some time, it had not been the object of widespread public debate. But when the Science Council coined the phrase "world product mandating" in its 1980 policy pronouncement, the concept suddenly came under much greater scrutiny.

Between 1980 and 1995 this strategy was the object of an intense debate, particularly among Canadian scholars of management science.

At the heart of this debate were two very different views of the multinational corporation's organizational structure and its decision-making process. As I have already noted, one view attributed the multinational's success to its highly centralized structure, with decision making a unidirectional process flowing from corporate headquarters to the subsidiaries. It was this "central brain" paradigm of the multinational corporation (MNC) that animated the Science Council's and the Canadian government's public policy thinking on WPMs. Public policy was thus aimed at persuading, cajoling, and arm-twisting the headquarters of MNCs to be good corporate citizens and hand over WPMs to Canadian subsidiaries. Using the same model of the MNC, critics pointed to weaknesses in the Science Council's articulation of a WPM strategy. The economist David Rutenberg argued that a parent firm would especially want to control new products because of their negative cash flows and uncertainty of success.[26] Along similar lines, Thomas Poynter and Alan Rugman emphasized that the parent firm's inherent desire to control all strategic R&D activity created high intra-organizational barriers to granting WPMs.[27] For many MNCs, Poynter and Rugman argued, granting a WPM "threatens the efficiency of its internal operations and hence its competitive advantage."[28]

By the mid 1980s, a new perspective of the MNC's structure and decision-making process had begun to displace the model of the centralized, all-powerful parent firm. The new model viewed the MNC as a networked institution with a complex set of internally competitive interactions.[29] This model of the MNC opened up the door to the potential opportunities for entrepreneurial initiative at the subsidiary level. In this new model, according to which the subsidiary was constrained rather than defined by the MNC's organizational structure, local management had the opportunity to make strategic decisions.[30] As a result, the center of the public policy debate shifted from the parent firm to the subsidiary. "The critical elements in achieving technological development in Canada," wrote the management scholars Norman McGuiness and Howard Conway, "are the willingness and ability of the Canadian subsidiary to engage in more innovative and search oriented behavior. The best product mandates will be earned, not allocated. . . . The challenge, then, is for the Canadian managers within these corporations to become more development, rather than short-term, profit, oriented."[31] This new scholarly interest in the issue of subsidiary initiative was fueled by the recession of the 1980s and by the crumbling of tariff barriers to international trade. These two long-term processes posed serious threats to the viability of Canadian subsidiaries as MNCs reevaluated their operations on a global basis. In this new economic world order, a Canadian subsidiary had to earn a WPM in order to survive and prosper within a multinational. The challenge for management theorists was to come up with models that could help subsidiary management formulate and execute WPM initiatives.

Despite the many pronouncements that have been made about the ability of foreign-owned subsidiaries to promote national technological self-reliance, and despite the considerable management theorizing on whether WPMs are given or earned, few detailed historical case studies have been carried out. In the area of digital technology, which certainly was crucial to postwar economic growth, this paucity is even more pronounced. The cases of Ferranti-Canada and Sperry Canada offer dramatic examples of the path-dependent manner in which subsidiaries can gain greater autonomy and forge their own design and manufacturing niches. The case of Control Data Canada illustrates the process, again path dependent, that can lead a parent firm to confer a significant mandate on a subsidiary where none existed before. But once a mandate is given, as the story of Control Data Canada reveals, local management must strive to reinvent its mandate if the subsidiary is to prove resilient in a rapidly changing competitive environment.

The development of electronic data processing equipment at Ferranti-Canada and of digital electronic machine tool automation at Sperry Canada suggests a Schumpeterian vitality in the behavior of foreign-owned branch plants that nationalist historians have greatly underestimated. Each of those subsidiaries showed a fiercely independent ambition to use domestically cultivated expertise to carve out a business niche in the fourth wave. After the collapse of the Korean War boom in defense spending, Ferranti-Canada set out to convert its DATAR experience into a commercial line of products. Sperry Canada, on the other hand, had no experience in digital electronics when it entered the field of machine tool automation. The pursuit of a self-reliant, sophisticated R&D program characterized these subsidiaries' business approach to the opportunities offered by the fourth wave. One wanted to be Canada's leading computer manufacturer; the other wanted to be a global player in digital automated machine tools. Neither Ferranti-Canada nor Sperry Canada was composed of the "terrified little men" that the nationalist historian J. J. Brown claimed populated Canadian industrial culture during the entire century surveyed in his 1967 history of Canadian technology.[32] Neither did the existence of foreign-owned branch plants in Canada preclude Schumpeterian corporate behavior, despite Kari Levitt's claims to the contrary in her oft-cited 1970 book *Silent Surrender*.

Ferranti-Canada's and Sperry Canada's experiences reveal that the relationship between the nationality of capital investment and Canada's industrial capacity to design and manufacture within a global innovation wave was far more nuanced than the black-and-white "de-industrialization" thesis raised by the nationalist discourse, which characterizes foreign-owned branch plants as agents of technological dependency. To be sure, it could be argued that Ferranti-Canada was an isolated case, and that, as a British-owned firm operating under what Alfred Chandler called a personal capitalism regime, this example does not directly contradict the nationalist argument that the domination of the Canadian economy by US-owned subsidiaries "arrested" Canadian development. But the story of Sperry Canada is more difficult to explain, for it showed that a US-owned branch plant run according to managerial capitalist principles could and did act as an agent of radical product innovation with its own ambitious business agenda.

At the time, Ferranti-Canada and Sperry Canada represented Canada's best hopes to produce a domestic capacity to design and manufacture electronic data processing equipment and machine tool automation technology. Their foreign parent firms did not exercise any iron grip over the capacity of these two subsidiaries to define and pursue their own R&D programs. Rather than a tale of passive dependency, the history of these firms is a tale of determined efforts by Canadian management and engineers to assert their own technological vision and product agenda.[33] In both cases, the subsidiary's autonomy to define, create, and sustain its own advanced digital electronics product development program resulted from a technology-driven corporate culture and a rather decentralized organizational structure that tended to be defined along geographic and product lines. Like their parent firms, both subsidiaries looked to radical technical innovation as the key to commercial success. Sperry Rand was a kind of federation of technology and product mandates; the Ferranti organization behaved like a collection of trading houses, each with its technological niche. As long as Sperry Canada and Ferranti-Canada did not lose much money, their parent organizations left them alone. To understand Sperry Canada's failure to turn its computer numerical control (CNC) innovation into a long-term business and Ferranti-Canada's inability to create a profitable business out of designing and building general-purpose computers, one must look to the inherent uncertainties in leading-edge experimental development, to the realities of the market, to the obstacles of capital formation, to the reluctance of civilian government to move in earlier and more aggressively to support this niche, and, most important, to the fact that by the late 1970s survival in the international NC market was extremely difficult for most North American companies.

Both subsidiaries had difficulties finding the financial resources to sustain their Schumpeterian ambitions. Sperry Canada had entered the nascent business of machine tool automation as a way to reduce its dependence on military contracts. However, its entire NC and CNC development program was constantly compromised by underfunding. Paradoxically, the very vulnerability to shrinking defense contracts from which Sperry Canada wanted to escape continually crippled its ability to diversify into the civilian sector. Ferranti-Canada, on the other hand was able to sustain much of its digital electronics R&D until 1964 through its opportunistic exploitation of public enterprise initiatives. Each of the one-off digital electronic systems it developed for the Canadian Post Office, Trans-Canada Air Lines, or the Federal Reserve Bank of New York allowed Ferranti-Canada to build incrementally on its DATAR experience and thus develop the design and manufacturing expertise required to move into the field of general-purpose computers. Though designing the FP6000 marked an important accomplishment for the company, marketing that computer was a greater and a far costlier challenge. Competing in a business dominated by IBM required far more money than Ferranti-Canada could raise.

For both Ferranti-Canada and Sperry Canada, along with the freedom to formulate and pursue aggressive R&D program came the crushing burden of self-financing. Though the Canadian government recognized that these two companies were its only hopes to establish a domestic hardware industry in electronic data processing equipment and machine tool automation technology, its financial support was timid and tardy. The federal government never really appreciated the fact that Sperry Canada and Ferranti-Canada, though subsidiaries of large foreign firms, essentially functioned as small Canadian companies with their own R&D programs and business strategies, and that they desperately needed the government's investment support. Canadian banks certainly were not an alternative source of support. After several years of exhaustive exploration of all the primary resource material available, I was struck by the conspicuous absence of any reference to banks' playing any role whatsoever in financing these companies' R&D, product development, or commercialization plans. The absence in the historical evidence of the slightest hint that these two companies ever approached the banks or even contemplated approaching them seems to reinforce the nationalist argument that banks have been an obstacle to the emergence of a diversified and sophisticated Canadian industrial sector.[34]

Sperry Canada's failure to turn UMAC-5 into a successful CNC business had less to do with its status as a subsidiary than with the uncertainties associated with any high-risk technological gamble and with the Canadian government's reluctance to act aggressively in support of the company's ambitions. Though there was some inter-divisional rivalry within Sperry Rand over the right to develop and commercialize NC, the parent firm was supportive of its Canadian subsidiary's overall product development strategy.[35] In adopting the novel strategy of using a programmable computer to breath new flexibility into digitally based machine tool automation, Sperry Canada's engineers never imagined that the development of software for the real-time control would lead them into a technical minefield. While applauding Sperry Canada's pioneering work, the Department of Industry, Trade, and Commerce declined to help the company stay in the vanguard of CNC. Its refusal, which had little to do with any assessment of Sperry Canada's capability to resolve future technological uncertainties, stemmed from the government's desire to use the company's unique Canadian expertise to promote the modernization of Canada's metalworking industry. But even when Sperry Canada decided to develop the simpler and less costly UMAC-6 system for the Canadian metalworking industry, the government was at first reluctant to move aggressively. DITC's unwillingness to put up any more money reflected its misunderstanding of intra-corporate realities that underlay Sperry Canada's world product mandate. DITC expected the parent firm to make up the shortfall. But as a profit center, Sperry Canada was responsible for paying its own way. In the end, DITC never quite grasped the fact that Sperry Canada was on its own. DITC's blindness to Sperry Canada's financial predicament was itself a manifestation of the "central brain" model of the multinational corporation; strategic product initiatives could originate only with the parent firm. As such, it was natural that the parent firm would also supply the financing.

In the case of Ferranti-Canada, public enterprise had played an important role in nurturing the company's new-found competence in digital electronics. The Canadian Post Office Department's pioneering electronic mail sorter gave Ferranti-Canada the opportunity to master the use of transistors in high-speed digital circuit design, as well as the opportunity to perfect its drum memory technology. The Trans-Canada Air Lines project to computerize reservations enabled Ferranti-Canada to design and build its first general-purpose computer. Though not Canadian, the Federal Reserve Bank of New York contract to build a digital electronic check-sorting machine allowed Ferranti-Canada to spawn the idea of endowing mid-size computers with multiprogrammable operating systems. One should not, however, confuse the Post Office and TCA contracts with any coherent government policy to nurture the growth of a Canadian computer industry—there was none.

Throughout the first 25 years of the fourth innovation wave, except for a brief flirtation with military self-reliance, the Canadian government never formulated any substantial policy to support the growth of a domestic computer hardware industry. Although there was considerable interest within the government in improving the scope and the efficiency of its own operations through greater use of EDP technology, there appears never to have been any discussion as to whether the state's purchasing power should be channeled to foster local equipment manufacture. Without an explicit "buy Canadian" policy backed up by some form of financial inducement, no government department could turn its gaze away from the aura of big names such as IBM and UNIVAC to consider Ferranti-Canada. Equal in importance to government procurement in fostering the growth of a domestic computer industry was the need to provide capital. Ferranti-Canada lacked the financial resources to establish the kind of marketing and service infrastructure that it needed to compete with the established players. Ferranti-Canada's inability to raise money—either internally or externally—made the Canadian government its only potential source of capital.

Lacking a clear industrial and technological policy related to the fourth wave, the DITC could not act quickly and aggressively to come to Ferranti-Canada's aid. The importance of the FP6000 to International Computer Limited's strategy of product development and commercialization attests to the technical worthiness of the Canadian-designed computer. The Digital Electronics Corporation's subsequent development of the minicomputer market suggests that Ferranti-Canada's attempts to commercialize a mid-size machine were the only means (before 1970) by which Ferranti-Canada might have penetrated IBM's market invincibility. Technical attractiveness, however, does not in itself guarantee market acceptance. Without a large infusion of capital, Ferranti-Canada's innovative computer had little chance of securing even a tiny toehold in the market. In the end, the FP6000's poor commercial showing in Canada made it easier for Ferranti UK to take the FP6000 design rights away from Ferranti-Canada and sell them to International Computers and

Tabulators (later to become International Computers Limited). One could argue that, even if the government had poured money into making the FP6000 the first flagship of Canada's computer industry, there were few guarantees that the product would have stolen a profitable niche away from IBM's 360. But without a vigorous and coordinated support program, Ferranti-Canada's failure was certain.

In contrast with the cases of Ferranti-Canada and Sperry Canada, the formation of technological competence at Control Data Canada had more to do with top-down strategic initiatives than with any locally driven ambition. This top-down initiative better fitted DITC's conception of how mandates should emerge in subsidiaries. The parent firm's decision to relocate the design, the development, and the manufacture of perhaps its most important mainframe to its Canadian subsidiary was not typical of the behavior of the other major US computer manufacturers of the day. Management scholars of the 1980s criticized the World Product Mandate strategy on the ground that it was not natural for parent firms to hand out these mandates easily. The historical trajectory that led to the PL-50 project and then to the Cyber 173 product offers one illustration of the path-dependent process that could lead a parent firm to confer a major product mandate on a Canadian subsidiary. DITC's strategy of moral suasion did not bring CDC to Canada. The availability of subsidies, although important, was not the deciding factor. By 1969, increases in the scale and scope of the R&D needed to compete with IBM, internal financial constraints, and the urgency of finding a new product line forced CDC to experiment with decentralizing its design and manufacturing competence.

CDC's efforts to cultivate international collaborative arrangements went beyond a one-shot attempt to exploit Canadian government funding. From CEO William Norris down to the company's senior executives, there was a strongly held belief that inter-firm and international collaborative alliances were the best way to move ahead in an industry dominated by IBM. The success of the joint venture with the Canadian government, as measured by the great success of the Cyber 173 series, provided Norris with the first and strongest reinforcement of his belief in international cooperation. In fact, Norris tried to use the Canadian experience as a model for collaboration with other national governments.

Was DITC's support of CDC-Canada a waste of money? DITC's support of the PL-50/Cyber 173 created a top-notch center of technological competence that otherwise would not have been readily available to Canada. The parent firm devoted considerable time, effort, know-how, and money to moving its Canadian subsidiary up the learning curve. The Canadian group did not work on bits and pieces; it designed and built entire systems. CDC-Canada had become CDC's most important computer development group and manufacturing facility outside the United States. Not only did CDC-Canada become Canada's only industrial center of excellence in mainframe design; the subsidiary also represented the parent firm's first and most ambitious experiment with moving R&D abroad. By 1980, Control Data was a bright light in the otherwise grim portrait of multinational behavior in Canada painted by DITC. "Where R&D is carried out under the auspices of the subsidiary in Canada," noted a DITC study already quoted in this volume, "it tends to consist of relatively unsophisticated items and a total computer systems capability has not developed in Canada with the exception of Control Data Canada Ltd."[36]

Perhaps the most noteworthy development at CDC-Canada was the clear sign of a move, albeit slow, toward greater autonomy. The reality of local strategic initiatives began to emerge. The entire superminicomputer project illustrates that CDC-Canada could articulate and win its own WPM. The superminicomputer initiative suggests that CDC-Canada may have become a profit center rather than a cost center had it not been for a disastrous reversal of the corporation's fortunes. Just because the Development Division at CDC-Canada was shut down in the late 1980s, one cannot conclude that DITC's financial support of CDC's Canadian subsidiary was a mistake. At the time, no one could have predict that continual advances in semi-conductor electronics would dramatically change the technological and corporate faces of the computer industry. No one could have foreseen that CDC would be a one of the casualties of this change.

In a 1971 Science Council report, Arthur Cordell offered up a typology of foreign-owned subsidiaries and R&D in Canada that reinforced the nationalist view.[37] He put all multinational R&D activities in Canada into just two categories: the "international interdependent laboratory" and the "support laboratory." "Research programs of an international interdependent type," Cordell argued, "may have little to do with the capacity for new product innovation in Canada. Specialized research of this type is often confined to a specific stage of the R&D process. . . . While innovation capability (both product and process types) is increased for the firm as whole, there is little obvious benefit which accrues to the Canadian economy. . . . An enclave operation of this type may be a form of 'brain drain.'"[38] The "support laboratory," on the other hand,

> acts as a technical service centre, i.e. to examine why a product may fail to operate in the Canadian market or to help with the adaptation of the product to the Canadian market, and/or acts a translator of foreign manufacturing technology,

i.e., to implement the process of 'technology transfer.' . . . However, one must distinguish the benefits which accrue when the R&D is undertaken in the process of transferring technology from those that accrue when R&D is undertaken in an indigenous firm. . . . The potential benefit to the Canadian economy arising from exploitation of either product or process is great in indigenous firms and almost non-existent in subsidiary operations.[39]

Ferranti-Canada, Sperry Canada, or CDC-Canada was neither an "international interdependent laboratory" nor a "support laboratory." Cordell's characterization of Canadian branch-plant R&D was not based on any detailed case studies of actual corporate practice over extended periods. Instead, Cordell used an industrial snapshot from around 1970 to abstract a "universal" typology of branch plants' R&D behavior in Canada.

The historian Graham Taylor, in contrast, tried to extrapolate a typology of the technology relationships between foreign parent firms from in-depth historical analyses. Taylor's assertion that the "parent firm's willingness to transfer resources, particularly advanced technology was closely linked to the degree of control they exercised over the affiliated companies" may prove to have wide validity. Although Taylor's model has some overlap with the CDC-Canada story, it has little relevance to Sperry Canada, and its relevance to Ferranti-Canada is questionable.[40] The creation of Sperry Canada and the creation of Ferranti-Canada's digital electronics program both entailed some initial technology transfer in the hope of winning Canadian defense contracts. But because of the decentralized management structure that defined the parent-subsidiary relationship of both firms, the direction and rhythm of the technical innovation process in the two subsidiaries was shaped more by the local managers' and engineers' ambitions and talents than by the parents' needs. As separate profit centers, both Sperry Canada and Ferranti-Canada were left to their own devices to formulate and implement their product development strategies. Paradoxically, the fate of those two strategies partially affirms Taylor's main point about the linkage between technology transfer and parental control. In both cases, the parent firm eventually exercised its authority to appropriate the Canadian-devised technology and to transfer it back to the home country. However, neither Ferranti nor Sperry Rand would have interfered with the autonomous operations of its Canadian subsidiary had the subsidiary found enough local funding—from the government or from the banks—to bring its R&D to commercial fruition.

My three corporate case studies suggest strongly that, when it comes to subsidiary behavior, the nationalist discourse of dependency and de-industrialization has been excessively one-sided. The history of RCA's Canadian subsidiary further supports this conclusion.[41] Recently, the scholarly view of the foreign-owned subsidiary has swung the other way. After an extensive analysis of the Canadian federal government's database, John Baldwin (an economist with the Micro-Economic Analysis Branch of Statistics Canada) and Petr Hanel (a professor of economics at the University of Sherbrooke) have concluded that the R&D done by subsidiaries in Canada is more sophisticated than that done by domestic companies.[42] In some respects, this conclusion brings us back to A. Edward Safarian's 1966 analysis. Though the nationalist argument accused foreign direct investment of depriving Canada of Schumpeterian innovators, Baldwin and Hanel argue that a subsidiary will tend to introduce more "original" innovations whereas a Canadian-owned firm will tend to introduce more "imitative" innovations. From a historian's perspective, Baldwin and Hanel's analysis does have an important limitation: It is a snapshot taken in 1993. It does not reveal the process by which subsidiaries gain the capacity for manufacturing and R&D. The three examples in this book not only show some examples of this process; they also highlight the diversity of subsidiary experiences.

Canadian Context and the Direction of Design

One of the challenges facing the historian of technology is to create a narrative that explains why and how technical change proceeded as it did. In writing such a story, historians must not only seek to portray design choices and the factors that shaped them; they must also elucidate the forces that set the design process in motion and explain how the process was maintained. The content and the conceptual structure of the tale that "technology's storytellers" weave will, however, depend on the factors they view as the determinants of change. The "internalist" narrative mistakenly treats dynamics of technology as essentially exogenous to society: Technology unfolds as some inexorable internal techno-scientific logic that transcends human will. The "externalist" narrative of technological change, however, falls into the trap of spinning a socio-economic and political tale devoid of technical contingency. The internalist narrative offers a technological-determinist view of history; the externalist narrative offers the equally mistaken view that only economic, social, and political categories are needed to account for the face of technological change. A more fruitful approach to the process of technological change is to see it as a simultaneous interplay involving the technical logic of design and a host of economic, political, and cultural forces that constitute the society within which science and engineering is practiced. As a result of this interplay, myriad of imaginable technical possibilities are reduced to what the German historian H. D. Hillige calls the "techno-scientific problem-solving horizon."[43] This horizon demarcates the socially, economically, politically, and/or culturally determined directions to which the knowledge and experience of a technological branch can extend.[44]

Despite considerable scholarship in other countries on how the interplay of technical and non-technical forces shapes design choices, this area of research remains relatively unexplored in Canadian historiography. I have presented considerable evidence that the Canadian context did shape the "techno-scientific problem-solving horizon." For example, the patronage of innovation in digital electronics by the Canadian Post Office Department and by Trans-Canada Air Lines illustrates that the design of technology in Canada resulted from the historically specific interplay of the practitioners and the government. In each of these cases, a monopoly public enterprise guided the direction of industrial innovation as a means of furthering political goals. The Post Office turned to the digital revolution to deflect growing political criticism of its efficiency. Whereas American Airlines pursued SABRE under a free-enterprise system, Trans-Canada Air Lines turned to technological efficiency and computerization in an attempt to preserve its monopoly status. But the most striking evidence that artifacts are the material images of technical logic refracted through a political economy can be found in the Royal Canadian Navy's development of DATAR and Sperry Canada's Schumpeterian entry into numerical control and computer numerical control.

The circumstances of World War II had thrust Canada into a position of geopolitical prominence that otherwise would have been inaccessible to it. Whereas at the end of World War I the Canadian military had returned to its former oblivion, at the end of World War II the armed services were determined to retain much of their wartime status. At the outbreak of World War II, the Royal Canadian Navy's senior officers had dreamed of leading a conventional naval task force into battle. Unable to win political approval for the program to build aircraft carriers, battleships, and destroyers, the RCN found itself thrust into the less glorious role of escorting shipping convoys. At the war's end, the RCN's senior staff again aspired to command a conventional task force of the sort launched by world naval powers, albeit on a smaller scale. When Prime

Minister William Lyon Mackenzie King's anti-militarism and the budgetary constraints of returning to a peacetime economy scuttled its postwar aspirations, the RCN returned to its anti-submarine/escort mission and set about using it to create a new role of importance within the North Atlantic Triangle alliance. But its wartime experience in this area had taught the RCN that the lack of a domestic R&D capacity in this area would surely compromise its postwar aspirations of self-reliance and sovereignty.[45] Accordingly, while the British and US navies devoted their R&D energies to the future battle requirements of large naval task forces, the RCN sponsored the creation of a strong national R&D capability to assert its preeminence in anti-submarine warfare.

This quest to transform Canada's limited naval resources into a respected military presence within the North Atlantic Triangle alliance determined the contents of the RCN's entire R&D program, of which the most daring element was DATAR. On a technical level, DATAR represented perhaps the earliest comprehensive fusion of communications, computation, and control into a unified digital paradigm. DATAR's digital, decentralized, modular design for processing and exchanging information was specifically developed for the needs of Canada's small fleet of frigates engaged in submarine hunting rather than for the large naval task forces of Canada's more powerful allies. From 1948 through much of 1954, the RCN, through its DATAR project, was the leading exponent of the naval use of digital electronics. In its search for a sovereign voice within the postwar military order, the RCN pushed the internal logic of the nascent digital technology into a new design paradigm for automated naval warfare.

Sperry Canada's pursuit of NC and subsequent "invention" of CNC also adds compelling evidence of the importance of national contexts in delimiting the "techno-scientific problem-solving horizon." The UMAC-5 story in particular suggests that the dominant interpretation in the US literature of the social construction of machine tool automation must become more nuanced if it is to explain developments north of the 49th parallel (and perhaps those south of it). David Noble, in his landmark study on the origins of the American NC program, argued that the eventual triumph of the NC design approach to machine tool automation was based more on the exercise of power than on the merits of technical and economic rationalization.[46] Support for NC over the alternative "record and playback" (RP) technology, Noble argued, arose because the former eliminated the machinist's monopoly over skill. Noble concluded that NC's very design embodied an alliance among the US Air Force, industrial management, and academia whose purpose was to strip machinists of their power over the labor process by de-skilling them. Though both RP and NC produced a tape of instructions, which could then be used to automate production, in RP the tape was produced by recording the skilled worker's motions during the actual machining process, whereas in NC the tape was produced by a programmer removed from the shop floor.[47] Noble emphasizes that management's superior control of machinists, not technical efficiency or economics, determined the success of the NC over RP. Showing that design choices in machine tool automation were not purely technical and neutral, Noble's work provided the first comprehensive validation of Langdon Winner's provocative assertion that artifacts have politics.[48]

Forty years have passed since the introduction of NC, yet evidence of widespread de-skilling remains inconclusive.[49] But the power of Noble's discourse stems less from the eventual ramifications of design per se than from the ideological intent that shapes design decisions. Even if NC did not de-skill workers, was the early engineering development motivated by a Machiavellian desire to de-skill the machinist? Put differently: Was the imperative to wrest control of production from the machinist the dominant factor that shaped "techno-scientific problem-solving horizon" within which NC developed? Noble answers with an unequivocal Yes. The comprehensiveness, cogency, and passion of Noble's study of NC's development at the Massachusetts Institute of Technology should not, however, blind us to the possibility that the social construction of NC proceeded differently in other nations.[50]

In 1955, with the global state of the art of NC technology still in its infancy, Sperry Canada decided to develop its own product line in this area. The corporate rationale for such a move was clear: Diversify or perish. Sharing the technology-driven ethos of its parent firm, Sperry Canada naturally chose "radical product innovation." But much of its analysis as to why NC was an attractive product to develop was animated by a belief that NC was the ideal technology for the small-batch, highly fragmented nature of Canada's metalworking industry. The dominance of Fordism, with its specialized tools and high setup costs amortized over large production runs of one product, had turned Canadian small-batch manufacturing into a competitive handicap. Sperry Canada consistently asserted that the flexibility of NC would level the playing field. All the evidence clearly suggests that promoting labor displacement and de-skilling was never a consideration in Sperry Canada's design, testing, or marketing decisions. Sperry Canada's view that digital electronic technology would empower machinists becomes more apparent in view of how it went about designing the world's first stand-alone computerized NC system for machine tools.

Though Noble assigned malevolent intentions to the design of NC, he was ambivalent about CNC. He even suggested that CNC could potentially enhance the worker's skill. Yet his ideological convictions led him to conclude that management would never bow to such worker-centered approach to automation. Harley Shaiken (in the United States) and Barry Wilkinson (in Great Britain) also argued that the process of CNC's implementation is still motivated by management's desire to enhance its control over the production process.[51] In all the discussions of CNC, scholars on the left are silent as to how CNC was socially constructed. What is clear, at least in the Canadian case, is that de-skilling, to which Harry Braverman attributed a leading role to technological change in the capitalist countries,[52] did not provoke the invention of CNC at Sperry Canada, a subsidiary of the very sort of corporation that according to the Braverman-Noble school should have been bent on de-skilling workers.

Sperry Canada's invention of CNC was sparked, first and foremost, by the need to escape from the swarm of NC imitators and to create a new Schumpeterian niche. Technological imperatives, rather than any de-skilling agenda, were paramount in the company's decision to embed computational power directly in the machine tool. The central issue for Sperry Canada was how to overcome the costly hard-wire inflexibility associated with trying to match one NC system to different configurations of machine tools. Solving this problem was the key to creating a more exclusive market niche. However, once the programmable-computer approach was adopted, Sperry Canada's management and engineers consciously sought to design UMAC-5 so that it would expand the machinist's skills.

Sperry Canada's design team put a great deal of effort into making the use of UMAC-5 transparent to the machinist. UMAC-5 could run an NC tape prepared elsewhere, or the machinist could program UMAC-5 directly. Using a menu-driven push-button panel, the machinist could program UMAC-5 to perform complex sets of point-to-point machining operations. The design logic of the interface was to make the programming task easy for the machinist.[53] Furthermore, UMAC-5 allowed the machinist to prepare the tape by actually performing each machining operation rather than through formal programming. UMAC-5's ability to store the machinist's operations was the digital version of the earlier analog RP method.[54]

Sperry Canada, which had been created to manufacture sophisticated aircraft instrumentation and inertial navigational systems, had its roots in high-precision complex machining. From the outset, Sperry Canada envisaged CNC as a technology to enhance the capabilities of skilled machinists. Despite persistent detective work, I have been unable to find a machinist who actually used UMAC-5. It still remains to be seen if UMAC-5 was used the way it had been designed to be used. Whether the customers shared Sperry Canada's philosophy on UMAC-5's use is impossible to tell. Certainly in Canada, Sperry Canada's inability to sell its technology reflected the entrenched attitude in the metalworking industries that NC and CNC were cost-cutting strategies. In the eyes of these industries, the high price of the technology made it an un-economical approach to cutting costs. In other words, Canadian manufacturers may have acted more like the archetypal capitalist firm, as depicted by radical political economists such as David Noble and Harry Braverman, than did Sperry Canada, an American branch plant.

The case studies add new meaning to Bruce Sinclair, Norman Ball, and James Paterson's observation that "nothing more characterizes the Canadian situation than the continual selection of techniques from other societies. . . . The ability to adapt is just as critical as the ability to originate."[55] Most if not all of the inventions—ENIAC, Colossus, the transistor, pulse-coded modulation, even radar—that set the digital electronics revolution in motion originated from just two vertices of the North Atlantic Triangle: Great Britain and the United States. Nevertheless, Canada's early participation in this fourth wave of global innovation was marked by creative adaptation that often merged seamlessly into invention. The Royal Canadian Navy's digitally mediated self-reliance, the Post Office's quest for digital automation, Trans-Canada Air Lines' use of the digital paradigm to consolidate its monopoly status, the Defence Telecommunications Research Establishment's exploration of transistor design to cope with the crippling problems of circuit complexity, Ferranti-Canada's ingenious search for a product niche in a market dominated by giants, and Sperry Canada's technology-driven strategy of diversification collectively demonstrate that the process of design in Canada did indeed refract the international state of the art through a historically specific national context.

Learning by Doing and Canada's First Community of Digital Practitioners

In the end, the reader may choose to interpret this story of public enterprise's flirtation with technological self-reliance, of the Schumpeterian vitality of foreign-owned branch plants, and of daring and creative technological adaptation in Canada as a litany of failure. UTEC was canceled.

DATAR failed to go beyond the prototype stage, as did the Post Office's mail sorter. Though Trans-Canada Air Lines' ReserVec system proved an operational success, the technology was never sold anywhere else in the world. The FP6000 never became the basis of a domestic computer hardware industry. Sperry Canada was unable to turn UMAC-5 into a profitable business in computer numerical control. CDC-Canada's Development Division had to be shut down. Despite considerable government funding there were no lasting commercial successes.

To focus on failure, however, would miss the crucial process that underlay Canada's articulation of the fourth innovation wave: moving people up the learning curve. Although measures of technical and market success are very important, they can obscure a fundamental historical issue: the timely formation of the national core of scientific and engineering expertise needed to keep a small and open political economy, such as Canada's, responsive to the longer-term opportunities arising from profound global shifts in technology. Emphasizing R&D as process over R&D as product, I have underscored the creation of expertise.[56] Whereas the R&D-as-product view measures the success of R&D efforts strictly through a cost-benefit perspective, the R&D-as-process view focuses attention on the important opportunities for skill formation and networking. R&D as result is tied to information production. R&D as process, on the other hand, is about knowledge creation. The French economists Patrick Cohendet, Jean-Alain Héraud, and Ehud Zuscovitch argue that the received view of innovation as basically an informational phenomenon fails to recognize that "the development of new technologies is fundamentally a localized and path dependent learning process to which access is quite limited even when the general characteristics are publicly known."[57] Information may be freely available "off the shelf"; technical knowledge is not. Chris Freeman argues that, unless there is the industrial capacity first to assimilate and then to improve upon this publicly available know-how, it is unlikely to be used efficiently.[58] That capacity is acquired through "learning by doing."[59] Michael Porter's work also reinforces the importance of learning by doing in forging a nation's technological capacity. Underlying a nation's competitive advantage among nations is its ability to foster the formation of industrial engineering and management skill and the accumulation of technical knowledge.[60] Similarly, Edward Constant, in his work on the origins of the "turbojet revolution," suggests that a nation's "technological knowledge comprises traditions of practice which are properties of communities of technological practitioners."[61] In other words, a nation's accumulated technical knowledge resides not in the material artifacts but in an existing community of technical practitioners. From the perspective of the present book, learning by doing is essential if the continual emergence of a community of technological practitioners is to be sustained.

Once one understands that modern prosperity depends on the creation and sustenance of a community of technologists, rather than on the fate of individual products, then one can understand the true tragedy engendered by the cancellation of Canada's most expensive R&D project in the period 1945–1960: the AVRO Arrow jet interceptor. The tragedy lay not in the cancellation (for Canada could not afford the Arrow) but in the subsequent departure from the country of an entire community of technologists. In the case of the digital revolution, thanks to all the actors who have appeared in this story, young Canadian talent, instead of going down the road to the United States, found a domestic community of technologists and a tradition of technical practice they could join.

The genealogical lineage of Canada's "high-tech" industry makes us aware of how much knowledge, and of how many careers, were fostered by the "failed" technologies described in the preceding chapters. Consider Denzil Doyle, the acknowledged "dean" of the "Silicon Valley North" that emerged in the Ottawa-Carleton region in the 1970s and the 1980s.[62] Upon his graduation from Queen's University (with the Governor General's Medal for academic excellence), Doyle first found employment at Computer Devices of Canada, then at the Defence Research Telecommunications Establishment. Later, Doyle was responsible for the creation of Digital Equipment Canada (a branch plant of the Digital Equipment Corporation), and he was its first Chief Executive Officer. In time, Digital Canada won the corporation's world product mandate for personal computers and workstations. CAE, a Montreal company known for sophisticated flight simulators, offered another Canadian, Stephen Dorsey, the opportunity to return home after his graduation from the Massachusetts Institute of Technology in 1958. Dorsey became a leader in the design and commercialization of the first generation of stand-alone word processors. The two companies he created in Montreal, AES Data Ltd. and Micom, which sold word-processing systems around the world, then became new nodes from which the network of Canadian technologists could grow. I. P. Sharp & Associates, a spinoff from the FP6000 software team, developed into one of the largest international time-sharing networks. Not only did this company provide new opportunities for young Canadian hardware and software engineers; it also attracted the renowned Harvard University mathematician and computer scientist Kenneth Iverson back to Canada.[63] From the Consolidated Computer failure flowed the Dynalogic Corporation and its aggressive campaign to win the US personal computer market in the early 1980s.[64]

Mitel, Newbridge, and Corel, Canada's high-tech stars of the 1990s, also owe much of their existence to a community of technologists—forged with the help of public enterprise—that a product-oriented analysis might dismiss as yet another failure." This story involves Northern Electric (Nortel). In the early 1950s, Northern Electric had little R&D capacity to speak of. But the search for a national capacity to resolve the potentially crippling problems of complexity and reliability associated with large-scale circuits led the military to enter into close cooperation with Northern Electric to explore applications and manufacture of transistor circuits. This collaboration contributed to the growth of the Advanced Devices Centre at Northern Electric and to the creation of Canada's first industrial training ground for the manufacture of microchips. In 1969, with the help of a $37 million government subsidy, the Advanced Devices Centre spun off into Microsystems International, the precursor of Mitel, an internationally successful telecommunications firm (subsequently acquired by British Telecom) and the training ground for Michael Copeland and Terrence Mathews. After leaving Mitel, Copeland started Corel. Mathews formed Newbridge, which continues to spin off new high-tech companies.

Thus, far from being tragic tales of lost opportunities, the case studies in this book underscore the invaluable skill formation that took place as Canadians pooled the resources of the military, the universities, business, and the federal government to finance the ascent up the learning curve of an entire generation of technologists. In the process, Canadians learned that they could make a living without having to pillage their natural resources or to compete ruthlessly on price with the world's poorest peoples in the global commodity markets.

Notes

Introduction

1. For a suggestion that Canada flirted with military self-reliance in the period 1945–1959, see Keith Krause, *Arms and the State* (Cambridge University Press, 1992). See also Robert Van Steenburg, "An Analysis of Canadian-American Defence Economic Co-operation," in *Canada's Industrial Base*, ed. D. Haglund (Ronald Frye, 1988).

2. See Norman R. Ball and John N. Vardalas, *Ferranti-Packard* (McGill-Queen's University Press, 1994). That narrative, however, stresses the history of the company's dominant business activity: the manufacture of capital goods for the electrical power sector, in which utilities were the dominant customers. As a result, the discussion of Ferranti-Packard's digital innovations, though it touches on all the key artifacts, fails to situate the process of technological change within the broader political economic context of postwar Canada. Reconstructed almost entirely from the personal recollections of former company engineers, *Ferranti-Packard* does not examine the extent to which public enterprise profoundly shaped the rhythm and direction of digital innovation at Ferranti-Canada, a theme that is central to the present volume. The Ferranti-Canada story, as told in *Ferranti-Packard*, is reconceptualized here thanks to new and extensive research into the records of the Defence Research Board, the former Department of Defence Production, the Royal Canadian Navy, Canada Post, Air Canada, the Federal Reserve Bank of New York, Industry Canada, and the Science Council of Canada.

3. Joseph Schumpeter, *Business Cycles* (McGraw-Hill, 1939).

4. Within this context, economic development is a process of reallocation of resources between industries. This process automatically leads to structural changes and disequilibria as labor, capital, and credit flood into the emerging leading sector.

5. For the most recent comprehensive overview of the long-run time series used and the relative merits of the various quantitative data analyses used, see Joshua Goldstein, *Long Waves* (Yale University Press, 1988), chapters 8–12 and appendixes A and B. Another problem with Schumpeter's model is its apparent determinism;

on that, see Arnold Pacey, *The Culture of Technology* (MIT Press, 1983). Carlo Cippola, on the other hand, underscores the creative response of history" that imbues the Schumpeterian model (*Between Two Cultures*, Norton, 1991, p. 71). Thus, contrary to Pacey, individual human action does play a key role in the Schumpeter model. In response to strong arguments made by Schumpeter's critics, recent interpretations of Schumpeter's work, particularly in Europe, have added new balance. See Andrew Tylecote, *The Long Wave in the World Economy* (Routledge, 1991); Christopher Freeman, John Clark, and Luc Soete, eds., *Unemployment and Technical Innovation* (Greenwood, 1982); Christopher Freeman and Luc Soete, *The Economics of Industrial Innovation* (MIT Press, 1997).

6. Simon Kuznets ("Technological Innovations and Economic Growth," in *Technological Innovation*, ed. P. Kelly and M. Kranzberg, San Francisco Press, 1978) took exception to the idea that innovations could appear clustered over time. Gerhard Mensch, on the other hand, produced an empirical study of innovation clustering along with mechanisms to explain this phenomenon (*Stalemate in Technology*, Ballinger, 1979). While criticizing Schumpeter's heroic innovator, Mensch asserted that major innovations had clustered around depressions because they offered the only way out of a "technological stalemate" (Gerhard Mensch, Charles Coutinho, and Klaus Kaasch, "Changing Capital Values and the Propensity to Innovate," *Futures* 13, no. 3 (1981): 276–292). Edwin Mansfield ("Long Waves and Technological Innovation," *American Economic Review* 73, no. 2, 1983, p. 144) challenged the empirical basis of Mensch's work: "[Mensch's] evidence does not persuade me that the number of major technological innovations conforms to long waves of the sort indicated by his data. . . . The hypothesis that severe depressions trigger and accelerate innovations is also questionable." Nathan Rosenberg and Claudio Frischtak ("Technological Innovation and Long Waves," in *Design, Innovation and Long Cycles in Economic Development*, ed. C. Freeman, St. Martins, 1986) also did not see why a depression should provide a push into new product lines that were rejected under earlier, more favorable economic circumstances.

7. Freeman et al., *Unemployment and Technical Change*; Freeman and Soete, *The Economics of Industrial Innovation*.

8. Rod Coombes and Alfred Kleinecht ("New Evidence on the Shift toward Process Innovation during the Long-Wave Upswing," in *Design, Innovation and Long Cycles in Economic Development*, ed. Freeman) have further disaggregated the innovation classification. Whereas some innovations can be classified easily as pure "product" or "process," other innovations can fit both depending on the perspective one takes. For example, a computer numerical control system may be a product innovation for a producer of capital goods, while the final user who adopts this technology sees it as a labor-saving improvement. Their work, however, still corroborates the basic assertion of Freeman et al. that product innovations dominate the upswing and process innovations dominate the downswing.

9. Jacob Schmookler, *Invention and Economic Growth* (Harvard University Press, 1966), p. 208.

10. Freeman et al., *Unemployment and Technical Innovation*. See also Tylecote, *The Long Wave in the World Economy*.

11. Freeman et al., *Unemployment and Technical Change*. Soete ("Long Cycles and the International Diffusion of Technology," in *Design, Innovation and Long Cycles in Economic Development*) offers an interesting discussion of the interplay between a nation's attempt to monopolize its technological advantages and the dynamics of international innovation diffusion.

12. Soete, "Long Cycles and the International Diffusion of Technology"; Tylecote, *The Long Wave in the World Economy*.

13. Ironically, Harold Innis, a contemporary of Mackintosh, became an intellectual cornerstone of the new political economy. Glen Williams ("On Determining Canada's Location within the International Political Economy," *Studies in Political Economy* 25, 1988: 107–140) points out that Innis was recast in the leftist ideological discourse of the 1970s and the 1980s as a Canadian nationalist, an anti-continentalist, an anti-imperialist, and a forerunner of the dependency school.

14. Clark-Jones (*A Staple State*, University of Toronto Press, 1987) argues that the US military's search for assured access to strategic natural resources during the Cold War increased Canada's dependence on staple exports. This view is also argued by Paul Phillips and Stephen Watson ("From Mobilization to Continentalism," in *Modern Canada, 1930–1980s*, ed. M. Cross and G. Kealy, McClelland and Stewart, 1984).

15. Jim Laxer, "Canadian Manufacturing and US Trade Policy," in *Canada Ltd.*, ed. R. Laxer (McClelland and Stewart, 1973). In the foreword to *Canada Ltd.*, Robert Laxer emphasizes the centrality of the "de-industrialization" thesis both as a framework within which to understand Canada's historical development in the 20th century and as "a rallying cry for Canadians to join the movement for independence though socialism and socialism through independence."

16. Kari Levitt (*Silent Surrender*, Macmillan, 1970, p. 58) argues that "Canada provided a dramatic illustration of the stultification of an indigenous entrepreneurial class" by the dominance of the foreign-owned branch plant.

17. In *The Canadian Corporate Elite* (McClelland and Stewart, 1975), Wallace Clement argues that, in exchange for control over the distribution and financial aspects of the Canadian economy, Canada's capitalist elite had ceded control of industrial development to US multinationals.

18. Mel Watkins, "Economic Development in Canada," in *World Inequality*, ed. I. Wallerstein (Black Rose, 1975).

19. R. T. Naylor, *The History of Canadian Business* (Lorimer, 1975).

20. Clement, *The Canadian Corporate Elite*. See also Glen Williams, *Not for Export* (McClelland and Stewart, 1983). For a more recent vintage of the debate over Canada's inability to break the staples trap, see *The New Canadian Political Economy*, ed. W. Clement and G. Williams (McGill-Queen's University Press,

1989); *Perspectives on Canadian Economic Development*, ed. G. Laxer (Oxford University Press, 1991).

21. After reviewing ten new books on the "New Political Economy," Jack McLeod ("Paradigms and Political Economy," *Journal of Canadian Studies* 26, 1991): 172–181) was struck by the conspicuous absence of any concrete discussion of how the process of technological change fits into political economy. For McLeod, this state of affairs merely corroborated observations made by Richard Simeon ("Inside the MacDonald Commission," *Studies in Political Economy* 22, 1987: 167–179), who, as research coordinator for the MacDonald Commission, observed that technology was simply not an issue for economists. During the commission's hearings, economists denied that there was any evidence of a technological revolution, or any speedup of technical change, and they did not present any models of economic change that encompassed technology as an explicit model.

22. J. J. Brown, *Ideas in Exile* (McClelland and Stewart, 1967), p. 339.

23. Ibid.

24. Ibid., p. 340; Science Council of Canada, Forging the Links, Report 29, 1979. For the in-depth analysis underlying this report, see John Britton and James Gilmour, *The Weakest Link* (Science Council of Canada, 1978).

25. Cristian DeBresson, "Have Canadians Failed to Innovate? The Brown Thesis Revisited," *History of Science and Technology in Canada Bulletin* 6 (1982), January: 10–23.

26. That adaptation has been central to Canadian technological development was the central theme of *Let Us Be Honest and Modest*, ed. B. Sinclair et al. (Oxford University Press, 1974). "Nothing more strongly characterizes the Canadian situation," Sinclair et al. assert (p. 2), "than the continual selection of techniques from other societies. . . . The ability to adapt is just as critical as the ability to originate."

27. Chris DeBresson and Brent Murray, *The Supply and Use of Technological Innovation in Canada* (Cooperative Research Unit on Science and Technology, 1983), p. 203.

28. Science Council of Canada, Forging the Links.

29. Ibid.

30. The Science Council of Canada first employed the term "technological sovereignty" in its Eleventh Annual Report, titled Technological Sovereignty (1977).

31. For a critique of the Science Council of Canada's position on "technological sovereignty," see Kristian Palda, *The Science Council's Weakest Link* (Fraser Institute, 1979).

32. Science Council of Canada, A Trans-Canada Computer Communications Network, Report 13, 1971; Strategies of Development for the Canadian Computer Industry, Report 21, 1973.

33. See Merritt Roe Smith, "Introduction," in *Military Enterprise and Technological Change*, ed. M. Smith (MIT Press, 1987). This introduction offers a valuable overview of the main themes that have emerged, in American historiography, in the study of military enterprise in its relation to economic technological, and social change. The entire collection of essays looks at various institutional aspects of military enterprise and technical change.

34. Merritt Roe Smith, *Harper's Ferry Armory and the New Technology* (Cornell University Press, 1977).

35. For an extensive historical analysis of interchangeable parts and specialized machine tools, and the rise of mass-production economy, see David Hounshell, *From American System to Mass Production, 1800–1932* (Johns Hopkins University Press, 1984).

36. Lewis Mumford (*Technics and Civilization*, Harcourt Brace and World, 1934; *The Myth of the Machine*, Harcourt Brace and World, 1966) argued that standardization, uniformity, mass consumption, the necessity for regulated order, and large-scale coordination—all characteristics assigned to mass-production culture—have been inherent dimensions of military organization since ancient times.

37. William McNeill, *The Pursuit of Power* (University of Chicago Press, 1982).

38. Gordon Laxer, "Class, Nationality, and the Roots of the Branch Plant Economy," in *Perspectives on Canadian Economic Development*, ed. Laxer, pp. 256–257.

39. The following military historiographies all deal in some way with the period 1945–1958, but none of them devotes much space to the question of how the military shaped Canada's capacity to design, adapt, build and use technology: B. J. C. McKercher and Lawrence Aronsen, eds., *The North Atlantic Triangle in a Changing World* (University of Toronto Press, 1996); Marc Milner, ed., *Canadian Military History* (Copp Clark Pitman, 1993); Desmond Morton, *A Military History of Canada* (McClelland and Stewart, 1992). See also the following articles by Lawrence Aronsen: "Planning Canada's Economic Mobilization for War," *American Review of Canadian Studies* 15 (1985): 38–58; "From World War to Limited War," *Revue Internationale d'Histoire Militaire*, no. 51 (1982): 208–245; *Canada's Postwar Re-armament* (Canadian Historical Association, 1981), pp. 175–196. In addition, see James Eayrs, *In Defence of Canada* (University of Toronto Press, 1972).

40. David Zimmerman, *The Great Naval Battle of Ottawa* (University of Toronto Press, 1989).

41. Eayrs, *In Defence of Canada*.

42. Lawrence Aronsen, "From World War to Cold War," in *The North Atlantic Triangle in Changing World*, ed. McKercher and Aronsen.

43. Ibid.

44. Aronsen, *Canada's Postwar Re-armament*, p. 195.

45. J. L. Granatstein, "The American Influence on the Canadian Military, 1939–1963," *Canadian Military History* 2 (1993): 63–73.

46. J. L. Granatstein, *Canada 1957–1967* (McClelland and Stewart, 1986). In particular, see chapter 5, "The Defence Debacle." Though Granatstein uses the word "innovation" in the subtitle of his book (*The Years of Uncertainty and Innovation*), he rarely brings up technological change.

47. On the Arrow, see James Dow, *The Arrow* (James Lorimer, 1979); Palmiro Campagna, *Storms of Controversy* (Stoddart, 1992). The only other frequently discussed examples used to illustrate the impact of the United States' Cold War imperatives on Canadian technological development are the "early warning lines." For the most extensive discussion of the construction of the Mid-Canada and DEW lines, see Joseph Jockel, *No Boundaries Upstairs* (University of British Columbia Press, 1987).

48. Other than a small number of publications on the AVRO Arrow and the DEW Line, one finds in the Canadian historical literature little evidence of interest in the relationship between Canada's peacetime military and the process of technological change. A review of *Scientia Canadensis*, the official journal of the Canadian Science and Technology Historical Association, reveals no published work on this subject. Even in the scholarly journal *Canadian Military History* one does not find any attempt to study military enterprise as a force shaping Canada's peacetime technological development.

49. Morton, *A Military History of Canada*.

50. Krause, *Arms and the State*, p. 131. See also Van Steenburg, "An Analysis of Canadian-American Defence Economic Co-operation."

51. Krause, *Arms and the State*, pp. 31–33. The relationship between the three tiers of suppliers lies in the generation and international diffusion of innovations. Krause has used Melvin Kranzberg's ("The Technical Elements in International Technology Transfer," in *The Political Economy of International Technology Transfer*, ed. J. McIntyre and D. Papp, Quorum, 1986) typology of technology transfer to characterize these levels.

52. Desmond Morton's account of Canada's military buildup during World War I (*A Military History of Canada*, Hurtig, 1990) illustrates the jump to the fourth tier.

53. W. A. Thomson (Defence Research Board), Memorandum to Dr. Meek (Defence Research Board), 16 May 1958, Records of Canada's Department of National Defence, RG 24, Accession 83-84/167, box 7369, file DRBS-170-80/E5, pt. 2, National Archives of Canada, p. 4.

54. Jockel relates how in 1947 (7 years before the Americans started thinking about the DEW Line) Canadians had already crystallized a technology popularly known as the "McGill Fence," which was used in the construction of the Mid-

Canada Line. Though the construction the Mid-Canada Line was initiated by the United States, Jockel suggests that it would be a mistake to think that Canada was coerced into building it. For Jockel, the Mid-Canada line served Canadian interests at least as much as it served US interests.

55. John Bartlet Brebner, *The North Atlantic Triangle* (Columbia University Press, 1945). Despite the United States' emergence from World War II as the dominant economic power as well as Canada's largest trading power, Brebner's prewar idea of the North Atlantic Triangle did continue to have some relevance in the early postwar years. After all, Canada still retained strong political and military ties with Britain. See Edgar McInnis, *The Atlantic Triangle and the Cold War* (University of Toronto Press, 1959).

Chapter 1

1. Morton, *A Military History of Canada*, p. 176.

2. J. L. Granatstein, *The Generals* (Stoddart, 1993), p. 258.

3. Air Marshal Robert Leackie, Vice Admiral Jones, and Lieutenant-General Foulkes, Postwar Policy for Scientific Research for Defence, brief to Cabinet Committee on Research for Defence, 31 October 1945, Records of Canada's Department of National Defence, RG 24, volume 11,997, file DRBS-1-0-181, part 1. *(Henceforth, Records of Department of National Defence, RG 24, will be cited as DND.)*

4. J. M. Hitsman, Canada's Postwar Defence Policy, 1945–50, Report No. 90, Historical Section, Army Headquarters, March 1, 1961, DND, volume 6928, file 90.

5. O. M. Solandt, Policy and Plans for Defence Research in Canada, May 1946, Cabinet Document D-61. Records of Office of Privy Council of Canada, RG2, volume 2750, file Cabinet Defence Committee Documents, volume II. *(Henceforth, Records of Office of Privy Council of Canada, RG2, will be cited as OPC.)*

6. The only published discussion of the early computer work at the University of Toronto is Michael Williams, "UTEC and FERUT," *IEEE Annals of the History of Computing* 16 (1994), summer: 4–12. Williams, a computer science professor at the University of Calgary, has long had an interest in the history of computer technology. Over the years, we have shared sources and exchanged ideas on the early history of computers in Canada. He started working on this article as I was preparing this chapter. The intended focus of his well-written account is internalist. Consequently, he did not attempt to place UTEC in the broader political and economic context of postwar Canada. On the internalist aspects of the UTEC computer there is considerable overlap between this chapter and Williams's account. But because he set out to write an overview, Williams had less opportunity to explore the broader technical issues that shaped some of the key design issues that I explore in this chapter.

7. Leackie et al., Postwar Policy for Scientific Research for Defence, 31 October 1945, DND, volume 11,997, file DRBS-1-0-181, part 1.

8. Ibid.

9. These principles were not original to the Canadian peacetime military mind. In fact they were taken directly from Vannevar Bush's 26 January 1945 presentation to a Select Committee on Postwar Military Policy of the US Congress.

10. To drive home the unassailable logic of their message to the Cabinet Committee on Research for Defense, the Canadian Chiefs of Staff even cited the views of the vanquished: "These principles wholly, or in part, have also been stated as prerequisite by Albert Speer, Reich Minister of Armaments and War Production and Col.-Gen. Alfred Jodl, Chief of the German Supreme General Staff, in recent interrogation carried out in Germany." (Leackie et al., Postwar Policy for Scientific Research for Defence, 31 October 1945, DND, volume 11,997, file DRBS-1-0-181, part 1)

11. In the midst of World War II, John Brebner (*The North Atlantic Triangle*, Columbia University Press, 1945) coined the metaphor of the North Atlantic Triangle, which cast Canadian history in the context of its interplay with the United States and Great Britain. The historian Edgar McInnis, president of the Canadian Institute of International Affairs, took up the "triangle" metaphor to explain the forces shaping Canadian foreign and military policy in the Cold War; see McInnis, *The Atlantic Triangle and the Cold War* (University of Toronto Press, 1959), p. 12.

12. Leackie et al., Postwar Policy for Scientific Research for Defence, 31 October 1945, DND, volume 11,997, file DRBS-1-0-181, part 1, p. 4.

13. O. M. Solandt, Policy and Plans for Defence Research in Canada, May 1946, OPC, volume 2750, file: Cabinet Defence Committee Documents, volume II.

14. Leackie et al., Postwar Policy for Scientific Research for Defence, 31 October 1945, DND, volume 11,997, file DRBS-1-0-181, part 1, p. 4.

15. Ibid.

16. Eayrs, *In Defence of Canada*.

17. MacGregor Dawson, *The Conscription Crisis of 1944* (University of Toronto Press, 1961), p. 27.

18. Eayrs, *In Defence of Canada*, pp. 55–56.

19. Minutes of Cabinet Defence Committee, August 14, 1945, OPC, volume 2748, file: Cabinet Defence Committee Conclusions, volume 1; Minutes of Cabinet Defence Committee, 24 September 1945, OPC, volume 2748, file Cabinet Defence Committee Conclusions, volume 1.

20. J. W. Pickersgill and D. F. Foster, *The Mackenzie King Record, III, 1945–1946* (University of Toronto Press, 1970), p. 394.

21. Brooke Claxton, memorandum for the Cabinet: Defence Estimates, 4 February 1947, OPC, box 70, file D-19 (1947), Cabinet Document 389.

22. In an unpublished autobiographical sketch, Claxton wrote: "Unless the free countries built up their defences together we would go down the drain one by one. About this I had no doubt and my experience at the Paris Conference... had confirmed my conviction." (Eayrs, *In Defence of Canada*, p. 22)

23. D. J. Goodspeed's *History of the Defense Research Board of Canada* (Queen's Printer, 1958) remains the best work on the early history of the Defence Research Board.

24. Goodspeed, *History*, p. 2.

25. In absolute terms, DRB appropriations rose from $3.9 million to $19.8 million. The first figure is based on DRB budget data (ibid., p. 110); and the second is deduced from appendix C of D. J. Goodspeed, The Canadian Army, 1950–55: Part 1, Canadian Defence Policy, Report No. 93, Historical Section, Army Headquarters, June 1961, DND, volume 6928, file 93.

26. B. G. Ballard, letter to P. W. Nasmyth, 2 January 1948, DND, volume 4233, file DRBS-3-640-43.

27. Ibid.

28. Minutes of Second Meeting of Electronics Advisory Committee, 8 July 1947, DND, volume 4233, file DRBS-3-640-43.

29. CARDE, headquartered in Valcartier, Quebec, was the military agency charged with carrying out the R&D associated with many of Canada's advanced weapons projects, including guided missiles, bomb ballistics, super-high-velocity field guns, and rockets.

30. B. G. Ballard, letter to P. W. Nasmyth, 2 January 1948, DND, volume 4233, file DRBS-3-640-43. Of the three services, the Navy was the most vocal proponent of the use of electronic computers in battle and in battle simulators.

31. Dean S. Beatty, a professor of mathematics at the University of Toronto, first raised the idea in late 1945. See Summary of Activities of Committee on Computing Machines, 10 October 1947, attachment to letter to O. M. Solandt from B. A. Griffith dated 14 October 1947 (Calvin Carl Gotlieb Papers, B88-0069, box 001, file UTEC, University of Toronto Archives). *(Henceforth, this collection will be cited as Gotlieb.)* The members of this committee were W. J. Webber (chairman), B. A. Griffith (secretary), A. F. Stevenson, C. Barnes, and V. G. Smith

32. From 10 June to 31 June 1946, a combined University of Toronto and National Research Council Canada delegation visited all the key research facilities in high-speed, large-scale computation: MIT, IBM, Harvard University, Bell Laboratories, RCA Laboratories, the University of Pennsylvania, and the Institute of Advanced Studies at Princeton. See J. W. Hopkins, memo to file: Report on Visits in Boston and New York Areas in Collaboration with University of Toronto Committee on Computing Machines, 5 July 1946 (Record of National Research Council of Canada, RG 77, volume 134, file 17-15-1-20, National Archives of Canada, *henceforth cited as NRC*).

33. Preliminary Report on Modern Computing Machines, August 1946,

34. J. W. Hopkins, memo to file: Report on Visits in Boston and New York Areas in Collaboration with University of Toronto Committee on Computing Machines, 5 July 1946, p. 7. NRC, volume 134, file 17-15-1-20.

35. Preliminary Plans for a Proposed Computing Centre at the University of Toronto, appendix 2 in a letter from B. A. Griffith to O. M. Solandt, 14 October 1947, Gotlieb, box 001, file UTECS.

36. The committee's views reflected the views of noted US scientists who argued that the advancement of the physical sciences had become contingent on high-speed computation. See William Aspray, *John von Neumann and the Origins of Modern Computing* (MIT Press, 1992), p. 59.

37. Preliminary Plans for a Proposed Computing Centre at the University of Toronto, 14 October 1947.

38. E. G. Cullwick, letter to B. A. Griffith, 27 November 1947, Gotlieb, box 001, file UTECS.

39. Originally, DRB and NRC agreed to give $15,000 each, and agreed that DRB would puts aside $100,000 for the actual construction of the computer center. Because of administrative constraints, DRB could not set aside funds for some possible future use. Instead, NRC and DRB each agreed to provide $20,000 per annum.

40. Progress Report on the Proposed Computing Centre at the University of Toronto, 22 March 1948, prepared by B. A. Griffith for Sidney Smith, Gotlieb, box 001, file UTECS.

41. Ibid.

42. Minutes of Second Meeting of Electronics Advisory Committee, 8 July 1947, DND, volume 4233, file DRBS-3-640-43, p. 5.

43. Ibid.

44. B. G. Ballard, letter to D. R. Hartree, 30 April 1948, NRC, volume 52. file 17-28B-1, part 1.

45. L. Stiles, letter to E. C. Bullard, 23 December 1948, Gotlieb, box 1, file UTECS. Northern Electric acted as Bell Telephone Laboratories' Canadian representative in this matter.

46. C. J. Mackenzie, letter to W. B. Lewis, 23 February 1949, NRC, volume 52, file 17-28B-1, part 1. Lewis was included in this meeting because he was the Director of Research of NRC's atomic energy research project at Chalk River, which was to be one of the university's main customers for computations.

47. W. H. Barton (secretary, Standing Committee on Extra-Mural Research, Defence Research Board), letter to E. C. Bullard, University of Toronto, 22 April 1949, Gotlieb, box 001, file UTECS.

48. This information was taken from "Existing and Practically Completed Digital Computing Machines Developed by or Presently on Loan or Contract to the Government," *Mathematical Tables and Other Aids to Computation* 2 (1946–47), p. 237. In Japan, relay computers played an important role into the early 1950s. See Ryota Suekane, "Early History of Computing in Japan," in *A History of Computing in the Twentieth Century*, ed. N. Metropolis et al. (Academic Press, 1980). At the same time that Canada canceled plans to build a relay computer, Sweden decided that its first large-scale computer should be based on relay technology. Called BARK (Binar Automatisk Rela Kalkylator), this computer was designed, built and tested in little more than a year (Goran Kjellberg and Gosta Neovius, "The BARK, A Swedish General Purpose Relay Computer," *Mathematical Tables and Other Aids to Computation* 5, 1951: 29–34).

49. *Digital Computer Newsletter* (Office of Naval Research, Mathematical Sciences Division) 3 (1951), no. 3, p. 2

50. Ibid. The Mark III's first set of computations concerned the development of fire-control systems.

51. L. Alt, "A Bell Telephone Laboratories' Computing Machine: Part I," *Mathematical Tables and Other Aids to Computation* (1948–49), p. 84. The Bell relay computer installed at the Aberdeen Proving Ground, however, was bigger than the one proposed for the University of Toronto. The former used 9000 relays; the latter called for 3000.

52. The university's confidence in its own abilities may have reinforced DRB's arguments. Two months earlier, the Committee on Computing Machines stated that only a year of experimentation in component design would be needed before its team's could start construction of a full-scale computer "without serious risk of major error in policy" (E. C. Bullard, Application for as Grant for Research, 27 January 1949, NRC, volume 52, file 17-28B-1, part 1).

53. N. L. Kusters had been one of the two representatives whom NRC selected to participate in the mission to the US organized, in the summer of 1946 by the University of Toronto Committee on Computing Machines, to learn about the state of computer developments.

54. N. L. Kusters, memorandum to B. G. Ballard, 13 June 1949. NRC, volume 134, file 17-15-1-20.

55. There is no evidence in either the University of Toronto, National Research Council of Canada, Defence Research Board, Institute of Advanced Studies, or in the John von Neumann archives that the University of Toronto team ever entered into any serious discussions with von Neumann over the question of copying the IAS computer.

56. Sometime in the 1950s Josef Katz had his name legally changed to Kates. In the text I will refer to him by his later name, even in the period before he changed his name. Only in the notes will the distinction be maintained when appropriate.

57. In time, students Casciato, Richardson, and Stein were added to the team.

58. Progress Report as of October 1, 1949, NRC, volume 52, file 17-28B-1, volume 1.

59. The notion of reverse salients plays a central role in Thomas Hughes's historical analysis of technological change "The Evolution of Large Technological Systems," in *The Social Construction of Technological Systems*, ed. W. Bijker et al. (MIT Press, 1990). See also Hughes, *Networks of Power* (Johns Hopkins University Press, 1983).

60. These figures for the IBM702 and the UNIVAC were taken from a detailed 1955 survey of commercially available computers. See John Carroll, "Electronic Computers for the Businessman," *Electronics* 28 (1955), June: 122–131.

61. The conventional approach of the day called for about six triodes for each one-bit adder circuit. Hence in a parallel machine, 240 triodes were needed just to add two 40-bit numbers. Kates argued that his proposed smaller tube could replace the six larger triodes. Josef Kates, UTECS Report No. 27: Special Tubes, 7 September 1949, Records of Canadian Patents and Development Limited, RG 121, volume 40, file L14-1220-1, part 1, National Archives of Canada. *(Henceforth, Records of Canadian Patents and Development Limited, RG 121, will be cited as CPDL.)*

62. V. G. Smith, letter to B. G. Ballard, 24 February 24, 1949, NRC, volume 52, file 17-28B-1, part 1.

63. V. G. Smith, letter to B. G. Ballard, 13 April 1949, NRC, volume 52, file 17-28B-1, part 1.

64. Josef Katz, A Binary Adder Tube, 21 May 1949, CPDL, volume 40, file L14-1220-1, part 1.

65. Redhead (letter to Josef Katz, 10 June 1949, CPDL, volume 40, file L14-1220-1, part 1) apologized that, because of "very unskilled labour" and the absence of jigs, "the concentricity of the digital plates leaves much to be desired." He also mentioned the great difficulty in "sealing the bulb to the stem. . . . It cracked all over the place; but we were able to save the tube."

66. In addition to sharing the same advantages of the binary adder tube, the binary subtractor tube allowed for direct subtraction without the additional time and circuitry needed for complementation. Josef Kates, UTECS Report No. 27: Special Tubes, 7 September 1949, CPDL, volume 40, file L14-1220-1, part 1.

67. Josef Kates, letter to P. A. Redhead, 15 June 1949, NRC, volume 52, file 17-28B-1, part 1. A month after his letter to Redhead, Kates produced a design for multiplier tube. Josef Kates, UTECS Report No. 23: Special Switch Tubes, 14 July 1949, CPDL, volume 40, file L14-1220-1, part 1.

68. The interplay of vacuum tube and transistor interests in the early days of computer technology is examined in chapter 3, where the role of the military in the promotion of solid-state technology is discussed in the context of the Defence Research Telecommunications Establishment Computer.

69. The companies were Northern Electric, Canadian Marconi, Radio Valve, Canadian General Electric, Rogers Majestic, and Canadian Westinghouse. B. G. Ballard, letter to P. A. Redhead, 22 June 1949, NRC, volume 52, file 17-28B-1, part 1.

70. J. W. MacDonald, letter to E. C. Bullard, 15 October 1949, CPDL, volume 40, file L14-1220-1, part 1.

71. Ruben Hadekel, Improvements in or relating to Telegraph Apparatus, British Patent no. 549,358, filed 1 April 1941, issued 18 November 1942.

72. Montford Morrison, High Frequency Circuit Selector, US Patent no. 1,977,398, filed 31 May 1930, issued 16 October 1934. Albert Skellett (assigned to Bell Telephone Laboratories), Electronic Discharge Apparatus, US Patent no. 2,217,774, filed 27 May 1939, issued 15 October 1940.

73. Alfred Stuart (assigned to Bendix Aviation Corporation), System of Frequency Conversion, US Patent 2,293,368, filed 20 June 1940, issued 18 August 1942.

74. George Morton and Leslie Flory (assigned to Radio Corporation of America), Electronic Computing Device, US Patent no. 2,446,945, filed 25 August 1942, issued August 10, 1948. Richard Snyder and Jan Rajchman (assigned to Radio Corporation of America), Calculating Device, US Patent no. 2,424,289, filed 30 July 1943, issued 22 July 1947. Jan Rajchman (assigned to Radio Corporation of America), Electronic Discharge Device, US Patent no. 2,494,670, filed 26 April 1946, granted 17 January 1950.

75. E. C. Bullard, letter to J. W. MacDonald, 7 November 1949, CPDL, volume 40, file L14-1229-1, part 1.

76. J. W. MacDonald, letter to E. C. Bullard, 18 November 1949, CPDL, volume 40, file L14-1220-1, part 1.

77. In documents sent to Smart & Biggar, from the US patent examiner, 17 January 1951 is given as the filing date of Kates' patent Electronic Vacuum Tube. See CPDL, volume 40, file L14-1220-1, part 2.

78. DATAR is discussed the subject of the next chapter.

79. The three types of Switching Tubes were a Double Beam Binary Adder Tube, with a rod construction T61/2 miniature envelop and 9 pin miniature base; a Double Beam Multipurpose Switch Tube, with a T61/2 envelop; and A Double Beam Binary Adder Tube with a T3 subminiature envelope and bottom base circular lead arrangement. See M. L. Card, ESSC memorandum: Information to DRB Liaison Officers concerning addition of Katz's Switch Tube to DRB Contract X-17, 30 March 1950, Gotlieb, box 001, file UTECS.

80. W. B. Lewis, memorandum to Director of Electrical Research, Defence Research Board, 3 January 1950, NRC, volume 52, file 17-28B-1, part 1.

81. Card, ESSC memorandum: concerning addition of Katz's Switch Tub, 30 March 1950.

82. Jan Rajchman, letter to J. Katz, 22 March 1949, Gotlieb, box 001, file UTECS. Ironically, the American patent examiner tried to use Rajchman's prior patents to disqualify certain claims in the Kates application.

83. Lewis, memorandum to Director of Electrical Research, 3 January 1950.

84. Minutes of 141st Meeting of Committee on Patents, 19 December 1956, 10, CPDL, volume 40, file L14-1220-3.

85. Rogers Majestic, Bertie the Brain (promotional pamphlet, circa 1950, in the possession of John Vardalas). On 13 March 1956, the *New York Times* noted: "The Bell Telephone Laboratories received a patent this week for a machine that plays tick-tack-toe. The machine was invented by William Kesiter, director of switching engineering at Bell."

86. During the early development work on the IAS computer many questioned the wisdom of pushing the existing technology by going parallel (Julian Bigelow, "Computer Development at the Institute for Advanced Study," in *A History of Computing in the Twentieth Century*, ed. Metropolis et al., p. 293).

87. For a period of time von Neumann worked as a consultant for IBM. See Aspray, *John von Neumann*.

88. Cuthbert Hurd, "Computer Development at IBM," in *A History of Computing in the Twentieth Century*, ed. Metropolis et al.

89. For list of the nineteen IBM 701 customers see Cuthbert Hurd, "Early IBM Computers," *Annals of the History of Computing* 3 (1981), April: 163–182. This article is an edited version of testimony that Cuthbert Hurd gave in US Federal Court, Southern New York District, in the 1979 antitrust trial of the US Department of Justice versus IBM.

90. Franklin Fisher et al., *IBM and the US Data Processing Industry* (Praeger, 1983), p. 17.

91. F. C. Williams and T. Kilburn, "A Storage for the Use With Digital Computing Machines," *Journal of the Institute of Electrical Engineers* 96 (1949), April: 183–200.

92. S. Lavington, *Early British Computers* (Manchester University Press, 1980).

93. Josef Katz, Storage by Means of Space Charge (UTECS Report 5, February 1950), Gotlieb, box 001, file UTECS.

94. Interview with C. C. Gotlieb, 29 June 1971, for Computer Oral History Project, Smithsonian Institution.

95. In Computation Centre Progress Report, April 1, 1951 to June 30, 1951, July 1950, one reads that "the theory of storage systems was further investigated; and a series of experiments, conducted by Mr. Kates and Mr. Richardson at Los Alamos, has confirmed the theory of space charge and pointed in the direction of more reliable storage" (Sidney Smith (president of University of Toronto)

Papers, A68-0007, box 110, file 04, University of Toronto Archives). *(Henceforth, these papers will be cited as Smith.)*

96. N. Metropolis, "The MANIAC," in *A History of Computing in the Twentieth Century*, ed. Metropolis et al., p. 461.

97. In the autumn of 1951, Gotlieb pointed to the success of the Los Alamos electrostatic memory as evidence that the University of Toronto computer group was on the right track (Computation Centre Progress report: October 1, 1950 to September 30, 1951, Gotlieb, box 001, file UTECS).

98. In all the documents John von Neumann's papers at the US Library of Congress, the Institute of Advanced Study archives in Princeton, and the University of Toronto there is not one shred of evidence to suggest that the UTEC group ever tried to establish a close working relationship with von Neumann's computer group. In fact, one cannot find any reference to the Canadian group in the papers of the first two collections. The converse is also true.

99. News that a one-digit stage memory had been built and tested with satisfactory results appeared in Computation Centre Progress Report: April 1 to June 30 1950, Smith, box 110, file 04. News that a "twelve digit memory for the model was assembled, along with the associated circuits consisting of the storage and location registers, the deflection generator, and reading and writing gates" was reported in Computation Centre Progress Report: October 1, 1950 to September 30, 1951 (Gotlieb, box 001, file UTECS).

100. Ibid.

101. Alfred Ratz, The University of Toronto Full Scale Computer, 20 September 1951, Gotlieb, box 001, file UTECS, p. 7. University of Toronto Archives. This report outlined the technical, manpower, and financial requirements needed to go on and build a full-scale computer. By the time the small experimental model, UTEC Mark I, had been built, the team had grown to six research engineers and four technicians. The engineers were L. Casciato, R. F. Johnston, J. Kates, A. Ratz, R. Stasior, and H. Stein.

102. Construction Schedule: U. of T. Computer, November 16, 1951, Gotlieb, box 001, file UTECS. This document is a graphical depiction of the timetable for the development of the various subsystems of the computer.

103. Lavington, Early British Computers.

104. Geoffrey Tweedale, "A Manchester Computer Pioneer," *IEEE Annals of the History of Computing* 15 (1993), no. 3: 37–43.

105. Lewis and Bowden had both worked in the Cavendish Laboratory, at Cambridge University, during the 1930s. During the war Bowden worked on radar at Britain's Telecommunication Research Establishment, where Lewis occupied the position of Chief Superintendent.

106. W. B. Lewis, letter to K. F. Tupper, 3 August 1951, Gotlieb, box 001, file UTECS. In this letter, Lewis gives several lengthy extracts from the letter he received from Bowden.

107. Little did Lewis know that Ferranti was desperate for a buyer for the Ferranti Mark I, because its first commercial sale to Britain's Atomic Energy Authority had already fallen through. During the computer's construction, the newly elected Conservative government put a freeze on all large capital expenditures. This left Ferranti's chief computer salesman, Bowden, with a computer and no buyer. As to the other impending sales, Ferranti only managed to deliver one computer in 1953 in addition to the first Ferranti Mark I which went to the University of Manchester.

108. C. J. Mackenzie, letter to K. F. Tupper, 26 December 1951, C. J. Mackenzie (president of National Research Council of Canada) Papers, MG 30 B 122, box 3, file: Correspondence 1951. *(Henceforth these papers will cited as Mackenzie.)*

109. By the end of 1951, the University of Toronto Computation Centre was still using IBM602-A relay-based multipliers. See item 3, Minutes of 5 October 1951 Meeting of Computation Centre Committee, Gotlieb, box 001, file UTECS

110. Cited in Robert Bothwell, *Nucleus* (University of Toronto Press, 1988).

111. See item 4, Minutes of 5 October 1951 Meeting of Computation Centre Committee.

112. C. J. Mackenzie, letter to K. F. Tupper, 5 December 1951, Mackenzie, box 3, file Correspondence 1951.

113. Ibid.

114. DRB had already approved $300,000 for design, construction, and testing of the full-scale computer.

115. This was how C. J. Mackenzie recounted his conversation with O. Solandt to K. F. Tupper in Mackenzie, letter to K. F. Tupper, 5 December 1951.

116. Minutes of Seventeenth Meeting of DRB Committee on Extra-mural Research, 8 December 1951, 2, DND, volume 11,996, file DRBS-1-0-43-2, part 5.

117. Minutes: Meeting of Computation Centre Committee, 11 January 1952, Gotlieb, box 001, file UTEC.

118. Ibid.

119. K. F. Tupper, letter to C. J. Mackenzie, 12 January 1952, Gotlieb, box 001, file UTECS.

120. A. G. Ratz, A Two-Year Plan for the Electronic Section, 20 February 1952, Gotlieb, box 001, file UTECS.

121. These were Tupper's words (memorandum—Conference: Computation Centre, 20 February 1952, Smith, box 110, file 04). At this conference were

Sidney Smith (president of the University of Toronto), K. F. Tupper (dean of engineering), Andrew Gordon (head of the chemistry department and a member of the DRB and NRC boards), and W. Watson.

122. W. B. Lewis, letter to Dean K. F. Tupper, 11 March 1952, Smith, box 110, file 04.

123. K. F. Tupper, letter to Sidney Smith, 13 March 1952, Smith, box 110, file 04.

124. Bothwell, *Nucleus*, p. 108.

125. The computer they proposed to build, called the Digitron Model-S, was to fit in two office desks. Taking their cue from the first UNIVAC computer, Casciato, Kates, and Johnston proposed in their business plan, to build a decimal-based machine because it would somehow seem less alien to business users than a binary one. As with UTEC they wanted to build a parallel machine using electrostatic memory. The internal memory was to hold 500 10-decimal-digit words, and mass storage was to consist of a 10,000-word memory drum. To simplify the arduous task of programming, they wanted to build in a large number of general-purpose registers. Taken from notes on a prospectus and business plan of the proposed company Digitronics. Josef Kates gave this document to me.

126. Ibid.

127. Kates was the aggressive marketer of the group and the man who shaped most of the business plan.

128. The British American Oil Company paid KCS $30,000 a month to design, install and run a computerized planning and scheduling allocation system (interview with Josef Kates, 15 August 1985, for National Museum of Science and Technology, Ottawa).

129. The technical documentation of this work can be found in two large studies done for Metropolitan Toronto. See KCS Data Control Ltd., A Centrally Controlled Traffic Signal System for Metropolitan Toronto (1959) and The Control of Traffic Signals in Metropolitan Toronto with an Electronic Computer (1961). I am in possession of copies of these studies. The originals still remain with Casciato. For a historical overview of the project, see John Vardalas and Ted Paull, "The Toronto Traffic Signal System," *Canadian Information Processing Society Review* 18 (1987), March-April: 34–39.

130. "The most sophisticated of Teleride's systems employ electronic odometers on each bus to measure its precise progress along the route. The bus' location is radioed every thirty seconds to the central computer, which calculates how long it will take for each vehicle to reach the next stop. . . . Now riders in ten North American cities can call the computer to learn the next scheduled arrival at their stop." (David Thomas, *Knights of the New Technology*, Key Porter Books, 1983, p. 84)

131. L. Casciato and H. White, Monitoring System for Vehicles, US Patent no. 4,009,375, issued 22 February 1977.

132. See also C. C. Gotlieb, "FERUT—The First Operational Electronic Computer in Canada," in Proceedings of the Canadian Information Processing

Society, 9–11 May 1984, Calgary. These papers as well as numerous others on FERUT can be found in the Gotlieb Collection at the University of Toronto Archives. Another source is Michael Williams's interviews with C. C. Gotlieb, 29–30 April 1992; these are available through Williams at the University of Calgary's Department of Computer Science.

133. Interview with Gotlieb, 29 July 1971. For a detailed retrospective of the calculations see H. W. Lea, "Hydraulic Problems In Connection With the Development of the St. Lawrence River," *Engineering Journal* 43, no. 2 (1960): 50–54; H. M. McFarlane, "Backwater Computations for the St. Lawrence Power Project," *Engineering Journal* 43, no. 2 (1960): 55–60; C. C. Gotlieb, "Backwater Computations for the St. Lawrence Power Project: Calculations on the Ferranti Digital Computer," *Engineering Journal* 43, no. 2 (1960): 61–66.

134. It appears that the spatial distribution of computational power later became a strategic concern in the event of war, because of the issue of vulnerability to a first strike. J. Jablonski (Operational Research Group, Defence Research Board), memorandum to Emergency Measures Organization, Privy Council Office: Location of Computers, 28 March 1960, DND, accession 83-84/167, box 7589, file DRB 9900-31, part 1.

Chapter 2

1. Zimmerman, *The Great Naval Battle of Ottawa*. ASDIC (named after the Anti-Submarine Detection Investigation Committee) used acoustical waves to detect submarines. The US Navy called it SONAR.

2. Gerhard Weinberg, *A World At Arms* (Cambridge University Press, 1994).

3. W. G. D. Lund, "The Royal Canadian Navy's Quest for Autonomy in the North West Atlantic, 1941–43," in *The RCN in Retrospect, 1910–1968*, ed. J. Boutilier (University of British Columbia, 1982), p. 39.

4. Weinberg, *A World At Arms*, p. 380.

5. Ibid.

6. For more details about Canada's heroic contribution to the protection of convoys against submarines see W. H. Pugsley, *Sailor Remembers* (Collins, 1948); G. N. Tucker, *The Naval Service of Canada*, volume 2 (Kings Printer, 1952); James Lamb, *The Corvette Navy* (Macmillan, 1977); John Swettenham, *Canada's Atlantic War* (Samuel-Stevens, 1979); Marc Milner, *North Atlantic Run* (University of Toronto Press, 1985); and Marc Milner, *The U-Boat Hunters* (University of Toronto Press, 1994).

7. Lund, *The Royal Canadian Navy's Quest*.

8. Ibid., p. 147.

9. Ibid., p. 157.

10. Minutes of 24th Meeting of Research Control Committee, item 24-6, The Future of A/S Research, item 24-6, 5 September 1946, DND, accession 83-84/167, box 139, file NSS-1279-33, part 2.

11. The chain started with the radar or sonar detection of a target. The range and bearing of the target was then passed on by telephone from the Reporter to the Plotter, who would then place the information, with a wax pencil, on the back side of a vertical transparent plotting board. The Air Plot Officer, with a wax pencil, would then filter this information onto the other side of the plotting board. The Direction Officer and Gunnery Liaison Officer then decided whether to engage the target with guns or fighters. The Gunnery Liaison Officer then passed the approximate bearing and speed to another person sitting in the Gun Direction Room. Then the Gunnery Direction Officer would pass on target identification information to the TIU Operator who then tells the director on what bearing to start pointing the gun.

12. Deputy Minister of Defence, memorandum to Minister of Defence, June 1950, DND, accession 83-84/167, box 3108, file NSS-7401-382-1.

13. Ibid.

14. Richard Snyder and Jan Rajchman, "Calculating Device," US Patent no. 2,424,289, filed 30 July 1943, issued 22 July 1947, p. 1.

15. Lieutenant-General Charles Foulkes (Chief of the General Staff) to Minister of National Defence, memorandum on Canadian Economic Mobilization, 10 February 1947, DND, volume 19,172, file 2130-30/3.

16. Ibid.

17. Ibid., p. 2.

18. Ibid., p. 6.

19. Minister of National Defence, memorandum to Chief of General Staff, 17 February, 1947, DND, volume 19,172, file 2130-30/3.

20. O. M. Solandt, Brief on Research, Defence and Industry, 28 March 1948, p. 5. DND, box 11995, file DRBS-1-0-43.

21. Of course there was no guarantee that Canada would be given an effective voice even if it did try to carry its fair share. World War II was ample proof of this possibility.

22. Solandt, Brief on Research, Defence and Industry, 28 March 1948, DND, box 11995, file DRBS-1-0-43.

23. Ibid.

24. Minutes of First Meeting of Electronics Advisory Committee, 22 May 1947, DND, box 4233, file DRBS-3-640-43.

25. Memorandum to secretary, Joint Communications Committee: Canadian Armed Services Electronic Research Requirements, June 19, 1947, 2, DND, box 4233, file DRBS-3-640-43.

26. Ibid., appendix B.

27. See item 3, Minutes of Seventh Meeting of Electronics Advisory Committee, 28 April 1948, DND, volume 4233, file DRBS-3-640-43.

28. The "primary" group were considered electronic and communications companies which had facilities for applied research, development and production. The secondary group embraced "those who are not primarily concerned with electronics and communications development and manufacturing. Minutes of Eight Meeting of Electronics Advisory Committee, 15 June 1948, appendix A, DND, box 4233, file DRBS-640-43.

29. Ibid., item 26.

30. Minutes of Ninth Meeting of Electronics Advisory Committee, 10 August 1948 and Minutes of Tenth Meeting of Electronics Advisory Committee, 14 September 1948, DND, volume 4233, file-3640-43.

31. Minutes of Ninth Meeting of Electronics Advisory Committee, 10 August 1948, DND, volume 4233, file DRBS-3-640-43.

32. Ball and Vardalas, *Ferranti-Packard*.

33. Minutes of Ninth Meeting of Electronics Advisory Committee, 10 August 1948, appendix A.

34. O. Solandt, letter to Vincent Ziani de Ferranti, 7 January 1949, Ferranti plc Archives, Manchester. E. G. Cullwick, however, continued to champion the Ferranti cause within DRB. E. G. Cullwick, letter to Sir Vincent Z. de Ferranti, 6 May 1949, Ferranti plc Archives, Manchester.

35. Letter from A. B. Cooper to Vincent Ziani de Ferranti, 7 February 1949 (Ferranti plc Archives, Manchester).

36. DRB only had $200,000 for the total industrial research program. See item 4. 9, Minutes of Tenth Meeting of Electronics Advisory Committee, 14 September 1948.

37. Letter from Cooper to Ferranti.

38. R. H. Davies (Ferranti Electric Inc., New York), letter to Vincent Ziani de Ferranti, 14 February 1949. Ferranti plc archives, Manchester.

39. Ibid., p. 4.

40. National economic policy also shaped DRB's rejection of the Ferranti UK offer. DRB was under considerable pressure from the Canadian electronics manufacturers to support an industry hit hard by demobilization. Canadian Business

Service Limited, The Electronics Industry in Canada, prepared for the Royal Commission on Canada's Economic Prospects, April 1956, p. 37. In this kind of economic climate, DRB felt that it would be counterproductive to encourage any R&D in companies that did not have an existing capacity to translate prototype design into actual production. Minutes of Ninth Meeting of Electronics Advisory Committee, Defence Research Board, 10 August 1948, item 2.3.5.

41. Minutes of Eleventh Meeting of Electronics Advisory Committee, 23 November 1948, p. 6.

42. Ibid.; "Present Government Development Work Coming Within The Scope Of The Canadian Defence Research Board's Requirements," September 1948, Ferranti plc Archives, Manchester.

43. Eric Grundy, memorandum to Sir Vincent de Ferranti: Research in Electronics Industry, Canada, 20 September 1948, Ferranti plc Archives, Manchester.

44. Ibid.

45. Lavington, *Early British Computers*. The contract was for an estimated £35,000 per annum over 5 years.

46. Kenyon Taylor, Autobiographical Notes, 15 February 1970 (Ferranti-Packard Archives, Provincial Archives of Ontario).

47. "The contract position has been somewhat confused on account of adequate funds not having been made available for this project during this financial year, we have been kept going on funds originally intended for other projects—this accounts for the 'bittyness' of the contracts." Arthur Porter, Preliminary Report, 15 October 1949. Ferranti plc Archives, Manchester England.

48. The first three contracts were issued on 18 November 1948 and were each worth $4000. The fourth was issued on 5 December 1948 and was for $5000. See Chairman, Defence Research Board, letter to R. B. Bryce, secretary of Treasury Board, 18 December 1952, DND, accession 83-84/167, box 3108, file NSS-7401-382-1.

49. The team consisted of M. K. Taylor, D. F. Walker, J. F. Harben and G. Ross. *Ferranti Journal* (March 1949), p. 25.

50. K. G. Ponting, "The Textile Innovations of Sebastian Ziani de Ferranti," *Textile History* 4. Exact year and issue number not indicated on reprint.

51. Kenyon Taylor, Dynamic Balancing While in Motion by Means of Weight Spirals, British Patent no. 399,845, application date January 1932.

52. Solandt expressed this view to Arthur Porter in October 1949 (ibid.).

53. See Chairman, DRB, letter to Bryce, 18 December 1952, DND, accession 83-84/167, box 3108, file NSS-7401-382-1.

54. Reference to date and money made in item 1 of attachment to memorandum to Minister of Defence Production, 25 July 1951, DND, accession 83-84/167, box 3108, file NSS-7401-382-1.

55. Arthur Porter, Preliminary Report, 15 October 1949.

56. Computer Devices of Canada was created expressly to design and build a real-time interactive digital simulator to train officers in anti-submarine warfare. The "Tactical Battle Simulator" was to be the first of its kind in the world.

57. Captain W. H. G. Roger (Electrical Engineer-in-Chief), memorandum to Chief of Naval Tactical Services, 5 April 1950, DND, accession 83-84/167, box 3108, file NSS-7401-382-1. The Research Control Committee was established in 1944 to coordinate the RCN's R&D programs. Funding approval for major R&D initiatives needed this Committee's approval.

58. Item 40-1, DATAR Development Contracts, in Minutes of Research Control Committee, 14 April 1950, DND, accession 83-84/167, box 139, file NSS-1279-33, part 2.

59. Minutes of Research Control Committee, 17 June 1950, DND, accession 83-84/167, box 139, file NSS-1279-33, part 2.

60. Deputy Minister of Defence, memorandum to Minister of Defence, June 1950, DND, accession 83-84/167, box 3108, file NSS-7401-382-1.

61. Captain W. H. G. Roger (Electrical Engineer-in-Chief), memorandum to Chief of Naval Tactical Services, 9 January 1950, DND, accession 83-84/167, box 3108, file NSS-7401-382-1. In addition, Ferranti-Canada wanted patent rights concessions similar to those granted by the British Ministry of Supply to the parent firm in order to pursue any future non-military commercial options.

62. Of the $279,000, only $189,000 was designated to Ferranti-Canada for the new 13-month engineering study.

63. By this point, Roger's rather unrealistic original estimate that $1.5 million was needed to get a production version of DATAR ready had jumped to $3.2 million.

64. In the end, the Tactical Battle Simulator turned out to be the RCN's single most expensive digital electronic project. Whether it deserves the label of pioneering effort or fiasco remains an unanswered question.

65. Vivian Bowden, letter to Arthur Porter, 4 January 1951, Ferranti-Packard Collection, Provincial Archives of Ontario.

66. Ironically, it was the sale of Ferranti UK computer to the University of Toronto that accelerated the demise of UTEC.

67. Taylor remained the Department's hands-on engineering mentor. In later years, Porter created the Department of Industrial Engineering at the University of Toronto.

68. R. Davies, letter to Arthur Porter, 30 June 1952, Ferranti-Packard Collection, Provincial Archives of Ontario.

69. Ibid.

70. Minutes of Research Control Committee, 24 October 1951, DND, accession 83-84/167, box 139, file NSS-1279-33, part 2.

71. Roger asked to double the money the Navy had planned to spend on DATAR from November 1951 to April 1953; i. e., from $400,000 to $800,000.

72. W. H. Roger, memorandum to Director of Scientific Services, 13 December 1951, DND, accession 83-84/167, box 3108, file NSS-7401-382-1. W. H. Roger, memorandum to Director of Weapons and Tactics, 15 January 1952, DND, accession 83-84/167, box 3169, file NSS-7407-382, part 1.

73. Minutes of Research Control Committee, 24 July 1953, DND, accession 83-84/167, box 139, file NSS-1279-33, part 2. To this total one should also add the over $400,000 worth of R&D work that preceded the decision to accelerate the DATAR program.

74. Goodspeed, "The Canadian Army, 1950–1955," part I, appendix C.

75. Ibid.

76. Ibid.

77. S. F. Knights, DATAR Contract F. E. 109,069 P33-40-881 With Ferranti Electric, 19 January 1953, DND, accession 83-84/167, box 3108, file NSS-7401-382-1.

78. A novel invention developed by Ferranti-Canada to allow the operator to quickly position the cursor over the moving targets on the CRT was the trackball, which today is an ubiquitous element in the information technology landscape. Sometimes it appears in its inverted form, as the "mouse." Surprisingly no patents were taken out on the "track ball." Whether this was due to military secrecy is not known. To my knowledge, this was the world's earliest implementation of the trackball.

79. This predictive ability required storing and using past data inputs.

80. Commander John Charles (Director of Naval Communications), memorandum: Report on Visit to DATAR Demonstration, 10–11 September, 15 September 1953, and Lieutenant Commander W. R. Howard, Report of Trip to DATAR Project in Toronto, 23 September, DND, accession 83-84/167, box 3169, file NSS-7407-382, part 2. Commander A. R. Hewitt (Director of Supplementary Radio Activities), Report of Visit to Scarboro Bluffs, NRC Establishment, 30 September 1953, DND, accession 83-84/167, box 3108 file NSS-7401-382-9. Commodore Lay (Assistant Chief of Naval Staff, Warfare), Conference of Demonstration of DATAR, 9 November 1953, Record of Proceedings, 13 November 1953, DND, accession 83-84/167, box 3169, file NSS-7407-382, part 2.

81. Minutes of Permanent Joint Board on Defence, September 1953, 19, DDP, volume 20,780, file CSC-6. 5, part 1. The PJBD was created in 1940 as a result of President Roosevelt's and Prime Minister Mackenzie King's desire to provide the necessary cooperation at the highest military levels to ensure the defense of North America. After World War II, the PJBD continued to play an important role in defense coordination. *(Henceforth, Records of the Department of Defence Production, RG49, National Archives, Ottawa, will be cited as DDP.)*

82. Appendix B in Lay, Conference of Demonstration of DATAR, 13 November 1953.

83. Minutes From the Project Committee on DATAR, 30 September 1953, DND, accession 83-84/167, box 139, file NSS-1279-33, part 1.

84. William Ford, secretary (RCN Research Control Committee), memorandum to secretary, Committee on Development, Defence Research Board: Naval comments on Army and RCAF 1955–56 R&D Estimates, 1 October 1954, DND, accession 83-84/167, box 139, file NSS-1279-33, part 2.

85. See item 1. 1. 4 in Electronics Branch, Department of Defence Production, Activity Report for the Month of June, 1955, DDP, volume 329, file 152-16-2, part 18.

86. The US system was based on the Badge prototype developed at the Bedford Laboratory, Cambridge Research Centre, Massachusetts. If Canada could not design its own system, the Department of Defence Production pushed to have Badge prototyped and manufactured in Canada. But the US pushed ahead with its own production engineering program. See page 7, Electronics Branch, Department of Defence Production, Monthly Report, December 1955, DDP, volume 329, file 152-16-2, part 20.

87. W. H. Roger, memorandum to Director of Scientific Services: Brief on DATAR, 12 January 1955, DND, accession 83-84/167, box 139, file NSS-1279-33, part 2.

88. RCN Proposal For International Automatic Radio Message, 7 February 1955, DND, accession 83-84/167, box 3108, file NSS-7401-382-6. The framework for the standardization of the data communication protocol was the CAN-UK-US Naval Data Transmission Working Group. The philosophy behind Canada's approach to communication protocol was shaped by the role the computer played in DATAR. The meanings conveyed by various groups of digits depended on values found elsewhere in the data stream. This technique resembles the idea of index registers in computer programming.

89. Canada's and Britain's very different technological choices in the area of naval tactical systems raises an important question. Did these two technical trajectories reflect important differences in the "cultures" of the RCN and the RN? Britain focused on battleship groups. The lack of capital ships, however, benefited Canada and spurred innovation.

90. Minutes of Research Control Committee, 4 May 1955, DND, accession 83-84/167, box 139, file NSS-1279-33, part 2.

91. The Research Control Committee had approved $1.95 million for this work; $910,000 to cover the data processing, communication, and display electronics; $1 million to cover the radar, sonar, navigational equipment, the installation, and testing. Royal Canadian Navy Research and Development Estimates, 1955–56, 28 September 1954, DND, 83-84/167, box 139, file NSS-1279-33, part 2. At 24% of total estimated expenditures, DATAR represented the single largest item in the RCN's R&D program for 1955–56.

92. RCA Victor was finally chosen in December 1955. S. I. Comach (Contracts Administrator, Electronics Branch, Department of Defence Production), letter to Deputy Minister of National Defence, 18 August 1955, DND, 83-84/167, box 3168, file NSS-7404-20698.

93. For example, the electronics in a destroyer contained some 3000 tubes; the navigational equipment in a large bomber used 2500. See C. I. Soucy, "A Survey of Electronics Equipment Failures Caused by Failures of Replacement Parts and Tubes and Improper Operation and Maintenance," 5 March 1954, internal Royal Canadian Air Force report, DND, volume 4185, file 260-640-43.

94. In a survey of 1050 USN fleet and shore station installations, it was found that one-third of the equipment was not operating properly, due to breakdowns. The US airborne AN/APQ-24 radar bombing and navigation systems experienced, on the average, one failure per 21.5 hour interval. Soucy, "A Survey of Electronic Equipment Failures."

95. This is based on anecdotal evidence from many of the engineers who worked on DATAR. Despite plans to use magnetic cores for the computer's random access memory, the vacuum tube remained the dominant device in DATAR's digital circuitry.

96. Wing Commander K. P. Likeness, memorandum to Chief of Air Staff, 26 May 1954, 1, DND,, volume 4185, file 260-640-43.

97. Ibid., p. 2. To dramatize his point, Likeness points out that by "statistical deduction," one can conclude that "the electronic equipment in fewer than four out of ten typical modern bombers will remain fully serviceable throughout a four hour mission."

98. The issue of transistor performance and manufacturing is discussed in greater detail in the next chapter.

99. Naval Secretary, letter to K. G. Thorne (Ferranti Electric Limited), 19 April 1955, DND, 83-84/167, box 3168, file NSS-7404-20698

100. This theme is taken up again with greater detail in the next chapter.

101. House of Commons, Debates, Second Session-Twenty Second Parliament 1955, p. 4862.

102. Wallace Nesbitt, Conservative member from Oxford (ibid., p. 4903).

103. Summary of Remarks Made by Minister of National Defence, R. O. Campney, to Cabinet Defence Committee, 6 February 1957, Office of the Privy Council, volume 2749, file Cabinet Defence Committee Conclusions, part XII.

104. D. P. Hoyt (Acting Director of Scientific Services), memorandum to Vice and Acting Chiefs of the Naval Staff: Naval Development Programme, 1956–57, 27 April 1956, DND, accession 83-84/167 box 139, file NSS-1279-33, part 2. Treasury Board of Canada told the RCN that if it wanted any approval for any other additional R&D expenditures, then it would have to go through the Cabinet. Treasury Board Minutes TB-499572, Records of Treasury Board, RG 55, National Archives of Canada. *(Hereafter, Records of Treasury Board, RG 55, will be cited as TB.)*

105. F. H. Sanders (RCN DATAR Sub-Committee), Report of the Sub-Committee Deputed to Prepare Recommendations to Research Control Committee on DATAR, 26 October 1956, DND, accession 83-84/167 box 139, file NSS-1279-33, part 4. This extensive report contains considerable information, given by the USN, on the estimated costs, schedules, contracts, and design specifications of its Naval Tactical Data System (NTDS).

106. Sanders, Report of the Sub-Committee, 26 October 1956.

107. For a detailed account of the NTDS project, see David Boslaugh, *When Computers Went to Sea* (IEEE Press, 1999). Boslaugh acknowledges the US Navy's intellectual debt to Canada's DATAR program. On UNIVAC's engineering role in the design and construction of the NTDS computer, see David Lundstrom, *A Few Good Men from Univac* (MIT Press, 1987).

108. Sanders, Report of the Sub-Committee, 26 October 1956.

109. Minutes of Research Control Committee, 20 December 1956, 3, DND, accession 83-84/167 box 139, file NSS-1279-33, part 4.

110. See Marc Milner, "Royal Canadian Navy Participation in the Battle of the Atlantic Crisis of 1943," in Boutilier, *RCN in Retrospect*; Milner, *North Atlantic Run*; Marc Milner, "The RCN and the Offensive against the U-Boats, 1943–45," unpublished narrative, History Directorate, National Defence Headquarters, June 1986; Marc Milner, "Inshore ASW," in *The RCN in Transition*, ed. W. Douglas (University of British Columbia Press, 1988); Marc Milner, "The Dawn of Modern ASW," *RUSI Journal* 134 (1989), spring: 61–68; Milner, The U-Boat Hunters.

111. The fact that by the end of the war 99 of the 104 qualified A/S officers in the fleet were reservists underscores this cleavage (Milner, The U-Boat Hunters, p. 257).

112. Ibid., p. 58.

113. Zimmerman, *The Great Naval Battle of Ottawa*, p. 162.

114. Minutes of PJBD Meeting, September 1953.

115. The drive to universalize screw threads presents a striking example of how Merritt Roe Smith's discussion of military enterprise can work within international alliances. Smith, has written quite extensively on how the military penchant for standardization and uniformity played a profound role in shaping early American industrialization. The US military's pursuit, in the early 19th century, to make standardized and interchangeable parts central to ordnance design and production was instrumental in the rise of the American System of Manufacture. See Smith, *Harper's Ferry*. For a comprehensive discussion of the historical process leading from early 19th century armory practice to Ford's mass-production see Hounshell, *From the American System to Mass Production*. For a valuable synthesis of the main themes in the American literature on the historic role of military enterprise in peacetime industrial development see Smith, Military Enterprise and Technological Change.

116. Arthur Porter was told of the Army's "soft-pedalling . . . on their development until the question of standardization of equipment between US, UK and Canada has been resolved," at a meeting with Army Technical Staff officers in Ottawa. See Porter, Preliminary Report, 15 October 1949.

117. The number of professional staff on DATAR was ascertained from a prospectus that Ferranti Electric's Research Division put out in 1956. At the back there are short biographical sketches of all the professional staff.

118. Electronics Branch, Department of Defence Production, Preliminary Statement of Engineering Facilities at Selected Plants in the Electronics and Telecommunications Industry, 7 March 1955, DDP, volume 298, file 152-4-3, part 1.

Chapter 3

1. This inseparability is reflected in the common periodization that technical, economic and social histories of the computer use. The introduction of discrete transistor components gave birth to the second generation of computers. The introduction of integrated circuits led to the third generation. The micro-miniaturization of transistors, such as Large-Scale Integration (LSI) and Very Large Scale Integration (VLSI), led to the rise of the personal computer and a decentralized, networked, paradigm of computation. For example see Gerald Brock, *The US Computer Industry* (Ballinger, 1975); John T. Soma, *The Computer Industry* (Lexington Books, 1976).

2. Robert Noyce, one of the co-founders of the Intel Corporation and co-inventor of the integrated circuit, presents a wonderful discussion of the demand-pull from the computer industry that drove the semiconductor industry and the supply-push from the semiconductor industry that simultaneously propelled the computer industry forward. "Microelectronics," *Scientific American* 237 (1977), September: 64–81.

3. Ernst Braun and Stuart MacDonald, *Revolution in Miniature* (Cambridge University Press, 1982). See also, by the same authors, "The Transistor and Attitude to Change," *American Journal of Physics* 45 (1977), November: 1061–1065.

4. Richard Levin, "The Semiconductor Industry," in *Government and Technical Progress*, ed. Richard Nelson (Pergamon, 1982), pp. 58–59. Thomas Misa's examination of efforts by the US Army Signal Corps to miniaturize field communications equipment offers a fascinating, more fine-grained detail of this process. Thomas Misa, "Military Needs, Commercial Realities, and the Development of the Transistor, 1948–58," in Smith, *Military Enterprise and Technological Change*.

5. Alfonso Molina, *The Social Basis for the Microelectronic Revolution* (Edinburgh University Press, 1989).

6. Cited in Arthur Van Dyck, "Aviation Electronics," *Proceedings of the Institute of Radio Engineers*, November 1953, p. 1578. This paper gives an interesting, but not very developed, discussion of the US military's historical relationship to issues of complexity and reliability in weapons systems from World War I.

7. Van Dyck, "Aviation Electronics," 1572.

8. Alfred Gray, "Guided Missile Reliability and Electronic Production Techniques," Institute of Radio Engineers, Convention Record, Part 8 (1956), p. 79. From 1950 through 1953, the US Navy, in collaboration with the US Bureau of Standards, spent $5 million on Project Tinkertoy (Levin, "The Semiconductor Industry").

9. On the role of the US Army in advancing high-precision machining see Smith, *Military Enterprise and Technological Change*.

10. In digital electronics, an element is either off or on. Because the amplification or gain of any stage is unity, any number of these elements can be cascaded and the final gain will still be only unity. In analog circuits, such as radio, the gain across any element is greater than one, typically of the order of ten. Thus if one cascades analog circuit elements, voltages will quickly climb to unacceptably high levels.

11. The "dynamic circuit" idea, for example, proposed cutting down the number of active vacuum tube elements required in computer circuits by shifting more of the computer's logic functions to passive gating, using much smaller and simpler crystal diodes. Similar proposals were first suggested during the EDVAC project at the Moore School at the University of Pennsylvania. See T. C. Chen, "Diode coincidence and mixing circuits in digital computers," *Proceedings of the Institute of Radio Engineers* 38 (1950), May: 511–514. Sidney Greenwald and R. C. Haueter, "SEAC," *Proceedings of the Institute of Radio Engineers* 41 (1953), October: 1300–1313; Robert Elbourn and Richard Witt, "Dynamic Circuit Techniques Used in SEAC and DYSEAC," *Proceedings of the Institute of Radio Engineers* 41 (1953), October: 1380–1387. It took US government financial support, of which a good portion was from the US Air Force, to build a computer based on "dynamic circuitry" techniques. Though there was a nearly 50% reduction in tube complexity, "dynamic circuitry" posed important limitations on performance. Another approach was to seek radically new types of vacuum tubes for digital circuits. In this regard, Josef Kates's invention of the "binary arithmetic

tube" is a striking example. The development and production engineering costs of Kates's invention were underwritten by Canada's military. Minutes of the First Meeting of the Electron Tube Development Committee, 18 September 1952, DND, box 4233, file DRBS-3-678-43.

12. Braun and MacDonald, *Revolution in Miniature*, p. 50. The demand pull from the commercial computer industry would not kick in until the late 1950s and the early 1960s. In 1957, tubes were still outselling transistors in the US by a ratio of more than 13 to 1. Only in 1959 did transistor sales start to match tube sales.

13. See Braun and MacDonald, "The Transistor and Attitude to Change," November 1977.

14. It is important to remember that the idea of a semiconductor replacement for the vacuum tube triode had been around for nearly 20 years before there was any interest to develop it. The first patent of a semiconductor type amplifier was filed by the physicist Julius Lilienfeld on 8 October 1928; "Method and Apparatus for Controlling Electronic Currents," US Patent 1,745,175, filed 8 October 1926, issued 28 January 1930. The physicist Virgil Bottom argues that Lilienfeld's patent bares a striking resemblance to the modern NPN type transistor. Virgil Bottom, "Invention of the Solid-State Amplifier," *Physics Today* 17 (1964), February: 24–26. See also J. B. Johnson, "More on the Solid-State Amplifier and Dr. Lilienfeld," *Physics Today* 17 (1964), May, p. 602. As Lilienfeld's work reveals, physicists were aware of the implications of the new quantum mechanics to semiconductor theory. But the absence of pure semiconductor crystals made any practical electronic applications impossible. World War II and the military's push to develop a pure semiconductor crystal material for radar first provided the material needed to exploit quantum mechanics for electronics.

15. "Equipment designed to perform complicated functions," wrote the chairman of DRB's Electronic Component Development Committee, "tends to become complex and bulky. On the other hand, in many military applications, size and weight must be kept at minimum. Consequently, a demand has been created for components not only of the highest quality but also of very small size." M. L. Card, "Electronic Component Development For The Canadian Armed Services," this is the final copy of an article prepared, during the latter part of 1954, for publication, sometime in 1955, in the journal *Electronics and Communications*. DND, accession 83-84/167, box 7368, file 170-80/E5, part 1. M. L. Card was both the deputy director of the Canadian Military Electronics Standards Agency and the chairman of the newly created Electronic Component Development Committee.

16. K. P. Likeness, memorandum to Chief of R. C. A. F Air Staff, 26 May 1954, DND, volume 4185, file 260-640-43.

17. Canadian Business Service Ltd., "The Electronics Industry In Canada," prepared for the Royal Commission on Canada's Economic Prospects, April 1956.

18. Ibid.

19. The total American defense expenditure to develop and bring the transistor into mass production has not been calculated, to my knowledge. However, it can be inferred from the fact that the US Army Signal Corps alone spent more than $50 million in the period 1956–1964 to do the production engineering work needed to bring the transistor to the level of standardization required for mass production (Levin, "The Semiconductor Industry").

20. After the war, the Department of National Defence started to form a comprehensive vision of Canada's northern flank. One facet of the military's study of northern defense was a re examination of the kind of weapons needed to operate in the Arctic's harsh physical environment Minutes of Eleventh Meeting of Defence Research Board Electronics Advisory Committee, 23 November 1948, item 7.3.4.3, DND, box 4233, file DRBS-3-640-43.

21. When the Canadian military wanted to start production of its own small field radios, it was faced with the problem that Canadian electronic component manufacturers did not know how to manufacture moisture-resistant diodes. Rogers Majestic had already started, of its own initiative, to develop diodes for the Canadian television market. But from the military's perspective, these devices were too bulky and susceptible to moisture problems. Rogers Majestic had attempted to develop subminiature, glass-enclosed diodes. The company quickly dropped its development program when it realized that there would be few commercial opportunities for these expensive diodes. The American firm Hughes Aircraft had solved the problem for the US Army, but the process was a trade secret and not available to the Canadian defense industry. Secretary of ECDC, memorandum to L. M. Chesley (Assistant Deputy Minister, Department of National Defence), 4 October 1955, DND, accession 83-84/167, box 7368, file 170-80/E5, part 1.

22. Minutes: Ninth Meeting of ECDC Transistor Panel, 3 January 1956, DND, accession 89-90/205, box 36, file 171-80/To.

23. [Services] Joint Telecommunications Committee Technical Meeting, 10 November 1952, DND box 4233, file DRBS-3-678-43.

24. Ibid.

25. DRB contributed $125,000 and $535,000 came from the Army, Navy and Air Force. Minutes of 13th Meeting of Joint Services Electrical and Electronics Committee, 17 September 1954, item 17, DND, volume 9285, file CSACS-2-70-67-1, part 1.

26. W. J. Eastwood (secretary of Joint Services Electrical and Electronics Committee), memorandum to secretary, Principal Supply Officers Committee, 19 August 1955, DND, box 9285, file SACS-2-70-67-1, part 1.

27. Whether this also proved to be the case over the broader range of electronic components still needs to be studied.

28. Concerns over the availability of transistors during an emergency even reached down to the question of ensuring a secure source of semiconductor

material. The Electronic Components Development Committee's special Transistor Panel underscored, to the Department of Defence Production, Canada's very limited ability to produce electronic grade germanium and silicon. Only Consolidated Smelters, from Trail, British Columbia, was in a position to undertake the production of an acceptable grade of germanium. No one in Canada could produce the more strategically important silicon. The sole source of this material was from DuPont in the United States. Minutes: Ninth Meeting of ECDC Transistor Panel, 3 January 3, 1956, DND, accession 89-90/205, box 36, file 171-80/To.

29. Minutes: Fifth Meeting of ECDC Transistor Panel, 12 April 12, 1955, DND, accession 89-90/205, box 36, file 171-80/To.

30. Ibid.

31. Recall that in the spring of 1955 the RCN sent representatives of Ferranti-Canada to the US to bring back several hundred surface-barrier transistors to study in the hope that they could be used to miniaturize DATAR.

32. L. M. Chesley, memorandum to chairman of Defence Research Board, 22 September 1955, DND, accession 83-84/167, box 7368, file 170-80/E5, part 1.

33. In 1955, Davies, a DRB Chief Scientist, wrote to the vice-president of the National Research Council about the need to eliminate this lacuna in the Canadian defense program (Frank T. Davies, letter to B. G. Ballard (Vice-President, National Research Council), 12 July 12, 1955, DND, accession 83-84/167, box 7368, file 170-80/E5, part 1).

34. N. F. Moody, "The Present State of the Transistor and its Associated Circuit Art," *Nuclear Instruments* 2 (1958), p. 182. Moody, as will be discussed below, came to lead a program to expand digital transistor circuit know-how within the Canadian defense research establishment.

35. E. Cullwick, memorandum to P. Nasmyth: Facilities for Applied Research and Experimental Development, 18 March 1948, DND, accession 83-84/167, box 7327, file DRBS-100-22/0, part 1. Invitations from the British and American armed services to cooperate in joint electronic weapons development projects first provoked concern over the need for centralized in-house electronic laboratories. The Canadian armed services inability to accept the invitation underscored "the problem of the absence of adequate facilities within the Department of National Defence for experimental development of electronic equipment and systems, [as well as facilities] to do the applied research which arises from such development." (ibid.)

36. "CF-105," DDP, volume 298, file 152-4-3, part 1. This document has no date, but it is reasonable to assume from its location within the file that it was prepared in March or April 1955.

37. D. B. Mundy, memorandum to D. A. Golden (Deputy Minister, Department of Defence Production), 6 December 1956, DDP, volume 298, file 152-4-3, part 2. F. T. Davies, memorandum: Shortage of Research Personnel in Canadian Electronics, 12 July 1955, DND, accession 83-84/167, box 7368, file 170-80/E5, part 1.

38. In the end the DDP had to recommend that Canada give the US Hughes Aircraft a 4-year, $15 million contract to develop the AVRO Arrow's electronic systems.

39. W. A. Thomson, letter to D. W. R. McKinley, 25 July 1955, DND, accession 83-84/167, box 7368, file 170-80/E5, part 1. Silicon-based semiconductors were initially more expensive to develop. But their greater resistance to heat and radiation made them far more attractive to the military.

40. F. W. Simpson's Electronic Material's Section undertook to develop a deeper understanding of silicon crystal growth methods and transistor behavior. D. W. McKinley, letter to (DRB) chairman, 26 July 1955, DND, accession 83-84/167, box 7368, file 170-80/E5, part 1.

41. Progress Report: Research and Development Project No. D48-99-31-03, 31 December 1960, DND accession 83-84/167, box 7593, file DRB-9931. 11, part 1.

42. Ibid.

43. Bothwell, *Nucleus*. According to Bothwell, it was Lewis's deep Christian faith that led him to preach the peaceful uses of nuclear energy. Though Moody did not start from the same deep religious convictions, he nevertheless felt uncomfortable contributing to such a murderous weapon.

44. N. F. Moody, "Electronic Counter," 10 April 1952, Canadian Patent Application 629,620, CPDL, volume 62, file L14-1495-2.

45. For example, see ibid. and N. F. Moody and W. D. Howell, "Electronic Counter," Canadian Patent Application no. 590,852, 4 August 1949, CPDL, box 63, file L14C-1508-2; William Battell and Norman Moody, "Discriminator Circuit for Kicksorter," Canadian Patent Application no. 619,148, filed 10 August 1951, CPDL, box 48, file L14C-1364-2; and Norman Moody, "Trigger Circuit," draft of patent application 28 April 1952, CPDL, box 50, file L14C-1364-2.

46. R. K. Richards, *Electronic Digital Components and Circuits* (Van Nostrand, 1967).

47. Cited in Joel Shurkin, *Engines of the Mind* (Washington Square Press, 1985), p. 91.

48. N. F. Moody, G. R. Maclusky, and M. O. Deighton, "Millimicrosecond Pulse Techniques," Lecture delivered at the Third Annual Joint AIEE-IRE Conference on Electronic Instrumentation in Nucleonics and Medicine, New York, October 25, 1950. Copy in CPDL, box 50, file L14C-1381-2.

49. Braun and MacDonald, *Revolution in Miniature*, p. 49. In 1953, the price of the best transistors was eight times that of a vacuum tube. One electronics expert recalled, with regard to consumer electronics, that "for a decade or so, from 1953 to 63, we had no choice but to go with vacuum tubes because they did a better job, and up until that time they were cheaper" (ibid., p. 50).

50. Attributed to Major-General C. Irvine, US Air Force, during an Electronic Component Conference in Los Angeles in May 1955, by CI Soucy, RCAF. See C. I.

Soucy, The Future Replacement of Tubes By Transistors and Magnetic Amplifiers, DND, accession 89-90/205, box 35, file 170-80/To, part 2.

51. E. F. Johnson, letter to Chairman, Defence Research Board, 5 April 1956, DND accession 89-90/205, box 35, file 170-80/To, part 1.

52. Levin, "The Semiconductor Industry."

53. The US Air Force spent $5 million on industrial R&D into the development of the silicon transistor. In 1956, the US Department of Defence committed $14 million to develop a large-scale capacity to manufacture silicon transistors. The development of this mass-production capacity, which became the engine of national economic growth, was not motivated by any "abstract desire to promote industry, but rather by concrete military needs" (Levin, "The Semiconductor Industry," pp. 67–68).

54. During the mid 1950s, the Japanese also decided to use a government laboratory, the Electro-Technical Laboratory (ETL), as a pole from which vitally needed transistor know-how could diffuse to industry. After some doubt as to how best to focus its energies, ETL shifted from studying the transistor itself to pursuing the circuit applications of the transistor. Like DRTE, ETL decided to design and build the nation's first solid-state computer as a vehicle for developing a domestic capacity to apply transistors. "At the time," the leader of the ETL transistor group later recalled, "none of us were interested in computer architecture; our goal was to build a computer as an application of transistors." Sigeru Takahashi, "Early Transistor Computers in Japan," *Annals of the History of Computing*, 8 (1986), April: 147. But the fact that ETL confined its initial investigations to the older and cheaper germanium transistor, whereas DRTE immediately explored the use of the embryonic and more costly silicon transistor, further illustrates the direct role that military enterprise plays in shaping both the rhythm and direction of technical change.

55. The information on this training program was gathered from correspondence contained in, to this writer's knowledge, the only file on the subject. See DDP, volume 303, file 152-14-5C3.

56. Only two companies turned down DRTE's training proposal: Ferranti-Canada and RCA Victor of Montreal. Each of these companies was already pursing its own R&D program. Canadian General Electric never responded.

57. The following patents resulted from the trigger circuit development work at DRTE: Norman F. Moody and David Florida, "Two-State Apparatus," Canadian Patent no. 584,379, filed on 24 January 1957, issued 6 October 6. Norman F. Moody, "Two-State Electronic Circuit," Canadian Patent no. 603,345, filed on 26 March 1957, issued 16 August.

58. In more technical terms, the PNPN trigger was able to run the transistors at saturation without sacrificing speed. Moody argued that the symmetric transistor adaptation of the conventional vacuum tube flip-flop was not well suited for heavy loads. As its current output increased, performance deteriorated. The PNPN

structure's regenerative loop minimized the storage time delay. N. F. Moody and C. D. Florida, "Some New Transistor Bistable Elements For Heavy Duty Application," *Institute of Radio Engineers Transactions on Circuit Theory* (1957).

59. The PNPN structure consisted of a complementary pair of PNP and NPN transistors joined to form a regenerative feedback loop. N. F. Moody, "The General Properties of the P-N-P-N Structure As Applied to a Trigger Circuit," 1954, DRTE Electronics Laboratory memorandum 5059-54. During the 1980s, a similar principle was used to develop CMOS (Complementary Metal Oxide Silicon) technology. Today, CMOS technology underlies a large segment of microchip design.

60. In 1957, Florida described how the "controlled saturation concept" had been tested on only a small sample of transistors of forty transistors. Florida points out that building a computer which uses saturated trigger circuits would provide the large statistical sample to support Moody's circuit technique. See Florida, *An Analysis of a P-N-P-N Structure*.

61. C. D. Florida, "Digital Computers (I)," Transcript of a Talk Given at DRTE(EL), February, 1956. Computer History Archives, University of Calgary. "To paraphrase Mr. Moody's remark that unless we learn about transistors we shall be extinct as electronic engineers," Florida once explained (p. 2), "I would say that unless we learn about computers, we shall be extinct period!"

62. Florida, An Analysis of a P-N-P-N Structure.

63. Florida's "passive routing" circuit architecture was not a new idea in computer technology. The American SEAC vacuum tube computer project had already pioneered similar principles between 1949 and 1953 under the name of "dynamic circuit techniques. The circuit design philosophy behind the SEAC computer, was "to minimize the number of tubes by performing all the logical operations of 'and', 'or' and 'not' between pulses in circuits comprised of germanium diodes and resistors." Elbourn and Witt, "Dynamic Circuit Techniques Used in SEAC and DYSEAC," October 1953, p. 1380. J. Felker adapted similar principles in his proposal to design of a germanium point-contact transistor computer in 1952. J. H. Felker, "The Transistor as a Digital Computer Component," *Review of Electronic Digital Computers* S-44 (1952), February: 5–109. The Japanese government laboratory, ETL, however did use the technique in its exploration of transistor application in order to reduce complexity. Takashi, "Early Transistor Computers in Japan," p. 146. But unlike other "dynamic circuit" approaches, the DRTE computer's passive circuitry did not sacrifice performance to reduce complexity.

64. C. D. Florida, "A Floating Point Arithmetic Unit," Defence Research Telecommunications Establishment Electronics Laboratory Report No. 5083-7, February 1959. The first commercial scientific computer to incorporate floating-point computation was the IBM 704. R. Moreau, *The Computer Comes of Age* (MIT Press, 1986), p. 66. First delivered in 1955, the design of this vacuum tube computer benefited from IBM's experience in the US Air Force's SAGE project.

65. The three-address architecture was not new, but its implementation was rare. The first commercial system in Europe to use a three-address system was Ferranti UK's ORION model.

66. David Florida, "The DRTE Solid State Digital Computer," in *Proceedings of the Second Canadian Conference for Computing and Data Processing, June 1960, Toronto, Ontario* (University of Toronto Press, 1960).

67. C. D. Florida, "A Non-restoring Method for the Extraction of the Square Root, and it Mechanization," Defence Research Telecommunications Establishment Electronics Laboratory Report No. 5083-6, February 1959. George Lake, "A Digital Decimal to Binary and Binary to Decimal Converter," Defence Research Telecommunications Establishment Electronics Laboratory Report No. 1044, 1960.

68. This converter was designed by George Lake. See Lake, "A Digital Decimal to Binary." Fred Longstaff, who at the time was Ferranti-Canada's chief computer designer, later recalled coming to Ottawa to see the DRTE computer and feeling that "Florida was trying to build more fanciness into the hardware than was really viable at the time."

69. Though the absence of a good programming environment contributed more to the burden of using the computer than hardware inadequacies, Florida gave scant attention to software development.

70. N. W. Morton, memorandum: Review of DND Computation Facilities, 30 November 1954, DND, accession 83-84/167, box 7589, file DRBS-9900-31, part 1. By 1956, Canada's aircraft manufacturer A. V. Roe had acquired a CRC102A computer as well, and about ten orders for computers had been placed by various Canadian universities, companies, and government agencies, some of which were again partially financed by the military. See Computation Centre, University of Toronto, "Submission to The Royal Commission on Canadian Economic Prospects," October 1955, No. 165. Library of Parliament. The world of computational technology was so small in 1955 that it was easy for Gotlieb, the director of the University of Toronto Computation Centre, to keep abreast of all activity in Canada.

71. E. W. Greenwood (for Chief Superintendent of CARDE), memorandum to chairman, Defence Research Board: Digital Computer, 20 October 1954, DND accession 83-84/167, box 7593, file DRBS-9931-05, part 1.

72. Ibid.

73. In a visit to CARDE in October 1954, G. D. Kaye, the head of DRB's Operations Research Group, observed: "The shortage of programmers is the critical point. CARDE are already employing outside programmers, which is expensive, and trying to train their own men as programmers, which is inefficient when the computer for which they are trained is a long way away. It seems to me that purchase of a machine at CARDE would be the best way to remedy this shortage. The mathematicians are there and so are the problems, and experience would be

rapidly gained on such a machine." G. D. Kaye, Visit to CARDE—October 15, 1954, DND, accession 83-84/167, box 7593, file DRBS-9931-05, part 1.

74. Greenwood, memorandum to chairman, 20 October 20, 1954. In addition to training more programmers, a computer on CARDE's premises was also needed to analyze the vast amounts of data generated directly by Doppler, radar, kinetheodolite, telemetring, wind tunnel, and V. T. fuze experiments.

75. A. S. Shore (Grants and Contracts), memorandum to chief superintendents, superintendents, and directors: Computation Requirements, 29 October 1956, DND, accession 83-84/167, box 7589, file DRBS-9900-31, part 1.

76. In a document to Treasury Board's Interdepartmental Committee one finds that DRTE accounted for 40% of computation time for defense R&D purposes. The next biggest user was the Operations Research Group, with 22%. DRB Annual Report 1956: Computation, TB, box 632, file 1501-22.

77. James Scott, letter to Chairman (Defence Research Board), 7 November 1956, DND, accession 83-84/167, box 7593, file DRBS-9931. 11, part 1.

78. For example, in the summer of 1958, David Florida convinced representatives from the Pacific Naval Laboratory (PNL) that copies of the DRTE computer could be made available to other Defence Research Board laboratories. Superintendent (Pacific Naval Laboratory), letter to Chief Superintendent, Defence Research Telecommunications Laboratory, 29 April 1959, DND accession 83-84/167, box 7593, file DRB-9931. 11, part 1. But the deal collapsed when DRTE refused to sell the computer at the low price PNL was calling for. O. A. Sandoz (for Chief Superintendent, D. R. T. E.), letter to Superintendent (Pacific Naval Laboratory), 8 June 1959, DND accession 83-83/167, box 7593, file DRB-9931. 11, volume 1. ; Frank Davies (Chief Superintendent), D. R. T. E., letter to Superintendent (Pacific Naval Laboratory), 14 August 1959, DND accession 83-84/167, box 7593, file DRB-9931. 11, part 1.

79. This view was expressed to G. D. Kaye, from DRB's Operations Research Group, on a visit to the DRTE computer group. G. D. Kaye, letter to Chairman, Defence Research Board, 11 December 1956, DND, accession 83-84/167, box 7593, file DRB-9931. 11, part 1. The public record does not allow one to discern if Florida put his group's work to wider application in order to ensure continued funding for his project.

80. G. T. Lake (for Chief Superintendent, DRTE), memorandum to Chairman, DRB: Future Research Programme for DRTE Computing Centre, 22 February 1963, DND, accession 83-84/167, box 7593, file DRB-9931. 11, part 1.

81. Progress Report: Research and Development Project D48-99-31-03, 31 December 1962.

82. In 1963, the administrative pretense ended. The DRTE's Chief Superintendent, Frank Davies, recommended to the chairman of DRB that funding be approved specifically "to provide a computing service for the establishment and

to encourage its use." Frank T. Davies, memorandum to Chairman of DRB, 28 January 1963, DND accession 83-84/167, box 7593, file DRB-9931. 11, part 1.

83. G. T. Lake, "Proposal for a New Computer in DRTE," 29 December 1964, John Chapman Papers, National Archives of Canada, MG 31, J43, box 12, file 17. (Hereafter, "Chapman" indicates MG 31 or J 43 in this set of papers.)

84. DRTE had started work on Alouette II. Canada had also committed itself to build four satellites for the ISIS program.

85. Linda Petiot, "Dirty Gertie," *IEEE Annals of the History of Computing* 16 (1994), summer: 43–52. This article offers the reader a good overview of the DRTE computer's architecture and operational specifications.

86. Ibid., p. 50.

87. Colossus, the first digital, electronic, information-processing machine built in Britain, was specifically conceived for breaking codes. Colossus's wartime success in breaking the German Enigma code provided the Allies with an invaluable strategic advantage. After the war, the computer became the essential tool for code breaking.

88. Despite an exhaustive examination of the primary sources, I was not able to uncover even the smallest shred of evidence to substantiate Petiot's suggestion.

89. Fred Longstaff, the central figure in Ferranti-Canada's computer design team and Florida's contemporary, later recalled thinking that DRTE group "was building too much fanciness into the hardware than was viable at that time. . . . For example [Florida] was building a square root as an instruction, not even a modern computer does that now. There is a sub-routine for it." (interview with Fred Longstaff, Les Wood, Gord Lang, and Ted Strain, 14 August 1984, for Computer Oral History Project, National Museum of Science and Technology, Ottawa)

90. Scott, letter to chairman of Defence Research Board, 7 November 1956, DND, accession 83-84/167, box 7593, file DRBS-9931. 11, part 1.

91. Doris Jelly, *Canada: 25 Years in Space* (Polyscience, 1988).

92. Norman Moody and Richard Cobbold, "Monostable Transistor Switching Circuit," Canadian Patent no. 636,768, filed 1959, issued 1962, p. 2. This patent built on the monostable device described in Norman Moody and David Florida, "Monostable Two-State Apparatus," Canadian Patent no. 595,885 filed 27 April 1957, issued 12 April 1960.

93. "The voltages induced in the read-out winding by the half-excited cores on the selected lines did present a problem, as these 'disturb' voltages are additive and may mask out the desired read-out signal, particularly in large arrays. Direct pick-up from the exciting windings also adds to the disturbs." J. A. Rajchman, "Computer Memories," *Proceedings of the Institute of Radio Engineers* 49 (1961), January: 104–127, 108. The most common approach to this problem was to thread the read-wire by zig-zagging back-and-forth diagonally through the core.

The idea was to have the small disturb voltages cancel one another rather than reinforce one another. In a survey of memory technology, Jan Rajchman concluded in 1961 that this "cancellation [was], in general, not a sufficient remedy to the problem of disturb voltages" (ibid., p. 100). The threading precision needed to minimize the disrupt voltage was, from the production side, very time consuming, and from the technical side, not completely satisfactory. Cobbold's idea to thread the read-wire to the drive wires eliminated the disrupt voltage problem completely. See Richard Cobbold, "Magnetic Core Memory," US Patent no. 2,995,733, filed 1959 issued 1961. See also "Discovery Improves Computers," *Financial Post*, October 1959. R. S. Cobbold, "Novel Magnetic Core Memory," *Electronic Equipment Engineering* 8 (1960), February: 73–79.

94. The same reasons that made silicon the semiconductor material of choice by the military—resistance to heat and radiation—also made it ideal for space electronics.

95. With a good portion of Canada's Arctic beyond radio range, the Canadian military was interested in developing self-contained navigational equipment for its long range fighter aircraft. Furthermore, it was reasoned that any self-contained system would not be vulnerable to enemy radio jamming. The development of a lightweight transistor system made deployment of such equipment feasible. Working with other sections of DRTE(EL), Moody's group played a leading role in the development of a self-contained airborne navigation system. A good deal of work went into applying transistors to Doppler navigational radar and analog computation. However, this hybrid system also integrated a digital-based positional computer to the analog electronics. A patent was taken out for the electronic analog computational sub-system. See R. K. Brown, N. F. Moody, P. M. Thompson, R. J. Bibby, C. A. Franklin, J. H. Ganton, and J. Mitchell, "A Lightweight and Self-contained Airborne Navigational System," *Proceedings of the Institute of Radio Engineers* 41 (1959), May: 778–807. Bibby, Franklin, and Florida were also key engineers in the development of Alouette II and the ISIS satellite program.

Chapter 4

1. In a corporate history of Ferranti-Canada, Ball and Vardalas briefly discussed the Post Office and Trans-Canada Air Lines computer projects Ball and Vardalas, Ferranti-Packard. Their study, however, did not explore in any detail the circumstances of why and how these two public enterprises supported new technological developments. Neither did they examine the interplay between Ferranti-Canada's own technological ambitions and those of these two public enterprises. Such a wider and deeper elaboration is the purpose of this chapter.

2. See comments from James Sinclair, the Liberal member of parliament from Coast-Capilano. Canada, House of Commons, Debates, 15 December 1951, 1990.

3. Work force reduction figure cited by the Postmaster General, Gabriel Rinfret. See House of Commons, Debates, 15 December 1951, p. 3899.

4. The total number of pieces of first-class mail that the Canadian Post Office handled for the years 1948–49, 1949–50, 1950–51, 1951–52, and 1952–53 is 1,287,556,498, 1,321,172,583, 1,412,252,979, 1,413,988,068, and 1,518,279,701 respectively. See Report of the Postmaster General for each of the years in question. Similar figures for earlier years are unavailable because volume of business was measured by weight. Joseph Noseworthy brings up the issue of the increasing resignations in the Toronto area. House of Commons, Debates, 30 May 1951, p. 3562.

5. Maurice Levy, "The Electronic Aspects of the Canadian Sorting of Mail System," reprinted from the *Proceedings of the National Electronics Conference* 10 (1955), February, p. 3. The breakdown lay at urban nodes on Canada's postal distribution system. In 1952, for example, Canada's 17 largest cities processed more than 70% of the nation's mail ("Thinking Machine Will Speed Mail," *Popular Science*, October 1953: 114–117). Of this volume, half was destined for distribution within each urban center. City sortation was thus the most time-consuming and costly phase of sorting mail. O. D. Lewis, memorandum to W. J. Turnbull, 29 January 1952. Records of Canada Post, RG 3, accession 90-91/257, box 274, file 78-2-10, part 1. National Archives of Canada. *(Henceforth, Records of Canada Post, RG 3, will be cited as PO.)*

6. "Electronic Mail Sortation," *The Postmark*, September-October 1960: 8–9, 15, 20.

7. Other than being a "Clerk, Grade 4" in 1951, very little is known about O. D. Lewis. In 1952, Lewis was promoted to "Administrative Assistant." Director of Organization and Classification (Civil Service Commission), memorandum: Reassignment and provision of a terminable allowance for Mr. O. D. Lewis, 16 October 1952. PO, accession 90-91/257, box 275, file 78-2-10, part 2.

8. At Adamson's request for written elaboration, Lewis prepared a four page memorandum outlining his ideas. Adamson was a superintendent in the Post Office Department and Lewis's superior. O. D. Lewis, memorandum for Mr. Adamson, 4 December 1951, PO, accession 90-91/257, box 275, file 78-2-10, part 2.

9. Lewis, memorandum for Adamson, 4 December 1951

10. This receptivity was no doubt shaped by public attacks on the post office's efficiency. For example, a few days after Christmas 1951, George Drew, the leader of the opposition, seized on the public's heightened sensitivity to mail delivery during the holiday season to attack what he saw to be the accelerated decline in the efficiency of Canada's postal service. Demanding to know why it took a week for mail delivery between Toronto, Ottawa, and Montreal, Drew reminded the government of the important place that efficient postal services had come to occupy in Canada's economic and social structure of the (House of Commons, Debates, 29 December 1951, p. 2511).

11. O. D. Lewis, memorandum to W. J. Turnbull, 9 February 1952, PO, accession 90-91/257, box 274, file 78-2-10, part 1. Since no electronic system could read handwriting at high speeds, someone was needed to read the addresses and type them into an apparatus that would then print the corresponding bar codes onto

the envelope. Typing required little if any decision making. The typist did not have to memorize complex geographic grids. For this redefinition of the labor process to be effective, the act of typing each address had to be a faster than the mental act of sorting it. Lewis suggested, after some rough calculations, that "150 lower paid typists could do in four hours, the work that 160 higher paid sorters did in eight hours" (ibid.).

12. Walter Turnbull, Deputy Postmaster General, Transcript of radio address, Recorded at CARTE Radio Bureau, on 28 February 1957, CP, accession 86-87/402, box 33, file Publicity 1953–57.

13. Turnbull, Radio Transcript, 1957. Criticism of postal services in Parliament reflected this crucial urban dimension. Criticism of the Post Office was undertaken by members of Parliament representing the fast-growing metropolitan areas, such as York West, York South, and Montreal.

14. At the age of 19, Turnbull joined the federal civil service as clerk in the post office. By the age of 27, Turnbull managed to get a new position created for him as the Post Office Department's first Head of Public Relations. Then in 1936, Turnbull came to the attention of the recently elected Prime Minister, William Lyon Mackenzie King. For the next 9 years, until the end of World War II, Turnbull worked as King's principle secretary. In 1945, Turnbull wanted to return to the Post Office. As a reward for loyal and excellent service, King saw to it that Turnbull was named Deputy Postmaster General.

15. After visiting with National Research Council officials, Lewis concluded that they could not take on the problem O. D. Lewis, memorandum to W. J. Turnbull, 29 January 1952, CP, accession 90-91/257, box 274, file 78-2-10, part 1.

16. Whether Levy found out about the Post Office's automation ambitions through the grapevine or through Lewis is not known. Furthermore, it is not clear whether Levy went directly to Turnbull or was introduced by Lewis.

17. Levy's recollection of the incident is taken from Levy, memorandum to Boyd, 19 March 1954. Levy wrote that he was surprised that Turnbull was already thinking along these lines. But he failed, or refused to recognize, that Lewis had already sparked interest in this issue some four months earlier.

18. IT&T had already designed and built the "Transorma," the mail sorter prototype for the Belgian postal authorities.

19. J. D. Stewart, letter to R. P. Morris, 8 July 1952, Maurice Moise Levy Papers (*henceforth cited as Levy Papers*). I hope to find a suitable public archival home for these papers.

20. In 1936, IBM had patented claims surrounding mail coding and sorting. See Ayres et al., US Patent 2056382. 6 October 1936. But these were quite narrow. IBM's bid to win very broad patentable rights to the idea had failed. J. D. Stewart, a Washington, D. C. based patent lawyer wrote back to IT&T representative, R. P. Morris, in New York, that IBM tried and failed to expand its 1936 patent to

include a "method of sorting mail in accordance with postal districts which comprise marking parcels of mail with code indications in accordance with various postal districts, analyzing the code markings with a light sensitive device and distributing the articles to various sorting machines under the control of an analyzing device." The claim, according to Stewart, was refused because the US Patent Office felt that a combination of older patents dealt with the general idea of marking and sorting. Stewart, letter to R. P. Morris, 8 July 1952. Westinghouse had already obtained marking and sorting patents which they used in a commercial billing system sold to electrical utilities. Electrical bills were sent as postcards, on which was also printed, in a coded form, the customer's address. When the customers mailed their payments, they enclosed the card. Then for the purposes of recording and statistical compilations, the utility passed the cards through special sorters. Friedman had a patent on bar-code scanning. US Patent no. 2,224,646. IBM had also obtained a wide range of patents dealing with the use of photoelectric techniques to sort cards. See US Patents 1853443, 2000403, 2000404, 285296, 2231494, 2268498. GE had patented the idea of scanning fluorescent markings, which were invisible to the human eye, by means of ultraviolet light for identification purposes. See Short, US Patent no. 2,593,206.

21. Maurice Levy, memorandum, 24 September 1952, Levy Papers.

22. The generic idea of coding and electronic sorting was not new. Nevertheless, there was still some latitude of originality because the Canadian letter sorting scheme did not infringe on any patents. Stewart, letter to Morris, 8 July 1952.

23. IT&T's American patent lawyer admitted the originality of the idea, but he felt it could never be patented.

24. It is ironic that the conversion of the code to binary data and the use of a digital memory, the two most novel features of Levy's proposal, were not in themselves patentable. See opinion in Stewart, letter to Morris, 8 July 1952.

25. T. P. Leddy (vice-president, Federal Electric Manufacturing Co.), letter to W. J. Turnbull, 15 May 1952, Levy Papers.

26. Cited in E. J. Underwood (Acting Deputy Postmaster General), letter to R. B. Bryce (Assistant Deputy Minister, Department of Finance), 9 June 1952, Levy Papers.

27. B. G. Ballard, letter to O. D. Lewis, 16 June 1952, Levy Papers. No copy exists of the Technical Proposal written by Levy and presented by the Federal Electric Manufacturing Co. to Turnbull. However, a detailed summary of this proposal can be found in OD Lewis's memorandum to E. J. Underwood. This summary accompanied the post office's request (See Underwood, letter to Bryce, 9 June 1952) to the Finance Department that a $100,000 be allocated, in 1952–53, toward the development of an electronic mail sorter prototype.

28. Electronics itself was not the issue. Rather, it was the importance of precisely controlling the aerodynamic forces which would affect letters as they moved at very high speeds.

29. Canada, House of Commons, Debates, Fourth Session—Twenty-First Parliament 1952 (Queen's Printer, 1952), 11 June 1952, p. 3901.

30. Ibid., 30 May 1952, p. 3562

31. Editorial, "Will Gadgets In Post Office Speed Delivery," *Calgary Herald*, 23 July 1952. Found in CP, accession 86-87/402, file Publicity 1953–57.

32. The editorial continued: "Electronic sorters are no doubt satisfying to dream about. But for ourselves, we should be more content if some of the Post Office's surplus energy were devoted to prosaic problems like trying to cut down local delivery time from the occasional four days to a maximum of say one."

33. T. P. Leddy, letter to W. J. Turnbull, 29 July 1952, Levy Papers.

34. The staffing requirements are given in O. D. Lewis, memorandum to W. J. Turnbull, 11 September 1952. Levy Papers.

35. In a letter to the Civil Service Commissions, the Deputy Postmaster General underscores the urgency of staffing Levy's R&D team as quickly as possible. W. J. Turnbull, letter to secretary, Civil Service Commission, 11 October 1952, Levy Papers.

36. W. J. Turnbull, letter to R. B. Bryce (Assistant Deputy Minister of Finance), 19 December 1952, Levy Papers. Turnbull asked that $300,000 be allotted for the 1953–54 fiscal year.

37. In the spring of 1952, when Levy had first articulated his vision of the electronic sorter to the post office, he referred to the idea of "cold tube" memory. Whether he was thinking of electrostatic memory, or some other tube variant is not known.

38. This reference to Arthur Porter was made in Maurice Levy, memorandum to W. J. Turnbull, 6 December 1952. PO, accession 86-86/402, box 6, file 71-2-1. Levy met Porter only in February 1953.

39. Canadian Marconi, however, actively campaigned to design and build a system using a different approach based on conductive ink scanning. This company argued that the use of memory unnecessarily complicated the design of the mail sorter. O. D. Lewis, memorandum to W. J. Turnbull, 9 January 1953, PO, accession 90-91/257, box 274, file 78-2-10, part 1. Canadian Marconi was confident that it could come up with a system more or less ready for manufacture. Levy had no interest in non-memory approaches, but others did. Lewis, the man who first put the ideas of electronic sorting in Turnbull's head, lobbied strongly for Canadian Marconi.

40. Ridenour proposed an approach to memory based on photographic storage that would place postal information on a glass disk 15.5 inches in diameter and ⅜ inch thick. The names of all of Canada's 10,757 postal stations and the nearly 6000 mail routing distinctions would be stored digitally as dark spots and empty spaces. To read this memory, a spot of light, analogous to a television tube beam,

would scan the appropriate track. Photographic Storage System for the Post Office of Canada, March 1953, prepared by the International Telemeter Corporation, Levy Papers. For further technical discussion of this memory see G. W. Brown and L. N. Ridenour, "Photographic Techniques for Information Storage," *Proceedings of the Institute of Radio Engineers* 41 (1953), October: 1421–1428.

41. Another weakness was the optical memory read-only capabilities. For Levy this was not a big disadvantage. Unlike a computer system, where one is constantly writing to memory, the postal computer needed only to read the memory.

42. Maurice Levy, letter to Louis Ridenour, 7 April, Levy Papers. This seven page single-spaced letter was accompanied by a whole set of hand-drawn illustrations.

43. See Turnbull's own handwritten comments which appear at the end of Levy, memorandum to W. J. Turnbull, 8 April 1953.

44. Levy, memorandum to W. J. Turnbull, 8 April 1953.

45. Turnbull understood the value of effective public relations. In 1923 Turnbull created the public relation's group at the post office. He stayed as Head of Public Relations until 1936. In his retirement years, Turnbull spoke very fondly of his success in promoting, in film and print, the Post Office's public image.

46. Postmaster General, Submission to Treasury Board, 26 June 26 1953, PO, accession 86-87/402, box 6, file 71-3-5E. Treasury Board approved the request on 14 July 1953. See Authority To Enter Into Contract, TB, volume 184, T. B. 454222.

47. Maurice Levy, memorandum to W. J. Turnbull: Visits to US, 3 July 1953, PO accession 86-87/402, box 4, file Visits (Reports).

48. W. J. Turnbull, memorandum to Dr. Levy, 24 July 1953. Original in Levy Papers. Copy can be found in PO, accession 86-87/402, box 6, file 71-2-1.

49. Ibid.

50. Levy, memorandum to W. J. Turnbull: Photographic Memory, 4 August 1953.

51. Ibid.

52. Levy, The Electronics Aspects of the Canadian Sorting of Mail System.

53. The overall management of the project was in disarray. Conflicts raged between the Post Office and Pitney Bowes over who should obtain exclusive rights on any patent developments surrounding the mechanical sorters. IBM refused to get involved in the coding desk and printer aspects of the project because of the post office's refusal to grant them exclusive ownership over any patents that arose during the project. The technical feasibility of fluorescent scanning was still unresolved. Visible light scanning was pursued as a temporary measure. Maurice Levy, Confidential memorandum to Turnbull, March 1955, Levy Papers. See also J. MacDonald (Comptroller), memorandum to Deputy Postmaster General, 13 April 1955, PO accession 90-91/257, box 275, file 78-2-10,

part 2; and Budgets, Costs and Estimates Division, Electronic Project, 19 May 1955, PO, accession 86-87/402, box 4, file (untitled).

54. Fred Longstaff, letter to Maurice Levy, 27 August 1956, PO accession 86-87/402, box 3, file—Ferranti Electric, 1956.

55. Maurice Levy, memorandum to Turnbull: Visit to DRB and Discussions With Moody and His Collaborators on Transistor Circuits for Our Computer, 19 September 1956, Levy Papers. This information was subsequently revealed to Levy in conversations with K. G. Thorne, Ferranti-Canada's former Business Manager.

56. Ibid.

57. For a detailed account of the delays and cost overruns see, for example, the handwritten table entitled "Successive Estimates By Ferranti," dated 20 November 1956, found in PO, accession 86-87/402, box 3, file—Ferranti Electric 1956.

58. Turnbull, memorandum to Maurice Levy, 7 March 1956.

59. Ibid.

60. House of Commons, Debates, 13 August 1956, p. 7530.

61. Ibid., p. 7547.

62. Ibid.

63. In the end, the cost of the computer had gone up nearly 400% from the original cost specified in the August 1955 contract.

64. Ibid. If one looks at operating deficits, the imperative to automate sorting was greater in the US than in Canada. Throughout the immediate postwar period the US Post Office's budget deficits climbed at an alarming rate. The first computerized mail-sorting system in the US went into operation in 1960 in Rhode Island. Great Britain followed in 1966. See P. Harpur (ed.), *The Timetable of Technology* (Hearst, 1982), p. 168.

65. Brown, Ideas In Exile, p. 339.

66. Ibid., p. 340.

67. Bill Hamilton, letter to W. J. Turnbull, 7 August 1957, PO, accession 90-91/257, box 275, file 78-2-10, part 3. Turnbull, who was on the verge of quitting, warned Levy that "the new Government [was] prepared to support this project if it [was] kept down to a practical level and the present not sacrificed for the remote future, that is, a search for perfection should not be carried to the point where long delays result." W. J. Turnbull, memorandum to Maurice Levy 15 August 1957, PO, accession 90-91/257, box 275, file 78-2-10, part 3.

68. Ibid. Director (Organization and Classification, Civil Service Commission), memorandum to Post Office: Fourteen Additional Positions, 30 September 1957, Levy Papers.

69. For example, the computer's reliability still had to be established. The coding desk had to be completely redesigned. The reader's performance was still unacceptable. Maurice Levy, memorandum to W. J. Turnbull: Report on the work at hand, 6 September 1957, PO, accession 86-87/402, box 4, file Untitled.

70. In the process, Levy was let go. Frustrated and disillusioned, Levy did a little consulting for the US Post Office and tried to build a business around some of the bits and pieces of the electronic sorter.

71. Very early in the project Turnbull had warned Levy about seeking advice from others in the department: "It could be taken as a general principle that you should not pick up suggestions made by anyone in the Postal Service other than this office. This is the real danger I mentioned to you. Various officers will want to tempt you into bypaths and, while this at the start may not involve any great loss of time, still the net result will, in the long run, be a dissipation of energy." W. J. Turnbull, memorandum to Maurice Levy, 30 September 1952, Levy Papers.

72. Alan Dornian's article "ReserVec (*IEEE Annals of the History of Computing* 16, 1994, summer: 32–42) added an internalist perspective to the previous work of Ball and Vardalas. Dornian's work, however, missed the fundamental importance of distributed on-line, real-time technology, within the TCA context, and the institutional forces that shaped its design.

73. House of Commons, Debates, 1937, p. 2041.

74. Ibid.

75. In 1964, TCA was renamed Air Canada. Only four books have been written on Air Canada: Gordon R. McGregor, *The Adolescence of an Airline* (Air Canada, 1970); David Collins, *Wings Across Time* (Griffin House, 1978); Philip Smith, *It Seems Like Only Yesterday* (McClelland and Stewart, 1986); Susan Goldenberg, *Troubled Skies* (McGraw-Hill Ryerson, 1994).

76. Gordon McGregor, letter to directors of Trans-Canada Air Lines, 2 March 1956, appendix B: Capacity Program, 1951–65, Records of Air Canada (formerly called Trans-Canada Air Lines), RG 70, volume 325, file 2-1C, part 1, National Archives of Canada. *(Henceforth, Records of Air Canada, RG 70, will be cited as TCA, the abbreviation standing for the former name "Trans-Canada Air Lines.")*

77. Gordon McGregor, memorandum to Honourable George Hees, 18 July 1957, TCA volume 325, file 2-1 C, part 1.

78. Ibid.

79. Ibid.

80. Gordon McGregor, letter to directors of Trans-Canada Air Lines, 2 March 1956, TCA, volume 325, file 2-1C, part 1. A sum this large required the board of directors to consider asking for a modification of the Trans-Canada Air Line Act in order to increase the company's capitalization.

81. Trans-Canada Air Lines, Annual Report, 1945.

82. Ibid.

83. C. E. Amann, "Airline Automation," *Computers and Automation* (1957), August: 10–14, 30.

84. In 1951, TCA had 31,378 no shows, with a resulting loss of nearly $1 million. Ronald Keith, "How TCA "Lost" $1 million," *Financial Post*, 20 September 1952. The difficulty was not in giving the passenger the empty seat on the departing flight but in confirming seat availability on the subsequent legs of the passengers trip, particularly those legs on the return trip.

85. Henry Mintzberg, J. Pierre Brunet, and John Waters, "Does Planning Impede Strategic Thinking?" *Advances in Strategic Management* 4 (1986), p. 14.

86. According to one senior TCA executive (cited in ibid.), "What Howe said, went in those years."

87. TCA's quest for autonomy from the government and the associated strategic management issues were first raised in Taïeb Hafsi's doctoral dissertation, The Strategic Decision-Making Process in State-Owned Enterprises (Harvard University School of Business Administration, 1981).

88. See McGregor, *The Adolescence of an Airline*.

89. Taïeb Hafsi, Moses Kiggundi, and Jan Jørgensen propose a management typology for government-owned enterprises in Structural Configurations in the Strategic Apex of State-Owned Enterprises (Report 85-07, Écoles Des Hautes Études Commerciales, University of Montreal, 1985).

90. Hafsi et al., "Structural Configurations in the Strategic Apex of State-Owned Enterprises," 15.

91. Mintzberg et al., "Does Planning Impede Strategic Thinking," 37.

92. Ibid.

93. Gordon McGregor, letter to Frank Ross, 4 November 1954, TCA, volume 325, file 2-1C.

94. Frank M. Ross, letter to Gordon McGregor, 9 November 1954, TCA, volume 325, file 2-1C, part 1.

95. Between 1940 and 1942, Canadian Pacific Railways bought Canadian Airways, and nine smaller airline airlines companies, then merged them to create Canadian Pacific Air Lines (CPAL). In 1949, CPAL moved its headquarters from Montreal to Vancouver. The shift of corporate power to the west no doubt reflected the strength of the western part of the operation and the CPAL's Pacific focus. Entry into the Pacific market occurred after TCA decided not fly those routes.

96. McGregor, letter to Frank Ross, 4 November 1954.

Notes to pp. 124–128   349

97. Ibid.

98. Donald Gordon, letter to Gordon McGregor.

99. House of Commons, Debates, 6 March 1956, p. 870.

100. Ibid., p. 1863. Not all westerners objected to TCA's monopoly status. Stanley Knowles, a leader of Canada's socialist Cooperative Commonwealth Federation (CCF) party, argued that a government-owned airline was essential to establish a first-rate and economically viable air service in Canada (ibid., p. 1851).

101. Goldenberg, *Troubled Skies*, p. 19.

102. Ibid.

103. G. R. McGregor, "Submission to the Royal Commission on Canada's Economic Prospects," No. 115, 19 January 1956. Copy in the Library of Parliament, Ottawa.

104. Ibid., p. 12.

105. C. J. Campbell (Director of Telecommunications, TCA), memorandum to W. H. Seagrim (General Manager of Operations, TCA), 27 July 1955, TCA, volume 100, file 1620-10.

106. See Amann, "Airline Automation."

107. Amann, "Airline Automation" and Jon Eklund, "The Reservisor Automated Airline Reservation System," *IEEE Annals of the History of Computing* 16 (1994), no. 1: 62–69.

108. Among those who attended the meeting were Gordon McGregor (president), W. F. English (vice-president of operations), C. J. Campbell (director of communications), the vice-president of traffic, and the general manager of operations. W. F. English, memorandum to Gordon McGregor: Automatic Reservations System, 16 December 1953, TCA, volume 100, file 1620-10.

109. Campbell, memorandum Seagrim, 27 July 1955.

110. The reader should recall that in 1953, Canada had only one general-purpose computer: FERUT. This was the machine that DRB and NRC bought and installed at the University of Toronto in 1952.

111. Richardson recommended a total of 300 "Agent Sets." No copies remain of Richardson's October 1953 report entitled "Automatic Reservations—Passenger Flow Systems." The description given here comes from another study, in which the main elements of Richardson's report are briefly recapitulated. See L. P. Casciato, R. F. Johnston, and J. Kates, "Preliminary Report of an Automatic Passenger Service for Trans-Canada Air Lines," circa spring 1954.

112. Associated with "Local Store," Richardson called for an additional input/output device that would function as a "Master Agent Set." The device had to be able to examine passenger continuity on any flight.

113. Richardson actually proposed two central computers working in parallel. Each would serve as a backup to the other.

114. English, memorandum to Gordon McGregor: Automatic Reservations System, 16 December 1953.

115. The others were Cossor Ltd., Halifax, Canadian General Electric, Toronto, Canadian Aviation Electronics and International Business Machines, both in Montreal. To TCA's solicitation for bids to do an engineering study they replied that their "organizational structure would not allow them to participate, at the present time, in this study" (ibid.). Computing Devices of Canada bid $24,000, Adalia bid $18,000, and Ferranti-Canada bid $10,000.

116. In fact, Ferranti-Canada had attempted an earlier unsolicited proposal the moment it learned of the contents of Richardson's report.

117. Ibid. For the details of what Adalia proposed to do in the study see J. A. Richards (Executive Vice-President, Adalia Ltd.), letter to C. J. Campbell (Director of Communications, TCA), 25 February 1954, TCA, volume 100, file 1620-10.

118. Casciato, Johnston, and Kates, "Preliminary Report of an Automatic Passenger Service for Trans-Canada Air Lines," spring 1954.

119. C. J. Campbell, memorandum to H. W. Seagrim (General Manager of Operations, TCA), undated, TCA volume 100, file 1620-10.

120. C. J. Campbell, memorandum to General Manager of Operations, 24 September 1956, TCA, volume 100, file 1620-10.

121. To these, one should add the Royal Canadian Navy's Tactical Battle Simulator and the US Air Force's flight simulators where the term man/machine interface was first coined.

122. Lyman Richardson, "Some Applications of Electronic Data Processing Equipment in the Air Transport Industry," Paper 4, delivered to Canadian Convention and Exposition of the Institute of Radio Engineers, Toronto, 1–3 October 1956. A copy of the paper was given to me by Lyman Richardson.

123. Campbell, memorandum to General Manager of Operations, 24 September 1956. According to one TCA contract, Ferranti-Canada presented a report on the design of the "Transactor" on 27 July 1956, in Ferranti Document No. L208, Terms of Contract between Ferranti Electric Limited and Trans-Canada Air Lines, 5 March 1957, TCA, volume 100, file 1620-10.

124. Although all the 300 sensor heads read the card in parallel, data from the heads was picked off serially. For a schematic of the signal pick-up mechanism see Interdepartmental Electronics Committee (Government of Canada), Report on the Visit of R. Ziola and N. G. Anderson to Ferranti Electric Limited, Toronto, for a Demonstration of the Transactor, 12 December 1957, TB, volume 632, file 1501-22.

125. The drawback to the Transactor technology was its inability to serve as an interface for batch processing. The manual process of inserting a card, the mechanical clamping and unclamping action of the plates, and the final ejection of the card was far too slow to read cards in bulk. But the booking transaction was not a batch process. One dealt with one customer at a time, in real time.

126. Another economic advantage was that agents could share a Transactor without impeding their work flow. Richardson also reasoned that the use of cards added flexibility. If the system went down, the ticket agents could nevertheless still make note of customer requests and easily process them in bulk once the system was back up again. Otherwise, the agents would have to write down the booking request information and then enter it again when the system came back on.

127. One way around the constraint was to design different pre-printed cards for the various facets of the airline's operations: reservations, scheduling, flight operations, etc.

128. C. J. Campbell, memorandum to H. W. Seagrim (General Manager of Operations, TCA), 27 July 1955, TCA volume 100, file 1620-10.

129. Lyman Richardson, memorandum to W. G. Rathborne, Automatic Reservations—Report Function, 9 June 1955, TCA, volume 100, file 1620-10.

130. Ferranti-Canada began to work officially on the $75,000 Transactor contract on 5 March 1957. Josef Kates, now president of his new consulting company called KCS, was hired as a consultant. J. C. Ogilvie was retained as a consultant on the "so called 'human engineering' aspects of the Transactor design and the layout of the necessary cards." As part of the contract, TCA let Ferranti-Canada retain all patent rights but reserved the right to use the technology free of any royalty fees. Terms of Contract Between Ferranti Electric Limited and Trans-Canada Air Lines, 5 March 1957, TCA, volume 100, file 1620-10. "Ticketing Handset 'Reads' Pencil Marks," *Aviation Week* 11 (1957), November, p. 47.

131. W. Gordon Wood, confidential memorandum to G. R. McGregor, 7 November 1958, TCA, volume 100, file 1620-10.

132. In the first year of operation, 1961, salaries would drop from $1,172,547 to $246,087. In 1963, Wood estimated a drop in salaries from $1,394,207 to $246,087 (ibid.). McGregor replied that these estimates needed more detailed elaboration (on such questions as depreciation, leasing vs. buying, cost of housing the computer etc.) before he could use them to convince the board of directors. G. R. McGregor, confidential memorandum to W. Gordon Wood, 17 November 1958, TCA, volume 100, file 1620-10. In his second estimate, Wood estimated the final total system cost to be $3,234,550. W. Gordon Wood, confidential memorandum to G. R. McGregor, 21 November 1958, TCA, volume 100, file 1620-10.

133. Minutes of Inter-Departmental Meeting in President's Office: Automatic Data Processing, 10 December 1958, TCA, volume 100, file 1620-10.

134. Ibid.

135. Some of the elements Reid wanted incorporated into the computer's design specifications were batch oriented input/output devices, greater bulk storage capacity (specifically, large magnetic tape units), an interrogate command, and enough auxiliary memory capacity to allow the Reservations program to be dumped and another program played in and vice versa (ibid.).

136. Ibid.

137. Ibid.

138. G. R. McGregor, memorandum to comptroller, vice-president of operations, vice-president of sales, director of operations research, and director of administrative procedures, 24 December 1958, TCA, volume 100, file 1620-10.

139. Minutes of Inter-Departmental Meeting in President's Office: Automatic Data Processing, 10 December 1958. The year 1961 coincided with TCA's next round of aircraft modernization, which was to bring TCA into the jet age.

140. Computer check sorting was a direct application of the post office project. Instead of routing letters, one used a special-purpose drum memory computer to route checks. This will be discussed in the next section of this chapter.

141. Committee on the Assessment of Electronic Data Processing Equipment, Minutes of Second Meeting, 19 June 1959, TCA, volume 230, file 1240-22-2. The companies that declined to bid were Canadair Ltd., Montreal; Canadian Aviation Electronics Ltd., Dorval; Canadian Curtiss Wright Ltd., Montreal; Canadian General Electric Co., Toronto; Computing Devices of Canada; Metropolitan Vickers Electrical Export Co., Montreal; Remington Rand Ltd., Montreal; Sperry Gyroscope Co. of Canada Ltd., Montreal; and Standard Telephones and Cables Manufacturing Co., Montreal. The RCA Victor Montreal bid, submitted after the 1 June deadline, was not accepted. The value of the bids were as follows: Burroughs, $1,640,833; Canadian Westinghouse, $1,189,875; EMI, $1,853,700; Ferranti-Canada, $1,299,277; Honeywell, $1,330,690; IBM 1,189,620; and Philco, $1,400,962. Abstract of Tenders for Central Computer: Automatic Reservation System, 23 July 1959, TCA, volume 230, file 1240-22-2.

142. This close working relationship was about to bear fruit in the second half of the reservation project. Ferranti-Canada had become the clear favorite to win the $1. 8 million contracts to build the 332 Transactors, and all the digital communication equipment linking all the Transactors to the central computer.

143. Abstract of Tenders for Central Computer: Automatic Reservation System, 23 July 1959.

144. Committee on the Assessment of Electronic Data Processing Equipment, Minutes of Fourth Meeting, 30 June 1959, TCA, volume 230, file 1240-22-2. McGregor had set up this inter-departmental committee, under the chairmanship of Director of Operations Research P. J. Sandiford to select the best bid. The Burroughs proposal was rejected almost immediately because it proposed a vacuum tube design. The EMI proposal was poor in both categories.

145. Abstract of Tenders for Central Computer: Automatic Reservation System, 23 July 1959. The Honeywell and Philco bids that did not offer any emergency backup. With three computers, EMI offered the best redundancy, but the computers were inadequate in all the other requirements.

146. These additions included doubling the core store, adding three index registers, and using radix conversion circuitry. Other batch-oriented improvements were four additional magnetic tape drives and buffers; a card reader, a card punch, and the requisite control circuitry; and a Samatronic printer. J. B. Reid, memorandum to W. S. Harvey, Comptroller, 23 July 1959, TCA, volume 230, file 1240-22-2. The Gemini system that finally emerged consisted of a core memory of 4096 words for each computer. A word was 25 bits long plus a parity bit. The Gemini computers used a one-address architecture. Programming was carried out through an assembly-type language with 90 mnemonic instructions. In this language, addresses could be given names which would logically apply to the program. In all each computer had 15 registers. The first eight served as intermediate storage in calculations and as index registers for instruction modification. The ninth and tenth registers allowed the Gemini computers to coordinate their access to mass storage by keeping track of which locations in drum memory both computers were accessing at any moment. The eleventh register recorded the settings to which the 25 switches on the console were set. The twelfth register recorded error in the output/input operations. The thirteenth register was used for paper tape operations. The remaining two registers were reserved for future expansion. Only two additional tape drives were added to the four already specified in the initial proposal.

147. TCA held a contest among its employees to name the new automated reservation system. "Harry J. Simper, passenger office manager at Lethbridge, won $100 for coming up with the name ReserVec—from Reservations Electrically Controlled." Smith, *It Seems Like Only Yesterday*, p. 213. One wonders, however why the peculiar typographic form, with a capital letter v, for ReserVec.

148. Reservations Division, "ReserVec Report," 15 May 1963, TCA, volume 100, file 1620-10, part 2.

149. Of course it was already a part of military enterprise.

150. TCA management's preoccupation, during the period leading up to the commitment of funds to an automated reservation system, is another illustration of Mintzberg's view that TCA was focused on "how to do things better" rather than "what kinds of activities to do." Mintzberg et al., "Does Planning Impede Strategic Thinking."

151. These details on the SABRE come from a report from a Canadian government official to Treasury Board, after his visit to American Airlines. E. O. Landry, Notes on Visits to US Commercial and Industrial Firms to Discuss Central Computer Operations, 13 December 1962. TB, accession 80-81/069, box 179, file 1501-22, part 4. Transactor price in Wood, memorandum to McGregor, 21 November 1958. As a comparison, the standard flexowriter interface to a computer cost about $6,600. Reid, memorandum to Harvey, 23 July 1959.

152. Two IBM 7090s cost about $7.2 million (excluding tax), not including the cost of mass storage facilities. See Richard Smith, "A Historical Overview of Computer Architecture," *Annals of the History of Computing* 10 (1989), no. 4: 277–303. The Ferranti-Canada's Gemini twin computers cost $0.9 million (excluding tax). Reid, memorandum to Harvey, 23 July 1959.

153. P. N. O'Hara, "Know Your Data Processing Machines," *Office Automation*, May 1960: 59–71. This article provides an extensive catalogue of the commercial computers of the day. Dornian ("ReserVec) also provides an interesting tabular synthesis of performance characteristics of the two computers.

154. ReserVec I was extremely limited in the kinds of information it could process. No information about the passengers, other than the first four letters of their surnames, was kept. Other details still had to be filed by hand. SABRE on the other hand stored the passenger's full name, telephone number, special meal requests, and hotel and automobile reservations. It was this dramatic reduction in the kinds of information that ReserVec had to process that allowed the Transactor to be a cheap, but effective, machine interface. ReserVec II greatly expanded the variety of passenger information handled by the system. ReserVec II was designed for much higher throughputs than ReserVec I, from 4000 transactions per hour to 40,000. Smith, *It Seems Like Only Yesterday*.

155. For a history of SAGE see George Valley Jr., "How the SAGE Development Began," *Annals of the History of Computing* 7 (1985), July: 196–226.

156. "In 1958, the US Air Force issued a request for proposals for computers to be used in the DEWLINE air defense system. In response, IBM proposed a computer designated 7090 in December 1958. . . . In late 1958 the Air Force ordered four 7090s, to be delivered starting in November 1959. Components being built for STRETCH . . . and engineers working on STRETCH were diverted to the 7090." Cuthbert Hurd, "Early IBM Computers (Edited Testimony Given at antitrust trial of US Justice Department versus IBM)," *Annals of the History of Computing* 3 (1981), April: 179.

157. Even Computing Devices of Canada (CDC) had given up on building its own general-purpose computer business. By 1960, CDC was content to sell the computers of known US manufacturers. In 1960, CDC was selling Bendix computers. See O'Hara, "Know Your Data Processing Machines."

158. A Woods, Gordon & Co. survey of commercial computers for sale in Canada listed Ferranti-Canada's Gemini computer, along with its parent firm's Serius and Pegasus computers (ibid.).

159. ReserVec II also used parallel computers.

160. By 1964, all the world's major airlines had some form of computerized reservation system. The problem now was how to share information between these independent computer networks. TCA tried to play a leading role in establishing international standards for inter-airline reservation data exchange. TCA's active role was again driven by the issue of operating efficiency. In this case, Air

Canada's TCA's problem was to improve the selling of tickets when its passengers had to link up with other airlines. See "Some Considerations in the Exchange of Reservation Interline Information," 1964. The thrust of TCA's efforts was to define a standard format for communicating digital information. "The Communication of Data Between Systems," draft outline of an Air Canada presentation to IATA, in New York, circa 1964; "The Communication of Message Formats by Means of Field Defining Characters Using the Baudot Code," presented by Air Canada to the ATA-IATA Interline Communications Systems Working Group, 23–25 June 1964, Munich, Germany.

Chapter 5

1. In 1949, the Research Department was specifically created to handle Ferranti-Canada's new digital electronics work. In 1958, after Ferranti Electric merged with Packard Electric to produce Ferranti-Packard, the Electronics Division was created to manage all digital electronics product development work. In the context of this chapter, the term "Electronics Division" is used to denote the computer group even in pre-1958 references.

2. Reflective of the US's regional development, twelve regional banks made up the new Federal Reserve System, with overall coordination held by a Board of Governors in Washington.

3. The reserve deposits, to which all members of the Federal Reserve System had to contribute, became the medium of adjustment during the daily clearing process.

4. The Federal Reserve System (Board of Governors of Federal Reserve System, July 1954), p. 157. The figure for 1939 is very approximate because it was read off a graph. For the 1956 figure see appendix B, "Selected Operating Data of the Federal Reserve Banks and Branches," in R. C. Amara and Bonnar Cox, A Study of Check Handling Operations in the Federal Reserve System, a Stanford Research Institute Study prepared for the Federal Reserve Banks, September 1957, Federal Reserve Bank of New York Archives. *(Henceforth this archive will be cited as FRBNY.)*

5. See "Banking Uses," in R. Ziola, Report on the Fourth Institute on Electronics in Management, American University (TB, volume 632, file 1501-22), pp. 5–6.

6. This Electronics Subcommittee reported to the FRBs' Committee on Miscellaneous Operations. The FRBs also participated in the US Treasury Department's efforts to promote the development of electronic equipment to process the paper checks issued by the Federal government. On 28 and 29 September 1953, the Treasury Department held a Symposium on "Utilization of Electronic Equipment in the Clearance, Payment and Reconciliation of Government Card Checks," which brought together its officials, those from the FRBs, and twelve leading electronics, computer, and business machine manufacturers. J. E. McCorvey (assistant vice-president, Federal Reserve Bank of Atlanta),

memorandum to Lowell Myrick (Assistant Director, Board of Governors of Federal Reserve System), 30 September 1953, FRBNY. After this symposium, the Treasury Board set up the Joint Government Committee on Electronics to pursue the same question. Once again the FRBs participated. E. F. Bartlett (fiscal assistant secretary, Treasury Department), memorandum to H. H. Kimball (vice-president, Federal Reserve Bank of New York), Agenda of Meeting, 1 April 1954, FRBNY.

7. C. Edgar Johnson (vice-president, First National Bank of Chicago), memorandum to John Hurts (vice-president, Federal Reserve Bank of New York), 19 February 1954, FRBNY.

8. The MICR method called for the use of black magnetic iron oxide ink to print Arabic numerals. This was a departure from encoding information as bar codes. Today, all checks have MICR printed numbers on the bottom.

9. Oliver Whitby (Division of Engineering Research, Stanford Research Institute), letter Proposal (E 56-39) to H. H. Kimball (Subcommittee on Electronics), 11 April 1957, FRBNY.

10. Amara and Cox, A Study of Check Handling Operations in the Federal Reserve System, September 1957, p. 1.

11. The American Bankers' Association had drawn up an eight-digit code, called the ABA number, which uniquely identified each of the 21,000 banks in the US. SRI's study reaffirmed that Arabic numerals printed in magnetic ink were the best approach to encoding the ABA number and amount. "Since dirt, overstamping, and other types of document defacement are a constant problem, reliance has not been placed on optical scanning methods, such as photocell pick-up or those using phosphorescent ink; consequently these and related condition have led to the choice of magnetic ink as the recording or printing medium. The early success with bar-code systems led to the investigation of means of reading Arabic numerals directly. Since such Arabic numerals must be printed in any case to convey information to the human reader, much less printing and duplication are required, and an enormous advantage is attained." (ibid., pp. 16–17)

12. Ibid., p. 54, table VII.

13. Ibid., p. 4.

14. Ibid.

15. H. H. Kimball, memorandum of Meeting of Subcommittee on Electronics in New York in November 6, 1957, 22 November 1957, FRBNY

16. Stanford Research Institute, "Operational Performance Specifications for Check Handling Operations of Federal Reserve System," March 1958; Ferranti-Packard Electric Ltd., "A Proposed Document Sorting System for the Federal Reserve Bank System," FRBNY.

17. Earlier electro-mechanical, office-automation technology, such as punched cards and pencil-marked cards, sorted one digit at a time. For example, if one were sorting cards by an encoded, three-digit, decimal number, the first pass through the sorter would arrange all the cards into ten pockets, one for each of the possible ten first digits. Then each of these ten piles would be sorted again by the second digit. The process would be repeated for the third digit. SRI's recommendation to use a special-purpose, drum-memory sorter was intended to eliminate the time-consuming sorting by digit. Although multiple-pass sorting may have been workable in check sorting, it was totally unacceptable for postal sorting. This was why Maurice Levy had from the beginning pushed for a stored-memory approach to sorting. In its contract proposal to the FRBs, Ferranti-Canada played up the fact that its computer for the Post Office had already demonstrated the idea that it was feasible to do reliable, high-speed sorting by the entire ABA number, as SRI had called for.

18. The reader will recall that in DATAR, many sensing devices (radar and sonar) on several ships were connected on-line to one computer, and that users of different ships all had on-line, real-time access, to the special-purpose computer.

19. For details on the terms of reference of this pilot project see Guideline to Pilot Planning, 21 December 1959, FRBNY.

20. A. J. Stanton, Check Mechanization Department (Federal Reserve Bank of New York), memorandum to Mr. MacInnes: Proposed Budget—1960, 5 August 1959, FRBNY.

21. M. A. Harris, memorandum to W. F. Treiber (First Vice-President, Federal Reserve Bank of New York), 7 September 1961, FRBNY.

22. W. F. Treiber (First Vice-President, Federal Reserve Bank of New York), letter to Board of Governors of Federal Reserve System, 5 December 1961, FRBNY.

23. Ferranti-Packard Electric, "A Proposal for a Modular Check Sorter Computer for High-Speed Sortation of Magnetic Ink Encoded Checks," Prepared for The Federal Reserve Bank of New York, August 1961, FRBNY.

24. Ferranti-Packard News Release, 15 December 1961 (original in my possession).

25. Ferranti-Canada's design approach was to make all aspects of the computer check-sorting system modular, to enable customers to expand the capacity of their sorting systems in an economical and flexible manner. These modules consisted of a central dictionary look-up reference unit, drum-load and print-out facilities, simulators, controls, and power supplies. Any one of these units could be replaced and upgraded, as needs changed. A fully expanded check-sorting system, Ferranti-Canada argued, could process up to 135,000 checks per hour.

26. Dominion Bureau of Statistics, Canada, Annual Report on Cheques Cashed, 1948 (Ottawa, Dominion Bureau of Statistics, April 1949), p. 13.

27. Ibid., p. 1.

28. "Our Bank Enters the Electronic Age," *Royal Bank Magazine*, May—June 1964.

29. Ibid.

30. Proposal for a General Purpose High-Speed Document Sorting System for the Handling of M. I. C. R. Encoded Documents for Federal Reserve Bank of New York, Ferranti-Packard Electric Proposal No. TY. 100, prepared by Paul J. Dixon, January 1962, FRBNY.

31. Ibid., p. 25.

32. Ibid.

33. Ibid., p. 31.

34. W. F. Treiber (First Vice-President, Federal Reserve Bank of New York), letter to Board of Governors of Federal Reserve System, 5 March 1962, FRBNY.

35. Ibid.

36. Ibid.

37. This was the price quoted in Dixon's January 1962 proposal.

38. Ferranti-Packard, "Draft Specification of the FP6000 High Speed Check Sorting System for the Federal Reserve Bank of New York," May 1962, FRBNY.

39. Like the Gemini computers, the FP6000 computer used a 24-bit word length, with an additional bit for parity, and one-address architecture. But the instruction formats were different. An instruction in Gemini was structured as follows: S, M, F, X, N; the Stop-Go, S, was 1 bit; reference to the register for indexing, M, was 3 bits; the operation code, F, was 6 bits; the accumulator registers, X, was 3 bits; and the address in core store, N, was 12 bits. FP6000 used only 2 bits for the index registers and but seven bits for a longer function code. The last difference was to accommodate a much larger order code. See Ferranti-Packard, Gemini Data Processing System (1963), in Canada's National Museum of Science and Technology collection.

40. Naval Research Establishment, Progress Report FP6000-2, 30 May 1963, DND, accession 83-84/167, box 7589, file DRBS 9900-31, part 1. The computer was delivered in June 1963.

41. This system was delivered in December 1963.

42. Peter Drucker, *Management* (Harper and Row, 1973), p. 61.

43. Throughout the period 1956–1964, Ferranti UK's computer operations were, from a profit perspective, marginal at best. See Ball and Vardalas, *Ferranti-Packard*, p. 255.

44. F. S. Beckman and F. P. Brooks, "Developments in the Logical Organization of Computer Arithmetic and Control Units," *Proceedings of the Institute of Radio Engineers* 49 (1961), January: 53–66.

45. M. J. Marcotty et al., "Time Sharing on the Ferranti-Packard FP6000 Computer System," Proceeding—Spring Joint Computer Conference, 1963, of the American Federation of Information Processing Society, p. 40. In the late 1950s and the early 1960s one finds the terms multiprogramming, concurrent programming, parallel programming and time-sharing often used interchangeably. However, the FP6000, by today's standards was not a time-sharing computer; i. e. there was no on-line sharing of the system's facilities by remote users. The FP6000 allowed many batch programs to run concurrently.

46. For a comprehensive technical review, as seen from the historical context of 1961, of the issues surrounding the optimal allocation of memory as well as other issues surrounding the implementation of multiprogramming see E. F. Codd, "Multiprogramming," in *Advances in Computers*, volume 3 (Academic Press, 1962).

47. The "datum" and "limit" points occurred in multiples of 64 words, where a word was 24 bits long. The FP6000 used a priority system to allocate the system's resources. The computer's supervisory system, called the Executive, assigned each program a priority code based on its expected use, as measured by duration and frequency, of the system's hardware. Before each transfer of control, the Executive scanned its priority list for the active program with the highest priority.

48. The task of dynamic memory allocation was even more complex because the FP6000's supervisory program, called the Executive, allowed the user to write a program in such a way that the different parts of the program could multiprogram among themselves.

49. Martin Campbell-Kelly (*ICL: A Business and Technical History*, Oxford University Press, 1990, p. 221) writes: "The FP6000 had originally been specified in England as an overtly 'commercial' computer by one of Ferranti's salesmen, Harry Johnson; and the design had much common ancestry with the Ferranti Pegasus. But because Ferranti had most of its resources committed to the ORION and ATLAS, it was not developed in England. The design was picked up by the Canadian subsidiary, where it was developed during 1962." A similar misunderstanding was expressed by Moreau in *The Computer Comes of Age*. Moreau suggests that the FP6000 was developed with "the aid of techniques developed in Manchester and used for the first time in Ferranti Packard 6000" (ibid., p. 123). In note 13 on page 123, Moreau further suggests mistakenly that the FP6000 was a Ferranti UK initiative.

50. S. Gill, "Parallel Programming," *Computer Journal* 1 (1958): 2–10.

51. Codd, "Multiprogramming," 1962

52. When Ferranti-Canada engineers first described FP6000 in the technical literature, they situated their achievement within the context of Gene Amdahl's criteria for multiprogramming, which were published in the early phases of the FP6000's design. See Marcotty et al., "Time sharing on the Ferranti-Packard FP6000 Computer System," 1963. For more on Amdahl's criteria see, Gene Amdahl, "New Concepts in Computing System Design," *Proceedings of the Institute of Radio Engineers* 50 (1962), May: 1073–1077.

53. The market need for a smaller and less costly alternative to large mainframe technology is apparent from the meteoric rise of DEC. In 1957, $70,000 of local venture capital set up DEC in an old woollen mill in the suburbs of Boston. In 1966, when DEC went public, the worth of the company had grown to $228. 6 million.

54. The resulting cost constraints naturally affected hardware and software design choices The choice of 24-bit words is one example of this compromise. One compromise which affected the design of the Executive was to use the central arithmetic unit to perform the addressing and counting during peripheral transfers. This was in contrast to the usual practice found on the large multiprogramming machines where these tasks were handled by independent logic in each peripheral control unit.

55. These systems differed in core memory, mass storage capacity and in input/output facilities. As a result all these systems differed greatly in price.

56. John Deutsch (secretary of Treasury Board), memorandum to all Departments, 31 May 1955, TB, volume 632, file 1501-22. The first meting of this Committee was held in March 1955; see Minutes of first Meeting of Committee to Study Developments in Electronic Computing Machines and Their Applications, 9 March 1955, volume 632, file 1501-22.

57. *Royal Commission on Government Organization*, volume 1 (Queen's Printers, 1962), p. 45.

58. Ibid.

59. Ibid., p. 589.

60. *Royal Commission on Government Organization*, volume 2 (Queen's Printers, 1962), p. 119. It is ironic that, the Canadian military having abandoned its earlier ambitions of technological and industrial self-reliance, a civilian government Commission would now champion the cause. The Canadian defense establishment, argued the Commission, showed a "marked lack of confidence in Canadian science and technology as a whole." *Royal Commission on Government Organization*, volume 4 (Queen's Printers, 1963), p. 208.

61. Industrial Development Group (Electrical and Electronics Branch, Department of Industry), Quarterly Report: 1 July to 30 September 1965, DITC, volume 1728, file P1410-07, part 1.

62. Ibid.

63. Michael Bliss, "Canadianizing American Business," in *Close the 49th Parallel*, ed. I. Lumsden (University of Toronto Press, 1974).

64. The word "encouraged" rather than "caused" is used because Canada's patent laws were also a crucial factor in the early rise of branch plants. This factor is not brought out in Bliss's article.

65. In this regard, Peter Drucker's observation of the relationship between "product" and "customer" is noteworthy. "What a customer thinks he is buying, what he considers value, is decisive—it determines what a business is, what it produces, and whether it will prosper. And what the customer thinks is buying is never a product. It is always utility, that is what a product or service does for him." Drucker, *Management*, p. 61.

66. Part of the Bank's sophistication came from the ongoing consulting role of the Stanford Research Institute.

67. Interview with Longstaff et al.

68. Expert Group on Electronic Computers (Directorate of Scientific Affairs, Organization for Economic Cooperation and Development), Gaps in Technology between Member Countries, Restricted Report-DAS/SPR/67. 93, 4 October 1967. Copy in DITC, volume 1790, file IRA-9000-320/T2-5

69. Ibid., p. 11.

70. Ibid.

71. The merger of Ferranti Electric and Packard Electric in 1958 was a direct result of the fierce competition ravaging the Canadian electrical industry. Ball and Vardalas, *Ferranti-Packard*.

72. Such a move, if successful, gave Ferranti-Canada an unusual, but rare, business advantage over its Canadian competitors. Since they were all subsidiaries of US electrical power manufacturers, they could not ship to the US and compete against the parent firm. Ferranti-Canada could sell domestically and to the US; its Canadian competitors could only sell locally.

73. Ball and Vardalas, *Ferranti-Packard*.

74. Summary of Twelfth Monthly Meeting with Directors, 5 November 1964, 8, DITC, volume 1727, file P. 1115-19.

75. Industrial Development (Group Electrical and Electronics Branch, Department of Industry), Quarterly Report: 1 January to 31 March 1964, DITC, volume 1728, file P1410-07, part 1.

76. J. H. Morgan, letter to Chief Superintendent, Naval Research Laboratory, 20 July 1964, DND, accession 83-84/167, box 7589, file DRBS9900-31, part 2.

77. The initial impetus for this was to export magnetic drums to the US defense industry. Thus the Department of Defence Production first started this initiative, then, after its creation, the Department of Industry took over. By the time the Department of Industry had decided to consider supporting Ferranti-Canada in its bid to sell general-purpose computers, "Ferranti had nearly completed the 3rd year of a 5 year 50/50 shared cost-development program. Items developed included tape readers and magnetic memory drums, some $3 million worth of which has already been sold to US companies for less than $1.0

million development effort." Industrial Development Group, Electrical and Electronics Branch, Department of Industry, Quarterly Report: 1 July to 30 September 1964, p. 4, DITC, volume 1728, file P1410-07, part 1.

78. E. A. Booth, letter to J. H, Morgan, 15 July 1964, DND, accession 83-84/167, box 7589, file DRBS9900-31, part 2.

79. Campbell-Kelly, *ICL*.

80. Ibid., p. 222.

81. Interview with John Picken, 11 May 1989.

82. Interview with Sir Donald MacCallum, 9 May 1989.

83. No documents have yet been found to corroborate or refute this claim. The $1 million compensation that Ferranti-Canada accepted from its parent firm was in exchange for the loss of the design rights to the FP6000.

84. The global news agency Reuters acquired I. P. Sharp Associates in 1987.

85. Bob Van der Wijst, "Saga of the FP6000," *Canadian Information Processing Society Computer Magazine* 3 (1972), November: 4.

86. "Letter to the Editor," *Canadian Information Processing Society Computer Magazine* 3 (1972), November: 7.

87. B. Bleackley and J. LaPrairie, *Entering the Computer Age* (Book Society of Canada, 1982), p. 52.

88. David Thomas, *Knights of the New Technology* (Key Porter Books, 1983), p. 110.

89. By the mid 1950s, the total number of computers in the non-communist world could be counted in the 100s. Ten years later, this number was over 50,000 and the global production in computers was about $4 billion per annum and growing rapidly (Gaps in Technology, OECD, 1969). It was estimated that the value of computer installations in the US alone would double from 1968 to 1972. Gilbert Burck, "Computer Industry's Expectations" *Fortune*, August 1968: 93–97, 142, 145–146. Another estimate depicted a ten-fold increase in the yearly amount of money that would spent in Europe on computers from 1969 to 1980. Philip Siekman, "Europe vs. IBM," *Fortune*, August 1969: 87–91, 174, 176, 178.

90. If one subtracts out the contribution of the very large US market, US firms still supplied 85% of the computers used by the rest of the industrialized world (Gaps in Technology, OECD, 1969).

91. By 1960, as the US military spending on computer R&D declined, a new infusion of money from NASA and the "space race" served sustain the technological lead of the US.

92. Many of these points were discussed in Kenneth Flamm's works. *Targeting the Computer* (Brookings Institution, 1987) and *Creating the Computer* (Brookings

Institution, 1988). On the role of State in the development of the US computer industry see *Government and Technical Progress*, ed. R. Nelson (Pergamon, 1982).

93. Gaps in Technology: Electronic Computers (OECD, 1969).

94. See chapter 2 of G. W. Brock, *The US Computer Industry* (Ballinger, 1975).

95. Institutional Research Department, New York Securities Co., Perspective on the Computer Industry: 1966–1970, August 1966. CBI 80, Series 2, box 4, Folder 9, Charles Babbage Institute, Minneapolis.

96. It is important to recall that few software applications were standardized and off-the-shelf. Companies devoted a great deal of resources to developing their own custom application software.

97. Robert Sobel, *IBM, Colossus in Transition* (Bantam, 1983), p. 213.

98. Ibid.

99. Ibid.

100. Gaps in Technology.

101. Harry Johnson, "Involvement of Ferranti Ltd. in the Development of the FP6000 Computer," February 1988, p. 4. Private communication to Basil Ziani de Ferranti. Location of original unknown. Copy was sent to former CEO of Ferranti-Packard, and is now in this writer's possession.

102. Ministry of Technology, Government of United Kingdom, "Strategic Importance of Computers," Brief to Select Committee on Science and Technology, Parliament of United Kingdom, 1 March 1970. DITC, accession 85-86/222, box 96, file IRA-9050-10, part 1. For a detailed historical analysis of ICL, see Campbell-Kelly, *ICL*.

103. Attaché, Scientific Affairs, Canadian Embassy, Paris France, The French "Plan Calcul," May 1970, Records of Science Council of Canada, RG 129, Vol. 5, file 700/3, National Archives of Canada. Also see the sections on CII in Alvin Harman, *The International Computer Industry* (Harvard University Press, 1971), Flamm, *Targeting the Computer*, and Flamm, *Creating the Computer*.

104. CII's immediate goal was to develop a range of computers that compete against the lower range of the IBM 360 series: the 360/20, /30, /40, and /50. Its longer term goals, for the early 1970s, was to develop large computers. "Summary of Foreign Government Incentives to Computer Industries," 8 February 1968, prepared by the Automation and Instrumentation Section, Electrical and Electronics Branch, Department of Industry, Trade and Commerce, Canada. DITC, Vol. 474, file P8001-7400/E1.

105. "Summary of Foreign Government Incentives to Computer Industries," 8 February 1968, prepared by the Automation and Instrumentation Section, Electrical and Electronics Branch, Department of Industry, Trade and Commerce, Canada. DITC, Vol. 474, file P8001-7400/E1.

106. Marcel Cote, Yvan Allaire, and Roger-Emile Miller, IBM Canada (Case Study 14, prepared for Canada's Royal Commission on Corporate Concentration, 1976). In setting the context for the Canadian subsidiary, this case study offers a brief analysis of IBM's worldwide operations and organizational structure.

107. In 1949, with more than 30% of all its employees in foreign operations, IBM merged all its international subsidiaries under one organizational umbrella, called IBM World Trade. These subsidiaries did not report to any functional units within the US operations but instead directly reported to Arthur Watson, Thomas Watson's younger son.

108. Cited in Siekman, "Europe vs. IBM."

109. Flamm, *Targeting the Computer*. Much of the discussion that follows on Japan is taken from Flamm, Targeting the Computer and Martin Fransman, *Japan's Computer and Communications Industry* (Oxford University Press, 1995)

110. "Summary of Foreign Government Incentives to Computer Industries," 8 February 1968, prepared by the Automation and Instrumentation Section, Electrical and Electronics Branch, Department of Industry, Trade and Commerce, Canada. DITC, Vol. 474, file P8001-7400/E1.

111. The collaborative ventures between Japanese and US companies included one between Hitachi and RCA, one between Mitsubishi Electric and TRW, one between NEC and Honeywell, one between Oki and Sperry Rand, and one between Toshiba and General Electric. Fujitsu was the only company that did not ally itself with a US firm for know-how.

112. The diversity and competitive structure of the Japanese computer industry rested on the existence of a strong user-supplier network that had been built up in telecommunications. Fransman argues that historians have failed to recognize the important contribution that this network made in the growth of the Japanese computer industry. A key reason for the existence of this network was the Japanese system of "controlled competition" that had started during the 1920s. "The essence of controlled competition," write Fransman, "is a co-operative relationship between a large procurer and user of equipment and a closed group of suppliers which compete in outside markets. Controlled competition," Fransman goes on to explain, "is an alternative form of organization to vertical integration on the one hand, and arm's-length market relationships on the other." Fransman, *Japan's Computer and Communications Industry*, p. 22.

113. Wayne J. Lee, ed., *The International Computer Industry* (Applied Library Resources, 1971).

114. Charles Drury, Minister of Industry, memorandum to Cabinet: Canadian Computer Industry, 24 January 1968. Records of Office of Privy Council of Canada, RG 2, Vol. 6331, Cabinet Document No. 102/68. National Archives of Canada.

115. Ibid., p. 4.

116. Mira Wilkins, *American Business Abroad from 1914 to 1970* (Harvard University Press, 1974).

117. R. A. Gordon, confidential memorandum to C. D. Quarterman, 28 December 1967. DITC, Vol. 474, file P. 8001-7400/E1

118. DITC expected IBM World Trade to allocate to Canada the same ratio of R&D and manufacturing to sales as it did in its overall operations in Europe.

119. Charles M. Drury, letter to J. E. Brent, 4 October 1967, DITC, volume 474, file P. 8001-74/E1.

120. By the end of the 1970s, as a recession set in and multinationals sought to cut costs, DITC explicitly tried to merge its idea of world products mandates with the growing trend in global rationalization.

121. Starting in 1970, computer services and software offered the best chances for the emergence of a strong domestically owned computer industry. Having already carried out considerable research in these two areas, I am preparing an in-depth historical treatment of the growth of Canada's computer-services and software firms.

122. The National Research Council of Canada's decision to buy an IBM computer over Ferranti-Canada's FP6000 best captures the depth problem of trying to use government procurement to support a domestic industry. If the National Research Council (NRC), whose mandate was to advance Canada's scientific and technological competence, refused to gamble on a Canadian made computer, then how could one expect other less technologically sophisticated departments to do otherwise. Believing that the FP6000 was comparable to the IBM machine, the E&E Branch was dismayed at NRC's decision.

123. D. H. Fullerton and H. A. Hampson, "Canadian Secondary Manufacturing Industry," for the Royal Commission on Canada's Economic Prospects, May 1957.

124. Ibid., p. 63.

125. Ibid., p. 72.

126. For detailed discussions of the Auto Pact, see R. Wonnacott and P. Wonnacott, *Free Trade Between the United states and Canada* (Harvard University Press, 1967); D. Wilton, *An Economic Analysis of the Canada-US Automobile Agreement* (Economic Council of Canada, 1976); Melvyn Fuss and Leonard Waverman, The Canada-US Auto Pact of 1965 (Working Paper 1953, National Bureau of Economic Research, Washington, 1986).

127. Over the years the framework of the Auto Pact greatly benefited Canada's trade surplus with the US In 1965, the ratio of the number cars produced in Canada by the Big Three to the number sold in Canada was one-to-one. In 1996 this ratio had grown to two-to-one, with the surplus all being shipped to the US market. Mark Heinzl, "Made in Canada," *Wall Street Journal*, 14 October 1996, A2.

Next to natural resources, automotive products have been the biggest contributor Canada's export trade.

128. David Crane, "Ottawa seeking active role in growth of Canadian computer industry," *Globe and Mail* (Toronto), 4 March 1968.

Chapter 6

1. *Henceforth "Sperry Canada."*

2. UMAC is an acronym for Universal Multi-Axis Controller. The 5 signifies that this was the company's fifth model of controller. The first four were purely NC systems.

3. In a speech before the House of Commons, on 8 February 1951, C. D. Howe, then the Minister of Trade and Commerce, described the extent of military spending during 1950. Military equipment, supplies and construction contracts of $56 million, $88 million, $132 million, and $318 million were issued in the first, second, third and fourth quarters of 1950 respectively. Of this amount, $267 million was for aircraft. Right Hon. C. D. Howe, "Speech from the Throne," House of Commons, Debates, 8 February 1951, pp. 179–187.

4. A capital stock of the new company was set at $50,000, 500 shares at $100 per share. letters Patent: Sperry Gyroscope Company of Canada Ltd., 8 November 1950, Sperry-Rand Corporation Administrative Records, accession 1910, Series 1, Minute Books, box 29, Sperry Gyroscope Co. of Canada Ltd., Hagley Museum and Library.

5. Sperry Canada also acted as an agent for the British subsidiary of Sperry in London and it very quickly became the Canadian representative for another US firm, Kollsman Instrument Corporation. R. E. Crawford, "Instruments to keep pace with today's aircraft," *Canadian Machinery and Manufacturing News* 66 (1955), September: 101–106.

6. *The Canadian Who's Who*, Volume X, 1964–66 (TransCanada Press, 1966), p. 579.

7. Minutes of Special Meeting of Board of Directors, 20 February 1951, Sperry-Rand, Minute Books, box 29, Sperry Gyroscope Co. of Canada Ltd. Except for a few Canadian representatives, the board of directors of the Canadian subsidiary was predominantly composed of Americans. All meetings of the board of directors took place in the American head office. Though any resolution passed by the Sperry Canada Board was, in effect, a decision of the Sperry Corporation's most senior executives, Wence King was the principal force steering the growth of the Canadian subsidiary.

8. B. W. King, "Instruments and Controls," *Canadian Aviation* 27 (1954), June: 82, 89.

9. Howe, "Speech from the Throne," 8 February 1951.

10. Minutes of Board of Directors Meeting, 21 May 1951, Sperry-Rand, Minute Books, box 29, Sperry Gyroscope Co. of Canada Ltd.

11. Minutes of Board Directors Meeting, 3 July 1951, at Sperry Gyroscope Co., Great Neck, Long Island, Sperry-Rand, Minute Books, box 29, Sperry Gyroscope Co. of Canada Ltd. One immediate consequence of this decision was the creation of new company, the Sperry Gyroscope Co. of Ottawa, which fell under the direct control of Sperry Canada in Montreal. This facility was retained as an instrument overhaul facility. By 1955, the Sperry Gyroscope Co. of Ottawa had grown to 250 employees "overhauling 500 different types of instruments for aircraft, marine and industrial service in all fields, and manufacturing parking metres, range switches, range timers and coin metres of many different patterns." Crawford, "Instruments to Keep Pace With Today's Aircraft," September 1955, p. 101. The latter manufacturing activities represented this company's efforts to survive the contraction in military spending after the end of the Korean War.

12. Until 1950, Côte de Liesse remained an undeveloped street. However, by the end of the Korean War considerable industrial location had taken place on the street, such as Canadian Aviation Electronics (CAE), Bepco Canada Ltd., RCA Victor, Hollinger-Hanna Shore and Labrador Railway Co. and the Standard Electric building, which housed the offices of Sylvania Electric, Chrysler, and Shell.

13. These figures came out in the Minutes of the Board Directors Meeting, 20 July 1954, Sperry-Rand, Minute Books, box 29, Sperry Gyroscope Co. of Canada Ltd. In 1954, Sperry Canada gained ownership of its manufacturing facilities. The sale price was half the original cost. In addition, the Department of Defence Production agreed to build in an accelerated depreciation rate into procurement contracts with Sperry Canada for an additional year.

14. This figure was mentioned in interviews that government representatives had with Sperry Canada's Employment Manager. Employment Forecast Survey Interview Report, 7 May 1956, DDP, accession 70/349, box 558, file 200-13-395.

15. The high-precision nature of the work led Sperry Canada to hire fifty watchmakers. "The work here," observed one Sperry Plant Superintendent, "is very meticulous involving production and assembly of fine gears, wheels and pinions and a lot of detailed wiring. For instance, the motor that powers the standard gyroscope is about two inches in diameter and turns over 2400 revs a minute." Employment Forecast Survey Interview Report, May 7, 1956, DDP, accession 70/349, box 558, file 200-13-395. Surprisingly, few women were hired to do the fine assembly work.

16. For more technical discussion of the design and manufacture of these instruments see Crawford, "Instruments to keep pace with today's aircraft."

17. King, "Instruments and Controls."

18. The machometers were manufactured under license from the Kollsmann Co.

19. The reader will recall that the second mission statement in Sperry Canada's founding charter was "to design, manufacture, purchase, sell, lease, import, export and deal in any gunfire, bomb, torpedo and rocket control equipment appliances and devices, for use on or under the sea, on land, or in the air."

20. The total contracts awarded for 1953, 1954, 1955, 1956, and 1957 were $1.3, $1.6, $1.2, $0.6, and $0.8 million, respectively. These figures were pieced together from monthly announcements, in *Canadian Machinery and Manufacturing News*, of Department of Defence Production contracts issued to Canadian companies. The data was limited because the contracts listed were at least $10,000 and did not include any secret military projects. The first limitation means that the totals for Sperry Canada were slightly underestimated. No evidence known to me suggests that Sperry Canada had obtained any substantial secret weapons contracts. The 1953 and 1955 values may be significantly undervalued because data for three months were unavailable.

21. The 50% figure was explicitly mentioned in Sperry Gyroscope Co. of Canada Employment Forecast Survey Interview Report, 7 May 1956, DDP, accession 70/349, box 558, file 200-13-395.

22. See the brief historical overview of the Remington Rand Corporation in M. N. Rand, Report To Operations Committee of Sperry Rand Corporation, 24 February, 1958, Sperry Rand, Sperry Gyroscope, Minute Books, box 30, file Operations Committee.

23. Eckert and Mauchly had formed this company to commercialize their pioneering work on the ENIAC computer.

24. Taken from balance sheet data found in Rand, Report to Operations Committee of Sperry Rand Corporation, 24 February 1958

25. Remington Rand manufactured its first electric razors in 1937. This business had a separate sales and service network of 1,800 wholesalers and 75,000 retailers.

26. This Committee was made up of Sperry Rand's president, executive vice-president, general counsel, treasurer, and secretary, and two additional executive vice-president from each of the two parallel lines of authority. Harry Vickers (President of Sperry Rand), memorandum on Corporate Reorganization, 8 January 1958, Sperry Rand, box 30, file Operations Committee Meeting January 8, 1958.

27. The Remington Rand Group also included the Remington Rand Division, the Electric Shaver Division, the Univac Division, and the International Division. See chart 2—Sperry Rand Organization in Countermeasures Division (Sperry Gyroscope Company), Presentation To The Operations Committee: Sperry Rand Corporation, 20 October 1958, Sperry Rand, box 30, file Operations Committee Meeting 20 October 1958.

28. The other nine divisions were Aero Equipment, Air Armament, Tubes, Marine, Microwave, Surface Armament, Wright, Semiconductor, and Engineering and Manufacturing Services. Carl Frische (President of Sperry

Gyroscope), The Gyro Group: A Progress Report, 7 February 1958, Sperry Rand, box 30, file Operations Committee Meeting 7 November 1958.

29. Carl Frische, The Gyro Group: A Progress Report, 7 November 1958

30. See ibid., figure 12, p. 27.

31. At one point, the downward trend in profits provoked the president of the Gyro Group to rethink the way Sperry Rand organized its R&D. Frische observed that "it might be difficult to maintain continuous long-term research programs within the framework of product-oriented operating Divisions and some consideration is being given to the feasibility of establishing an R&D organization and facility for the common benefit of the Gyro Group." Frische, The Gyro Group: A Progress Report, 7 November 1958.

32. This anecdote is from my 18 February 1985 interview with Burton Wenceley King, founder of Sperry Canada, for the Computer Oral History Project of the National Museum of Science and Technology.

33. J. J. Childs, *Numerical Control* (Industrial Press, 1973), p. 5.

34. Interview with Peter Herzl, 7 February 1995.

35. Herzl recalls that several Sperry Canada NC systems were installed at Great Neck. He suggests that the successful operation of the Canadian numerical control systems played an important role in increasing the Gyro Group head office's confidence in the technical merits of the Canadian system.

36. These innovations were formalized in a series of patents. In order to get the most protection of a commercial development, Sperry Canada tended to delay filing patents until the product was ready for commercial sale. This meant that the filing dates on Herzl's patents came at least a year after the idea was designed, tested and implemented. Peter Herzl, Fluid Valve Apparatus, Canadian Patent no. 646,723, filed 16 July 1959, issued 14 August 1962. Peter Herzl, Positional Control System and Pick off Therefor, Canadian Patent no. 668,938, filed 21 January 1960, issued 20 August 1963. Peter Herzl, Digital-to-Analog Converter, Canadian Patent no. 625,838, filed 4 February 1960, issued 15 August 1961. Peter Herzl, Reliability Check for Programmed Multi-Channel Position Control, Canadian Patent no. 624,649, filed 4 February 1960, issued 25 July 1961. Peter Herzl, Coarse-Fine Positioning System, Canadian Patent no. 714,200, filed 12 January 1962, issued 30 July 1965.

37. Among the company's first innovations were a balanced vane type hydraulic pump for powering hydraulic control systems and a hydraulic power-steering device for automobiles. See "Sperry Rand Corporation Chronology," Inventory for accession 1910, Hagley Museum and Library.

38. Sperry, Ford and Vickers Committee on Fields of Endeavour, Participation of the Sperry Group in FIELDS OF ENDEAVOR, March 1959, p. 27, Sperry Rand, Harry Vickers Papers, box 13, Stockholder Relations, file 1959.

39. The total cumulative error, over any number of operations, remained less than 0.00025 inch. See "Positioning control holds .00025 limits; rugged enough for use 'out in the shop'," *Canadian Machinery and Manufacturing News* 70 (1959), June: 98–100.

40. See chart A1, "Group Activities Showing Customer and Company Funds," in Sperry, Ford and Vickers Committee on Fields of Endeavour, Participation of the Sperry Group in FIELDS OF ENDVEAVOR, March 1959. The major sources of military funding was for the development of missile systems and subsystems, radar systems and bombing-navigation systems. Of the massive missile program launched by President Eisenhower in 1957, Sperry had a role, either as a prime contractor or subcontractor, in 35 missile projects including ICBM Atlas and Titan, the IRBM Navy Polaris, the Redstone, the Bomarc, the Jupiter, the Nike, and Sparrow missiles. See the table entitled, The Breadth and Depth of Sperry Rand Participation In The Nation's Major Missile Programs, Sperry Rand, Gyroscope Co., box 31, file Operations Committee Meeting 26 January 1959.

41. "Return on Total Assets," Sperry Rand, Sperry Gyroscope Co., Minute Books, box 31, file Operations Committee Meeting, 26 January 1959.

42. The definition of the return on capital invested used here is the ratio of net-income to total assets, and not to capital employed. The 1959 value was a forecasted figure. E. M. Brown, Presentation on Cost Plus Fixed Fee Contracts, 1 April 1959, Sperry Rand, Gyroscope, Minute Books, box 31, file Operations Committee Meeting, 25 August 1958. In an attempt to underscore that this decline was not due to excessive profits in the earlier years, Brown predicted that the Gyro Group's 1959 after-tax return on invested capital would be nearly, as a percentage, half the average of the 100 largest US manufacturers. Brown's argumentation on this point is weak. He uses data from "The One Hundred largest Manufacturing Corporations in 1956," *The National Industrial Conference Board's Business Record* 15 (1958), February: 68–69, to substantiate his comparison. Despite the 1958 publication data, this data is for 1956. If one looks at the Gyro Group's after tax percent return on invested capital during fiscal 1956 and 1957, the company comes close to the average. However, its performance still falls far below that of the top 50 companies. In actual fact Sperry Gyroscope's percent return on investment had declined steadily since 1940. Sperry Gyroscope's most profitable years, in terms of percent of return on investment, were 1935–1939. Surprisingly, the very sharp increase in sales during both World War II and the Korean War did little to alter this trend. Profits as a percent of sales, however, did show considerable movement. Graphs depicting changes in profits, profits as percent of sales and return on investment, as well as total sales were compiled by the president of the Sperry Gyroscope Co. in a report to the Sperry Rand board of directors. Frische, "The Gyro Group," 7 November 1958.

43. Employment Forecast Survey Interview Report, DDP, accession, box 555, file 200-13-395.

44. D. A. Golden (Deputy Minister of Defence Production), letter to secretary of Treasury Board, 19 December 1961, DITC, box 1784, file P8270-270/S18-10, part 1.

45. Ibid.

46. In late 1961, Sperry Canada received an order of $2 million from Sperry Rand. Within a few months, Sperry Canada was to receive an additional order, or a comparable amount, from the parent firm. Taken together $4 million dollars of purchases from the parent firm represented a large proportion of Sperry Canada's $7–8 million sales (ibid.).

47. To my knowledge, there have been no historical studies of how US multinationals with branch plants in Canada used the Defence Sharing Agreement.

48. The figure of 60 was reported in *Metal Working News*, 28 August 28 1961.

49. This amount was stated in Golden, letter to secretary of Treasury Board, 19 December 1961

50. *Canada Month* 1 (1961), July, p. 12.

51. *Industrial Canada* 62 (1963), April, p. 17.

52. *Canadian Controls and Instrumentation* 2 (1963), April, p. 7.

53. Irwin Wolfe, *Canadian Machinery and Metalworking* 73 (1962), April: 36–38.

54. Sperry Canada was awarded the SIP contract in the spring of 1961. Brief announcement in *Metalworking News* (New York), 8 May 1961.

55. The key innovation here was to develop digital-to-analog and analog-to-digital interfaces to SIP's rather unique measuring system. SIP designed and manufactured its own lead screw measuring systems. For each lead screw, SIP would custom make a cam that would correctly compensate for any small errors in the lead screw's production. SIP prided itself on the accuracy of this solution. Sperry Canada had to adapt its NC technology to SIP's unusual approach.

56. Interview with Bob Cox, 25 January 1995.

57. SIP's single order for 15 NC systems from Sperry Canada prompted one trade journal to call the sale "probably the largest single order for numerical units ever placed by a European company." After this initial order, the Italian firm San Georgio signed a licensing agreement to produce and sell Canadian NC systems in Europe. For many years after, San Georgio continued to sell a transistorized version of the original Sperry Canada NC system to SIP.

58. H. G. Hauck (Manager of Industrial Marketing, Sperry Canada), Numerical Control Newsletter 6, 13 February 1963, accession 1910, Sperry Rand Records, Harry Vickers Papers, box 13, file—Sperry Gyroscope Company Group, 1962–64, Hagley Museum and Library

59. Rocketdyne was a division of California-based North American Aviation.

60. Golden, letter to secretary of Treasury Board, 19 December 1961.

61. The rest of the $400,000 would go to the machine tool builder who would build the machine-bed for the inspection unit.

62. This information was a contained in one of Sperry Canada's proposal submissions to Canada's Department of Industry. Sperry Gyroscope Company of Canada, Limited, Proposal to Department of Industry Concerning the Development of Numerical Control Systems, DITC, volume 1784, file P8270-270/S18-10, part 1.

63. The purpose of this R&D contract was to develop a highly precise machine and digital adaptive control system for the production of small bore holes to extremely tight tolerances in aerospace materials (ibid.).

64. D. H. Fullerton and H. A. Hampson, "Canadian Secondary Manufacturing Industry," for Royal Commission on Canada's Economic Prospects, May 1957.

65. Alfred Chandler, *The Visible Hand* (Belknap, 1977)

66. Fullerton and Hampson, "Canadian Secondary Manufacturing Industry," 63.

67. Ibid., p. 72.

68. Ibid. Diversification was itself a Canadian manufacturing strategy "to reduce the handicap for short runs in a relatively small and fragmented market." Whereas a US tire manufacturer produced for a large orders of only a few types of tires, its Canadian counterpart manufactured 600 different types of tires in much smaller batches. Some Canadian wire and cable plants produced more than 1000 kinds of wires.

69. Hugh G. Hauck, "Are We Snubbing a Profit Maker," *Canadian Machinery and Metalworking* 73 (1962), January: 48.

70. One approach did not exclude the other. One could have free trade in some products based on mass production and a technological solution for the domestic market.

71. Ibid. Sperry Canada's espousal of the advantages of NC for the Canadian context was picked up by other observers. "A possible answer to the problem of applying automation to short-run production is numerical control of machine tools," wrote the journal *Plant Management*. "Available for some time, it is surprising that more companies have not taken advantage of it. There are probably no more than half-dozen installations in Canada." "Numerical Control Equipment," *Plant Management* 7 (1962), February: 33.

72. The figure of nine NC systems in Canada was extracted from a special issue of *Canadian Machinery and Metalworking* (77, April 1966). From the table of all NC installations on pp. 130–137 one can select out those installed on or before 1962. The US data comes from "Machine Tools with Numerical and Prerecorded Motion Program Controls," *Current Industrial Reports Series BDSAF-630(63)-1* (US Bureau of the Census, 1964).

73. The proportion of the US GDP arising from manufacturing was 18 times Canada's (Bernard Mueller, *A Statistical Handbook of the North Atlantic Area*, Twentieth Century Fund, 1965). As an internal report within Canada's DITC noted, in 1965, the US was investing about $1.5 billion per year on machine tools and $250 million in NC technology. In Canada, the annual investment in machine tools was $75 million and $5 million in NC technology. "This means," the brief underscored, "that our demand was one-twentieth of the USA and the numerical content one-fiftieth." Advanced Machine Tool Development, 6 May 1965, DITC, volume 1784, file P8270-270/S18-10, part 1. In 1962, with only nine NC systems in Canada, this disproportion was probably even more exaggerated.

74. "The acceptance of [NC] is still agonizingly slow," observed one writer for Canadian Machinery and Metalworking. He then added: ". . . the view that it is best suited to a large volume still lingers—along with other misconceptions." T. W. Weissmann, "Can You Really Afford to Be Without Tape Control?" *Canadian Machinery and Metalworking* 75 (1964), April: 110. According to a 1966 survey on the state of NC in Canada, the increase to 120 NC systems in Canadian machine shops demonstrated that the misconception of NC as a mass-production strategy was less widespread than in the early 1960s. The fact that the number of NC installations was not higher indicated that this misconception was nevertheless still present. Richard M. Dyke, "Numerical Control In Canada," *Canadian Machinery and Metalworking* 77 (1966), April: 75–127.

75. Interview with Cox, 25 January 1995.

76. Barker Industrial Equipment reasoned that Canada's import tariff rates, which taxed machines tools at 7.5% and numerical control equipment at 22.5%, would privilege made-in-Canada NC technology. The company imported American machine tools and then fitted them with Sperry Canada numerical control units. This strategy was only of marginal benefit to Sperry Canada. Weissmann, "Can You Really Afford to Be Without Tape Control?"

77. Herbert E. Klein, "Numerical Control," *Dunn's Review and Modern Industry* (1965), August: 34–35, 63–66.

78. Employment Forecast Survey Interview Report, 2 May 1963, DDP, volume 555, file 200-13-395.

79. William Stocker, "Building Adaptability into N/C," *American Machinist/Metalworking Manufacturing* 107 (1963), 24 June: 64–65.

80. "Machining centers," which started to appear in the early 1960s in the most advanced machine environments, were capable of a multiplicity of operations in one setup. Unlike the traditional special-purpose machine tool characteristic of mass-production, a single machining center could face-mill, end-mill, profile, contour, drill, ream, tap, bore, and counter-bore in five or more axes.

81. The Velvet Glove project to design a short range guided missile arose from Canada's postwar determination to be an active contributor to the military technology pool of the North Atlantic Triangle alliance. See chapters 1–3 of this volume.

82. Work on the Argus simulator had introduced Caden to the design of electronic digital systems. While still at CAE, he also designed a small relatively simple, solid-state, special-purpose, digital computer to process hydrographic data.

83. Interview with Cox, 25 January 1995.

84. W. S. Kendall, memorandum to Inter-departmental Committee, 26 May 1965, DITC, volume 1784, file P8270-270/S18-10, part 1.

85. "Canadian Firm Develops New Control System," *Montreal Gazette*, 4 July 1963.

86. Stocker, "Building Versatility Into NC," 24 June 1963.

87. "Sperry Has Order Backlog Before First Delivery Made," *Montreal Gazette* 15 November 1963, p. 31.

88. For a list of the orders see B. W. King (President, Sperry Gyroscope Co. of Canada Ltd.), letter to Department of Industry, 6 November 1964, DITC, volume 1784, file P8270-270/S18-10, part 1.

89. "Parts programming" should not be confused with the UMAC-5 software development. "Part programming" entailed preparing the tape with the machining instructions for each job. Part programming was carried out by the machine shop. UMAC-5's software then translated the very general instructions into all the detailed measuring and control functions needed to carry out the work. UMAC-5's software can be likened to the operating system that oversees a computer's operation. Part programming was an application that ran within the operating system.

90. Paul Caden remembers two other women but is unable to recall their full names.

91. Interview with Cox, 25 January 1995.

92. All those with whom I spoke made reference to this incident. In Paul Caden's version, the machine tool's table fell one story and landed on a parked car!

93. W. Wayt Gibbs, "Software's Chronic Crisis," *Scientific American*, September 1994, p. 87.

94. "It's like musket making. Before the industrial revolution, there was a no specialized approach to manufacturing goods that involved very little interchangeability and maximum of craftsmanship. If we are going to lick the software crisis, we're going to have to stop this hand-to-mouth, every-programmer-builds-everything-from-the-ground-up, pre-industrial approach." (ibid.)

95. Bob Cox, personal communication.

96. Sperry Gyroscope Company of Canada Limited, UMAC 5 Development, 1 November 1965, 2, DITC, volume 1784, file P8270-270/S18-10, part 1.

97. Problems with software also made the installation far more time consuming and costly than was necessary. The first UMAC-5 controller took 90 days to install (ibid., p. 8).

98. Ibid., p. 2.

99. "General purpose compilers, such as used with conventional data processing facilities," explained one company document, "are not suitable for our purpose due to the fact that the internal organization of the UMAC-5 computer is specifically directed at machine control functions rather than general purpose data processing." Sperry Gyroscope Co. of Canada Ltd., Proposal to Department of Industry Concerning the Development of Numerical Control System, 7 April 1965, 13-4, DITC, volume 1784, file P8270-270/S18-10, part 1.

100. Both Bob Cox and Paul Caden mentioned this figure in interviews. Unfortunately, neither can remember the details of how the $5000 figure was arrived at.

101. Sperry Canada used Bryant drum which was an aluminum cylinder ten inches in diameter and five inches long. Memory was distributed along 96 of the drum's 100 tracks. The other tracks were for timing purposes. Each track could hold a maximum of 4096 bits of information. The drum rotated at sixty revolutions per second.

102. "Canadian NC system prunes tooling costs in US plant," *Canadian Machinery and Metalworking* 74 (1963), April: 170.

103. "An automatic wire-wrap machine fitted with numerical control designed and produced by Sperry Gyroscope Co. of Canada, Ltd., Montreal, is handling 2000-wire panels at less than 7% of the cost of hand wiring." "Automatic Wire Wrap Proves Out," *Canadian Controls and Instrumentation* 2 (1963), February: 36.

104. Interview with Paul Caden, December 1994.

105. Ibid.

106. Sperry Gyroscope Company of Canada Limited, UMAC 5 Development, 1 November 1965, DITC, volume 1784, file P8270-270/S18-10, part 1. p. 2

107. House of Commons, Debates, 24 October 1963, p. 3937.

108. Ibid., p. 3998. There was a certain irony in this political debate, which was being fought under the shadow of the canceled AVRO Arrow program. In Opposition, Hellyer had condemned the Diefenbaker government's destruction of Canada's defense industry. Now it was the Conservative party's turn to decry the Liberal party's dismantling of the Canadian defense industry.

109. Sperry Rand's various US divisions, explained one company official to a Canadian government official, "were now experiencing a period of greatly reduced volume and [were] unable to provide subcontract work" to the Canadian division. R. H. Littlefield, letter to C. R. Nixon, 11 January 1965, DITC, box 1784, file P8270-270/S18-10, part 1.

110. See chart 3 in B. W. King, letter to Department of Industry, DITC, box 1784, file P8270-270/S18-10, part 1.

111. Bendix, Hughes, TRW, and GE had the largest share of the high-performance NC business. Sperry Canada concluded that these four companies would maintain approximately 63% of this market over the foreseeable future. But the Canadian company hoped that it would eat into the market share of the other smaller competitors. According to Sperry Canada's scenario, its share would grow from 6% in 1964, to 8% in 1965, 14% in 1966 and then taper off to 17% over 1967 and 1968. With firm orders adding to $1.2 million before even delivering its first system, the company believed that by 1968 its UMAC-5 sales could total as much as $4.5 million. (ibid., chart 4)

112. Among the advantages cited were an internal computing ability to shorten program tapes; easy set-up features; high adaptability; a capacity to combine point-to-point and contour capabilities without loss of accuracy; a high order of interpolation in contour cutting; easy installation; and the ability to reenter a cut without retraversing (ibid., p. 4).

113. Ibid., chart 3.

114. See chart entitled "Continuous Path Numerical Control System Schedule," and graph entitled "Proposed Rate of Expenditure for Continuous path N/C System," in Littlefield, letter to C. R. Nixon, 11 January 1965.

115. Ibid.

116. Ibid.

117. "CN machine tool census sounded industry warning," *Canadian Machinery and Manufacturing* 69 (1958), December: 190. In the same issue, see also "Canada has 262,000 machine tools in the metalworking industry" (126–127) and "How the census was prepared and what it tells and how" (130–150).

118. Economic Council of Canada, First Annual Review: Economic Goals for Canada to 1970 (1964).

119. In 1962, there were only nine NC-equipped machine tools in Canada. By the end of 1964, only 51 machine tools with NC existed in Canada. "Internal Report: Advanced Machine Tool Development," 6 May 1965, DITC, volume 1784, file P8270-270/S18-10, part 1. Over a year later, the number had climbed to 121, eleven of which had been made by Sperry Canada. Though considerable in absolute terms, in relative terms, Canada still lagged far behind its biggest competitor, the US While the ratio of machine tools between the US and Canada was 17 to 1, the ratio for NC systems deployed was 62 to 1. Simon Reisman (Deputy Minister of Department of Industry), Draft of memorandum to C. M. Drury (Minister of Department of Industry), 21 April 1966, DITC, volume 1784, file P8270-270-/S18-10, part 2.

120. "Canada must catch up or crumble," *Canadian Metalworking/Machine Production*, February 1965: 31–32.

121. Reisman, Draft of memorandum to Drury, 21 April 1966.

122. Ibid.

123. Ibid.

124. King, letter to Department of Industry, 6 November 1964.

125. The first order for an American Kearney & Trecker machining center had been placed in Canada in 1965. The purchase was funded by the Defence Modernization Vote. See Advanced Machine Tool Development, 6 May 1965.

126. Very little data are available showing both the number and value of machine tools purchased by Canadian industry. From the study referred to here, it was impossible to find any category of machine tool for which the average price was more than $20,000. "Metalworking Machinery Imports," *Canadian Machinery and Metalworking* 75 (1964), August: 67–69. By contrast, the average cost of a machine tool found in the US was considerably higher: $73,000 for a boring machine, $94,000 for milling machine, $40,000 for a lathe, and up to $140,000 for other more exotic types of machine tools. See US, Department of Commerce, Current Industrial Reports: Machine Tools With Numerical and Prerecorded Motion Program Controls—1963, Series BDSAF-630-(63)-1, 1964, table 1. In these average prices are for the machine tool alone. The cost of the NC unit has been subtracted.

127. F. Dugal (Machinery Branch, Department of Industry), memorandum to W. H. Huck, 23 April 1965, DITC, volume 1784, file P8270-270/S18-10, part 1.

128. Instead of a 60-40 split of $1.2 million, Sperry Canada now asked for a 50-50 split of $557,000. The amount of money needed to complete UMAC-5 was increased from $125,000 to $277,000. The total cost to develop the new NC system was $280,000. R. H. Littlefield, letter to C. R. Nixon, 7 April 1965, DITC, volume 1784, file P8270-270/S18-10, part 1.

129. Sperry Gyroscope Company of Canada Ltd., "Proposal to the Department of Industry Concerning the Development of Numerical Control Systems," 7 April 1965, pp. 11–12. DITC's official request for approval from the Treasury Board reaffirms the government's view that a cheaper NC technology was not only needed to promote Canadian defense production requirements, but that it was also an important step in meeting the Economic Council of Canada's call for ways to improve national productivity. Assistant Deputy Minister (Department of Industry), Details of Request to the Honourable Minister of the Treasury Board, 18 June 1965, DITC, volume 1784, file P8270-270/S18-10, part 1.

130. Assistant secretary (Treasury Board), letter to S. S. Reisman (Deputy Minister of Industry), 5 July 1965, DITC, volume 1784, file P8270-270/S18-10, part 1.

131. The cost sharing agreement was dated retroactively to November 1964, when Sperry had made its first proposal.

132. This entailed the computer's assembler language, all the needed subroutines, and the programming aids for debugging. The structure of the memory

drum was also improved with the inclusion of "fast tracks," which helped to optimize the software. The assembler had one fundamental flaw. It was written in a non-standard language that ran on the commercial Bendix G20 computer. This choice was purely utilitarian. Located at the federal government's Meteorological Department at Dorval Airport, the computer was accessible and only ten minutes from the plant. However, over the course of writing the assembler, it became apparent that Fortran had become the language of choice in software development. With a Fortran compiler, UMAC-5 software development could be carried out on any mainframe. After a competitive bid, Sperry Canada chose the Service Bureau Corporation, in Boston, over two Canadian firms to rewrite UMAC-5's assembler in Fortran IV.

133. For example, Sperry Canada sold a number of UMAC-5s to Europe in which they were used to make side press frames for printing presses. This point-to-point application required that a lot of precise holes be bored in a very precise relationship to one another. The Burghardt boring machine used for this was fitted with over 160 different tools. (interview with Burton Wencely King, 18 February 1985) The ability to program all the tool offsets into a computer made Sperry Canada's CNC approach quite advantageous.

134. G. J. House, Notes on Sperry, 7 August 1968, DITC, accession 85-86/666, box 96, file P. 8001-270/S. 18, part 2.

135. R. H. Littlefield (Director of Planning, Sperry Canada), letter to J. R. Killick (Chief, Avionics and Electronic Data Processing, Department of Defence Production), 8 April 1968, DITC, accession 85-86/666, box 96, file P. 8001-270/S. 18, part 2.

136. House, Notes on Sperry, 7 August 1968.

137. Littlefield, letter to Killick, 8 April 1968.

138. D. B. Mundy, Draft of memorandum to J. S. Glassford, 29 May 1968, DITC, accession 85-86/666, box 96, file P. 8001-270/S. 18, part 2.

139. Littlefield, letter to Killick, 8 April 1968.

140. To the Department of Defence Production, Sperry Canada wrote that the cancellation of the Primary Display Unit carried "grave implications as to the future of the Company. . . . Despite our best efforts in search of business, we know of no other program to take its place. Without it or some alternative source of volume we would be operating below our break-even point for factory operations. To add to this loss, the substantial investment required to insure success on the UMAC-6 program is cause for great concern. It is questionable whether as responsible management, we should proceed with UMAC-6 investment under these circumstances." Littlefield, letter to Killick, April 8, 1968. p. 3

141. House, Notes on Sperry, 7 August 1968.

142. R. A. Gordon, memorandum To file: A Visit to Sperry, Montreal—May 8, 1968, 10 May 1968, DITC, accession 85-86/666, box 96, file P. 8001-270/S. 18, part 2.

143. On the question of the impact of the tariff on the Canadian manufacture of computers see preceding chapter on Ferranti-Canada.

144. E. A. McIntyre (Deputy Director of Electrical and Electronics Branch, Department of Industry), memorandum to D. B. Mundy on Meeting with Sperry Officials, 8 August 1968, DITC, accession 85-86/666, box 96, file P. 8001-270/S. 18, part 2.

145. The circumstances of King's resignation are still unclear. In his eighties, King currently lives in Montreal. Still bitter, he adamantly refuses to discuss these events. His only comment was that the leadership of Sperry Rand had changed for the worse during those years. This P. Caden, who came from the head office in New York, should not be confused with the Paul Caden, the engineer who designed the computer in UMAC-5. The primary sources do not reveal what the initial "P" in P. Caden stands for.

146. J. Marshall, letter to E. McIntyre (Deputy Director, Electrical and Electronics Branch, Department of Industry), 7 October 1968, accession 85-86/666 box 96, file P. 8001-270/S. 18, part 2.

147. E. A. McIntyre, memorandum to D. B. Mundy (Assistant Deputy Minister, The Department of Industry), 4 September 1968, DITC, accession 85-86/666, box 96, file P. 8001-270/S. 18, part 2.

148. Ibid.

149. R. A. Gordon, memorandum to file: UMAC-6, 27 August 1968, DITC, accession 85-86/666, box 96, file P. 8001-270/S. 18, part 2.

150. R. A. Gordon, memorandum to file: Meeting with Sperry and Honeywell on September 3, 1968, 5 September 1968. E. A. McIntyre, memorandum to D. B. Bundy: UMAC-6, 6 September 1968, DITC, accession 85-86/666, box 96, file P. 8001-270/S. 18, part 2.

151. E. A. McIntyre, letter to J. Marshall, 23 October 1968, DITC, accession 85-86/666, box 96, file P. 8001-270/S. 18, part 2.

152. The programs cited were the Program for Advancement of Industrial Technology (PAIT), the Industrial Research and Development Incentives Act (IRDIA) and Industry Modernization for Defence Exports (IMDE).

153. The name Sperry Canada will also refer Vickers UMAC Division.

154. "Sperry Assigns NC to Vickers," *Financial Post*, 7 December 1968.

155. Ibid.

156. Daniel B. Dallas, "Industry at War," *Manufacturing Engineering* 88 (1982), January, p. 79.

157. See Gary Vasilash and Glenn Hartwig, "The Boom Begins: 1946–1950," *Manufacturing Engineering* 88 (1982), January: 119–140; Gary Vasilash, "The

Advent of Numerical Control," *Manufacturing Engineering* 88 (1982), January: 143–172.

158. The Strippit Division of Houdaille Industries Inc., of Akron, New York, Hartwig explained, "started out just trying to bring the flexibility of digital computing to machine controls." But in the end, Hartwig added, the company "wound up creating an entire new field, computer numerical control, or CNC." Glenn Hartwig, "Electronic Control Moves Into Ascendancy," *Manufacturing Engineering* 88 (1982), January: 181–202.

159. Ted Crozier, memorandum to Dan Dallas, 13 October 1982. Copies of this memorandum and other letters between Crozier and Dallas were made available to me by Jack Scrimgeour, who was a Senior Advisor for Advanced Manufacturing Technologies and Industrial Automation, at the National Research Council. *(Henceforth, this source will be cited as Scrimgeour.)* Crozier had sent Scrimgeour a copies of all the correspondence. At the time, Scrimgeour was responsible for a small newsletter called *CAD/CAM and Canada*.

160. Daniel Dallas, letter to J. E. Crozier, 21 October 1982, Scrimgeour.

161. J. E. Crozier, letter to Daniel Dallas, 29 October 1982, Scrimgeour.

162. Hartwig, "Electronic Control Moves Into Ascendancy, 1960–69," p. 199.

163. Daniel Dallas, letter to J. E. Crozier, 5 November 1982, Scrimgeour.

164. Ibid. A retraction was never printed in *Manufacturing Engineering*. In view of all the coverage that UMAC-5 received in the trade literature, Dallas's inability to find evidence of Sperry Canada's achievements in the historical record is quite surprising.

165. "Where Canada's NC machines are located," *Canadian Machinery and Metalworking*, April 1966: 130–137; "The NC picture in Canada today," *Canadian Machinery and Metalworking*, July 1968: 96–104; J. A. Weller, "Where Canada stands in NC," *Canadian Machinery and Metalworking*, March 1970: 69–72; "Canada's NC installations," *Canadian Machinery and Metalworking*, March 1974: 60–69.

166. Walter Jones, "Government support put focus on NC machines," *Canadian Machinery and Metalworking*, March 1970: 74–75, 132.

167. "Opinion," *Canadian Machinery and Metalworking*, March 1977, p. 3.

168. One survey, carried out by the Canadian Manufacturer's Association and the CAD/CAM Technology Advancement Council, found that only 38% of the metal fabricating industry used NC. "Information Needs of Canadian Companies on Computer Aided Design and Computer Aided Manufacture," February 1982, Results of Questionnaire Prepared by and Distributed by the Canadian Manufacturer's Association in Cooperation with the CAD/CAM Technology Advancement Council.

169. Canadian CAD/CAM Council, Management in Crisis (1982), p. 2.

170. From 1969 to 1977, Sperry Canada sold about 80 NC systems in Canada, most of the sales occurring between 1972 and 1976.

171. These figures represented total cumulative presence of Sperry Canada NC systems up to the given year and not the percentage NC systems introduced that year. Calculated from data from the trade journal *Canadian Machinery and Metal Working*. Owing to the absence of any corporate records, one cannot assess the success that the new NC-based UMAC strategy had in the international arena. But it is safe to assume that its share of the international market also steadily declined.

Chapter 7

1. A detailed institutional history of Control Data is not the goal of this work. The company's history is far too complex and rich to fit within one chapter. This narrative will confine itself only those elements of Control Data's history that set the stage for the company's entry into Canada and explain the ensuing relationship.

2. All financial data on Control Data used in this chapter have been compiled from the annual *Moody's Industrial Manual* (Moody's Investors Service); B. R. Eng, M. G. Rogers, and J. J. Karnowski, "Control Data Financial History: July 8, 1957 to Dec. 31, 1979," Control Data Corporation Collection, CBI, Series 4, box 5, Folder 2, Charles Babbage Institute Archives.

3. Gregory H. Wierzynski, "Control Data's Newest Cliffhanger," *Fortune*, February 1968, p. 126.

4. For example, in 1965 Sperry Rand, Honeywell, Burroughs, RCA, NCR, and GE derived respectively 24%, 12%, 25%, 5%, 7%, and 1% of their revenues from their computer businesses. Institutional Research Department, New York Securities Co., Perspective on the Computer Industry, August 1966, CBI, Series 2, box 4, Folder 9.

5. It is of interest to note that much of the Digital Equipment Company's early success came from selling to technically sophisticated scientists and engineers. While CDC pursued the top-down supercomputer approach, DEC pursued a bottom-up approach, marketing the minicomputer for smaller computational jobs.

6. The CDC 1604 used lower circuitry than the IBM 7090, but regained performance through a "clever design of the processor." See R. W. Allard, "Product Line History: Control Data EDP Systems" (attachment to memorandum to William Norris, 21 July 1980), CBI, Series 4, box 1, Folder 4.

7. These prices include the cost of peripheral equipment (Report: Control Data, 15 December 1961, CBI, Series 2, box 4, Folder 8).

8. Thornton was another important member of the 6600 design team. For a history of 6600 computer see James Thornton, *Design of a Computer* (Scott, Foresman, 1970).

9. The long and perilous R&D journey of the 6600 nearly sank the company.

10. Wierzynski, "Control Data's Newest Cliffhanger."

11. Unsatisfied with the quality of the peripheral equipment, CDC entered the peripheral-equipment in order to supply its own needs. But it soon became apparent that internal demand alone could not cover the growing R&D costs of developing new peripheral equipment. In a bold move, CDC became an "original equipment manufacturer" (OEM) that supplied peripheral equipment to all computer manufacturers. Its role as an OEM supplier was enhanced when it started producing IBM plug-compatible peripherals.

12. Wierzynski, "Control Data's Newest Cliffhanger," 126.

13. Developing supercomputers for large time-sharing systems also reinforced Control Data's computer services business, the company's other important axis of diversification. By means of time-sharing technology and large, centrally located computers, Control Data's computer services business sold access to powerful computational facilities to small businesses.

14. The 3600 project was led by E. Zimmer and Chuck Casale.

15. Derrel Slais, Private Communication, 8 January 2000. Slais worked on the design of the 3600 later rose to become VP of Research and Development at Control Data.

16. Allard, "Product Line History: Control Data EDP Systems," 1980.

17. Interview with Derrel Slais, 19 May 1999.

18. Allard, "Product Line History."

19. Systems prices ranged between $2.5 and $4 million (ibid.).

20. John Eastling was the product manager for the 3200.

21. Yearly figures are not available; however, according to Allard, "Product Line History: Control Data EDP Systems," the total revenues to 1974 arising from the entire 3000 series of computers was about $0.77 billion. The 6000 and 7000 series, which came from Cray's group, yielded sales of $1.09 billion. However, Although the revenue from the 3000 computers is less than that from the 6000 and 7000 series, the profits before taxes are higher: $49 million for the 3000L computers alone; $43 million for the 6000 series; and a $53 million loss for the 7000.

22. T. D. Rowan, memorandum to W. C. Norris, 18 June 1980. CBI, Series 4, box 1, Folder 1.

23. These discussions took place between F. C. Mullaney, ; T. N. Kisch, general manager of Computer Division (CD), T. D. Rowan, head of Product Planning, CD; C. E. Miller, head of Applications, CD; E. D. Larson, head of Manufacturing, CD; and E. D. Zimmer, head of Engineering, CD.

24. Computer Division, Control Data Corporation, "Analysis of Product Strategy," 9 June 1965. CBI, Series 4, box 1, Folder 3.

25. The new generation of the 3000L line was to match the Model 30 to Model 70 range of the 360 series.

26. Emmanuel Otis, email to John Vardalas, 7 May 1999. In 1965, Emmanuel Otis was the Chief Engineer of the 3000 and 6000 lines. Mr. Otis was in charge of taking Seymour Cray's 6600 design into production. In 1967, he became a vice president and stayed in the engineering until 1972 when he then moved into sales. Since Cray was working on the next generation of supercomputer, the 7600, there was an idea to extend the range to include the 6600.

27. Interview with Slais, 19 May 1999.

28. CDC-Canada, Third Quarterly Project Report, Canadian Development Division, 11 June 1971. RG 20, accession 83-84/229/ box 186, file P8803-EA-44, part 5. These marketing objectives appeared in the context of a computer project that was subsequently pursued in Canada, which is the subject of the next section. Although the Canadian project, as the reader will learn, took a completely different technical approach than the MPL, it inherited the identical marketing objectives.

29. Tom Rowan, "STAR-100," 31 October 1974. CBI, Series 4, box 1 Folder 5.

30. The Army Research Projects Agency then took up the Solomon concept and awarded a contract to the University of Illinois to build the machine. The resulting computer was instead called Illiac IV.

31. Other key figures in the design and development of the STAR-100 are Neil Lincoln, D. H. Toth, and C. L. Hawley.

32. Thornton had already articulated the string-array concept around the time the 66000 was being completed

33. The STAR also made use of extensive pipelining techniques. Because the STAR-100 used memory-to-memory approach, there were no limitations, apart from the size of the RAM, as to how big the vectors of operands could be. Vectors with thousands upon thousands of operands could be "streamed" into the processing unit. Registers-to-register could have never accommodated large vectors. A register –to-register operation was faster than a memory-to-memory one. So for small vectors, the overhead needed to set up the vector made the STAR-100 slower than a Cray's 7600. But as the vector got longer the overhead became quite small. For long vectors the STAR-100 was at last least ten times faster than Cray's 7600.

34. On a US Weather Bureau benchmark test program, the Texas Instruments Advanced Scientific Computer could only accomplish the task in 33 minutes. The STAR 100 later ran through the program in 12 minutes and 30 seconds. Rowan, "STAR-100."

35. Control Data Corporation press Release, 9 November 1970. DITC, accession 83-84/229, box 186, file P. 8803-EA-44, Pt. 4.

36. Ibid.

37. The number "50" in PL-50 signified the mid-price point within the entire product line. STAR-65 was the name under which the product would be marketed.

38. Interview with Emmanuel Otis, 17 September. CDC actively sought to establish an alliance with another computer manufacturer, either in the US or in Europe, who would develop a line of smaller computers that were upwardly compatible with the PL series. The alliance would ensure common standards and compatibility. In the mid 1970s, this search would produce an ambitious R&D collaboration between National Cash Register and CDC.

39. William Norris, memorandum to file: Meeting with Philips in Holland, 12 November 1962. CBI, box 3C6B, Folder 28.

40. Sumantra Ghoshal and Christopher Bartlett, "The Multinational Corporation as an Interorganizational Network," *Academy of Management Review* 15 (1990), no. 4: 603–625.

41. Interview with Robert Price, 16 March 1999. Price joined CDC in 1961 and quickly rose through the company. He succeeded Norris as CEO of Control Data.

42. Norris, memorandum to file, 12 November 1962.

43. Wilkins, *American Business Abroad from 1914 to 1970*.

44. By the end of 1969, CDC had 24 subsidiaries around the world (Control Data Corporation, "Dates of Incorporation of Foreign Subsidiaries," 1980. CBI, Series 4, box 2, Folder 20).

45. R. D. Sirrs, Senior Trade Commissioner, Canadian Consulate in the US, letter to Ralph Risch, Director of Planning, Control Data Corporation, 29 October 1969. DITC, accession 83-84/229, box 186, file P. 8803-EA-44, part 1.

46. Interview with William Norris, 18 March 1999. CDC's Peripheral Division operated as an OEM (original equipment manufacturer) business supplying peripherals to other computer manufacturers. To finance the R&D requirements of this Division, CDC had entered into several joint ventures for product development with a variety of US and European computer manufacturers. For example there were "Computer Peripherals" which included CDC, National Cash Register (NCR), and the British firm International Computer Limited (ICL); and "Magnetic Peripherals," which also included the preceding companies and the German firm Nixdorf. The importance of cooperation and collaboration in Control Data's corporate culture was reinforced in the many other interviews I have conducted with former company executives.

47. There were also logistical advantages to Canada's proximity. The PL-50 would entail moving a good number of the engineers and families abroad for several years. The Canadian option was cheaper and far less disruptive to all the families with children.

48. The reader will recall that, in 1949, the Royal Canadian Navy had created Computing Devices of Canada (ComDev) and Ferranti-Canada as the industrial

centers of excellence to spearhead its plans to create DATAR—the world's first computer automated naval warfare system. Although ComDev withdrew from DATAR, it went on to do pioneering work in digitally based battle simulators for the Navy. While Ferranti-Canada took its DATAR experience into the commercial computer sector, ComDev continued to specialize in the military sector.

49. Computing Devices of Canada, Looking Back, Reaching Forward, 1978.

50. Interview with Norris, 18 March 1999.

51. Before closing the deal on ComDev, a group of CDC vice-presidents had come to Ottawa in July 1968 for a closer inspection of the company and to meet with DITC officials to discuss ComDev's future. During these meetings, CDC executives then learned of DITC's efforts to promote Canada's computer industry.

52. Sirrs, letter to Ralph Risch, 29 October 1969.

53. Representing CDC were Paul Miller (vice-president and group general manager, Computer Systems), Robert Chinn (vice-president. Manufacturing), Jim Thornton (vice-president. Development), Ronald Manning (director, Product Planning), and J. J. Doyle (assistant controller). Representing ComDev were E. B. Daubney (president) and Doug Bassett. The Canadian delegates from DITC's Electrical and Electronics Branch were Dave Quarterman, Ray Koski, and F. Hermann. (J. J. Doyle, Assistant Controller, memorandum to file, 13 February 1970. DITC, accession 83-84/229, box 186, file P8803-EA-44, part 1)

54. John Doyle, Controller, Control Data Corporation, letter to David Quarterman, DITC, 4 March 1970; and R. C. Hall, vice-president and Group Executive EDP Systems, letter to E. A. Booth, General Director, Electrical and Electronics Branch, DITC. DITC, accession 83-84/229, box 186, file P8803-EA-44, part 1.

55. Ibid.

56. John Doyle, letter to David Quarterman, 25 March 1970; and E. A. Booth, General Director, Electrical and Electronics Branch, letter to John Doyle, 16 April 1970. DITC, accession 83-84/229, box 186, file P8803-EA-44, part 1.

57. In the 1970s, CDC pursued cooperative ventures with the nations of the Socialist Bloc ("Control Data Corp.," 8 October 1973, CBI, Series 2, box 4, Folder 11; Control Data Corp., "16th Annual meeting: Report to the Stockholders," 1 May 1974, CBI, Series 2, box 1, Folder 18).

58. See CDC-91, Control Data Corporation Collection, Charles Babbage Institute, Minneapolis.

59. Wilkins, *American Business Abroad from 1914 to 1970.*

60. Control Data had acquired Commercial Credit in order to match IBM's financial capacity to lease computers.

61. See data in *Moody's Industrial Manual* for relevant years.

62. Dain, Kalman & Quail Inc., "An Upper Midwest Research Report on Control Data Corporation," April 1972. CBI, Series 2, box 4, Folder 11. Dain, Kalman & Quail was investment firm located in Minneapolis.

63. Kendal Windeyer, "Big computer plant for Quebec," *Montreal Gazette*, 3 August 1970.

64. Remarks by the Honourable Jean-Luc Pepin, Minister of Industry, Trade, and Commerce, at the Control Data Corporation Press Conference, 14 August 1970. DITC, accession 83-84/229, box 186, file P8803-EA-44, part 3. A CDC market analysis had predicted that the PL-50 alone would account for over 700 sales worldwide over its lifetime, of which over 50% would be produced in Canada. R. A. Manning, Director, Advanced Product Planning, CDC, letter to Raymond Koski, Project Officer, 24 June 1970. In this letter, Manning, provides numerous tables sales, value-added and research cost estimates for the PL-50.

65. W. C. Norris, Remarks at Press Conference, 14 August 1970, Ottawa. DITC, accession 83-84/229, box 186, file P8803-EA-44, part 3

66. R. A. Manning, Director of Advanced Product Planning, CDC, letter to Raymond Koski, DITC, 24 June 1970. DITC, accession 85-86/229, box 186, file P8803-EA-44, part 1. Attached to this letter are extensive tables of R&D and sales estimates.

67. *Moody's Industrial Manual*, 1965, 1966, 1967, 1968, 1969.

68. Manning, letter to Koski, 24 June 1970.

69. CDC-Canada Ltd., "The PL-50 System," 16 July 1970. DITC, accession 83-84/229, box 186, file P8803-EA-44, part 4.

70. The military-civilian interaction had, from the start, been an important element in the parent firm's business and technical strategy. See James C. Worthy, *William C. Norris: Portrait of a Maverick* (Ballinger, 1987).

71. These contracts, which were for $5 million, accounted for an important share of CDC total revenues. J. M. Dain & Co., "Report on Control Data Corporation."

72. CDC finally abandoned commercialization of the 904 computer when it became apparent that the company lacked the financial resources to support the massive software development and marketing efforts needed to launch yet another product line.

73. One analysis of CDC noted: "Control Data's objective is to concentrate its military efforts in areas that provide for cross-application of software expertise or technology with the EDP group such as the emulation work that it is presently doing for the military that will have application in its commercial effort." Dain, Kalman & Quail Inc., "Focus: An Upper Midwest Research Report on Control Data Corporation," 19 April 1972, CBI, Series 2, box 4, Folder 11.

74. Jean Marchand, Minister of Forestry and Rural Development, memorandum to cabinet: Philosophy and Policy of Regional Development, 29 January 1969. Office of Privy Council of Canada, Cabinet Document 741/69.

75. "The industrial centers will be chosen by agreement with the provinces according to guidelines identifying (1) where industry is already well established; (2) likely locations for new industry; (3) existing transport, power, water, housing, etc.—in short infrastructure; (4) ease of development of new or additional infrastructure." (ibid., p. 5)

76. Infrastructure development and social adjustment and rural development were the other two pillars. Before the Trudeau government, federal efforts to encourage regional development focused on infrastructure programs and rural development. Under DREE, industrial incentives became the strongest of the three. R. W. Phidd, "Regional Development Policy," in *Issues in Canadian Public Policy*, G. Bruce Doern and V. Seymour Wilson eds. (Macmillan Canada, 1974).

77. Donald J. Savoie, *Regional Economic Development* (University of Toronto Press, 1992).

78. W. J. Lavigne, Assistant Deputy Minister, DREE, letter to E. A. Booth, General Director, E&E branch, 17 August 1970. DITC, accession 83-84/229, box 186, file P8803-EA-44, part 4.

79. When DREE was formed in late 1969, the total projected budget for industrial incentives over the fiscal year 1972–73 was $65 million. P. M. Pitfield, secretary, Cabinet Committee on Priorities and Planning, memorandum to chairman of Joint Meeting of Cabinet Committees on Economic Policy and Programs and Social Policy and Cultural Affairs, 10 March 1969. Office of Privy Council of Canada, Cabinet Document 219/69.

80. IBM had also just received federal government incentives to build a component plant in the Eastern Townships of Quebec. Ivor Boggiss, "Confirm Plans by CDC To Build Canadian Plants," *Electronic News* 17 (1970), August: 2. But scope and significance of the IBM plant paled in comparison to the proposed CDC facility.

81. "Control Data Unit Said to Plan a Plant Near Quebec City," *Wall Street Journal*, 4 August 1970.

82. The remaining 160 positions would be taken up by administrative and technical support personnel and by material control and supervision personnel (R. Hall, Remarks at Press Conference, Ottawa, 14 August 1970, DITC, accession 83-84/229, box 186, file P8803-EA-44, part 3). Jean Labonté, The Director- General of Commercial and Industrial Development for the province of Quebec, explained that Quebec City was chosen for the size of its female labor pool.

83. This group's name was changed many times over the life span of the company.

84. For example, in 1969, against the accepted wisdom of the white business community, Norris established a new plant to build peripheral controllers in the poor black community of Northside in Minneapolis. Norris did not want the plant to be a mindless assembly operation. "Manufacturing [peripheral controllers] involved an array of skills ranging form entry-level to fairly high, providing incentives and opportunities for advancement" (Worthy, *Norris*, p. 110).

85. CDC-Canada Ltd., "The PL-50 System." The PL-50 was to have a monthly lease in the range of $30,000–$60,000. The price of the system was to be in the order of $2.5 million.

86. Other compilers were to be also to be developed in Canada for the PL line: APL, ALGOL, and PL/1. CDC-Canada, Third Quarterly Project Report, Canadian Development Division, 11 June 1971. DITC, accession 83-84/229/ box 186, file P8803-EA-44, part 5.

87. Progress in the project was documented in a series of quarterly progress reports that CDC-Canada had to provide to the Canadian government.

88. The central processing unit incorporated "a large IC register file of 256 64-bit words with a cycle time of 50 nanoseconds; a 'look behind' association file of 16 64-bit words; a large virtual paging mechanism, invisible to the user, providing storage accessibility of approximately 35 x 1012 by virtue of a 48-bit address; capability of a single instruction to address up to 65,000 elements; and data streaming allowing several instructions to be in various stages of execution simultaneously."

89. Oral History with Bill Richardson, 8 August 1999, conducted by John Vardalas. University of Waterloo was, and still is, one of Canada's top, if not the leading, engineering universities.

90. Northern Electric evolved into Northern Telecom (Nortel).

91. Thomas, *Knights of the New Technology*.

92. Electronics Advisory Group, DITC, Minutes of Meeting, 6 September 1968. DITC, accession 83-84/297, box 2, file P. 1106-230/E1.

93. Thomas, *Knights of the New Technology*. After spending $91 million, the Advanced Devices Centre developed the SP1, an electronic switchboard. Earning Northern Electric a ten-fold return, the SP1 sold well to the small US telephone companies who were independent of AT&T (ibid.).

94. Electronics Advisory Group, DITC, Minutes of Meeting, 12 September 1968. DITC, accession 83-84/297, box 2, file P. 1106-230/E1.

95. Microsystems International, *The Financial Post Corporation Service*, 25 October 1972.

96. Electronics Advisory Group, DITC, Minutes of Meeting, 15 June 1973. DITC, accession 83-84/297, box 2, file P. 1106-230/E1, part 2.

97. Tim Jackson, *Inside Intel* (Plume, 1997). Second sourcing meant giving another company the right to manufacture one's chips. In the context of the late 1960s and the early 1970s, "second sourcing" was a necessary marketing tool. Users would not buy a company's microelectronic devices if there wasn't the security of second source. So companies, like Intel, that brought a new product on the market would also give the technology, under license, to a weaker partner to act as a second source. Second sourcing was always a risk for companies like Intel

because it would allow another company the opportunity to move up the learning curve and become a possible future competitor. Jackson argues that MIL was given second sourcing rights because Intel had no fears that MIL could ever become a threat.

98. CDC-Canada, "An Indigenous Computer Industry," attachment to Jean-Luc Pepin, Minister of DITC, letter to Edgar J. Benson, Minister of Finance, 19 January 1972. Records of Department of Finance, Government of Canada, RG 19, Volume 4934, file 3710-02, part 1, National Archives of Canada, Ottawa. *(Henceforth the Department of Finance, RG 19, record group will be cited as DOF.)*

99. Ibid.

100. Nothing is known about the internal decision-making process that led to the submission. Clearly people in CDC-Canada provided the political intelligence, but did CDC-Canada initiate this submission? How much parent-subsidiary interaction was behind this submission? Nor is there any information that can shed light on who participated in framing the submission conceptually, or on who actually penned it. The document, however, does divide itself neatly into two sections, each with a different voice.

101. The Honourable John Turner, Minister of Finance, letter to the Honourable Jean-Luc Pepin, Minister of Industry, Trade and Commerce, 4 February, 1972. DOF, volume 4934, file 3710-02, part 1.

102. By autonomy Tuner is essentially arguing that CDC-Canada change from a cost center to a profit center and reorganize its reporting structure to the parent firm so as to allow strategic planning across the entire subsidiary.

103. Raymond Koski, memo to file: Marketing Discussions with CDC-Canada, 25 November 1971. DITC, accession 83-84/229, box 186, file P8803-EA-44, part 6.

104. "We never would have gone to Canada if weren't sure" was the view expressed by all these men when asked about the project many years later. (I interviewed Ray Nienburg on 25 May 1999, Gerald Shumacher on 8 July 1999, Michael Sherck on 7 July 1999, Bob Olson on 19 July 1999, and Derrel Slais on 19 May and 17 September 1999.)

105. Interview with Slais, 19 May 1999.

106. Examples where the STAR excelled are in modeling thermonuclear detonations, which was the underlying rationale for the Lawrence Livermore Laboratory STAR, and large-scale, complex modeling of weather systems.

107. The team used "kernels" of different code, which mainly involved scalar processing, that came from various CDC customers.

108. The PL-50's target was to surpass the performance of the IBM 360/65 but sell a the same price, i. e. about $30,000–$40,000 per month lease. In terms of the newer IBM 370 line, the PL-50 was to compete against the 370/165. But in trying to overcome the overhead problem, the Canadian team thought it was heading

toward a machine that was more comparable to the IBM 370/165 which was in the $45,000–$100,000 monthly lease range. Raymond Koski, memo to file: Marketing Discussions with CDC-Canada, 25 November 1971.

109. The Cyber 72, 73, and 74 models corresponded to the 6200, the 64000, and the 6600, respectively. Soon after the 7600 was updated with the Cyber 76.

110. The average purchase prices for the Cyber 72, 73, 74, and 76 were $1.6 million, $2.2 million, $4.5 million, and $7 million. Monthly leasing rates were $32,000, $47,000, $105,000, and $165,000 respectively.

111. Dain, Kalman & Quail, "An Upper Midwest Research Report on Control Data Corporation." The first of the Cyber 70 series was delivered in October 1971. In the 2 years that followed, CDC had received 88 orders for various Cyber 70 models. A. G. Becker & Co., "Control Data Report."

112. Interview with Larry Jodsaas, 2 March 2000.

113. Interview with Slais, 7 February 2000.

114. The first Cyber 173 was delivered in early 1975. Offering about two and a half times the processing speed of comparable Cyber 70 computers, the Cyber 170 sold for $1 to $4 million and leased for $18,000 –$75,000 per month, depending on the system size. "Control Data launched Canada-made computer line," *Canadian Datasystems*, April 1974: 42–43.

115. Interview with Terry Kirsch, 8 November 1999. "The peripheral processor subsystem [in the Cyber 170 line] consisted of ten functionally independent computers which provided 4096 12-bit words(plus 1 parity of MOS memory and a repertoire of 64 instructions. Due to a multiplexing system organization, the peripheral processors share common hardware for arithmetic, logic and I/O operations without a loss of speed and independence. The Cyber 170 mainframe can be expanded to include up to 20 peripheral processors." Control Data Canada, *INFO*, April 1974, inserted in *Canadian Datasystems*, April 1974.

116. In September 1973, CDC sent Terry Kirsch to manage the Cyber 170 program in Canada.

117. The growth of CDC peripheral business stemmed from the opposite approach. Unable to support the high R&D costs of developing new peripherals from internal demand alone, CDC became an Original Equipment Manufacturer (OEM) and sold its peripherals to all its competitors. In effect, CDC produced standardized components for the industry.

118. See the brief interview with Bob Olson in "Control Data launches Canadian-made computer line," *Canadian Datasystems*, April 1974: 42–43.

119. Interview with Bob Olson, 19 July 1999.

120. Ibid.

121. Ibid.

122. Ibid.

123. Norris, address to shareholders, in Control Data Corporation, 17th Annual Meeting: Report to Stockholders, 7 May 1975.

124. Initially, the 6000 line began with two operating system developments: SIPROS and COS (Chippewa Operating System). COS was a minimal system that was used to get the 6600 tested and ready for production. SIPROS was to have been the more advanced operating system. Improperly managed, the SIPROS development ran into serious problems and was dropped. COS, which then became the basis for all 6000 and Cyber 70 computers, evolved along two lines: SCOPE and KRONOS.

125. CDC was to develop a customized operating systems for the Union Bank of Switzerland, the US Air Force's data management systems, and for NASA Ames. CDC was also trying to develop an Advanced Timer Sharing (ATS) operating system in partnership with a French company.

126. The adoption of KRONOS over SCOPE 3. 4 brought forth considerable protests from the users of SCOPE. CDC was forced to enhance NOS by adding elements of SCOPE.

127. See Norris, address to shareholders, Control Data Corporation, 17th Annual Meeting: Report to Stockholders, 7 May 1975, CBI Series 2, box 1, Folder 19. Thinking about unbundling went back to 1970 after IBM had already introduced its "unbundling" pricing. But with the introduction of NOS, Price instituted "Separate Element Pricing" (SEP) in an attempt to factor in the true cost of software. (source: interview with Robert Price, 18 September 1999)

128. Control Data Corporation, 1978 Annual Stockholders' Meeting and First Quarter Report, 12, CBI Series 2 box 1 Folder 22.

129. Allard, "Product Line History."

130. The strategy was produced much better result in CDC' Computer Services Division.

131. Interview with John Titsworth, 15 September 1999.

132. Ibid. The Union Bank of Switzerland had launched a $30 million dollar lawsuit against CDC for failure to meet contract obligations.

133. In its budget forecast for the PL-50 project CDC had allocated $7.5 million for hardware development and $27. 9 million for software over the first 6 years of the project. Ron Manning, Director of Advanced Product Planning, letter to Raymond Koski, 24 June 1970, DITC, accession 85-86/229, box 186, file P8803-EA-44, part 1.

134. Interview with Amnon Zohar, 10 June 1999.

135. Ibid.

136. Bill Richardson believes that the Canadian software group also held onto the responsibility, which had started with the PL-50, to develop the "implementation language." For the PL-50 the implementation language was called SWIL— software writers implementation language. (interview with Richardson, 8 August 1999)

137. Interview with Schumacher, 8 July 1999.

138. The Cyber 174 became a dual processor version of the 173. The old discrete component 7600 was made 170 compatible, and called the Cyber 176. The Cyber 175 entailed a complete circuit redesign of the 7600's architecture.

139. The two companies had formed an Advanced Systems Laboratory (ASL). Derrel Slais, who had returned to the US from Canada to take up his normal duties as the general manager of CDC's advanced computer development program, became the ASL's director. The ASL was split between two locations: at CDC in Minneapolis and near NCR's Data Processing Division in San Diego. An NCR man R. O. Gunderson, headed up the ASL in Minneapolis, and a CDC man, T. H. Elrod, headed up the ASL in California.

140. Interview with Sheldon Fewer, 18 March 1999; interviews with Derrel Slais, 2 August and 19 May 1999.

141. Interview with Robert Price, 19 September 1999.

142. As Price recalled it in an interview (19 September 1999), NCR executives came back and told CDC: "'We do not need a mainframe. . . . All we need is a mini.' And so they did not see the point of the collaboration if they did not need a mainframe. They changed all their products. They never had a mainframe again and what came out of that for them was the "Tower" series. . . . It was a tremendously successful product line."

143. The 64 bit machine featured a virtual memory system. It offered a direct addressable address base large than any of its competitors. The Cyber 180 marked the first time CDC had moved from the 6-bit to the 8-bit world.

144. Soon after the appearance of the Cyber 180s, Larry Jodsaas, vice-president of CDC's Computer Systems Division, proclaimed that the computer "perhaps represent[ed] the only new major operating system for mainframes in the 80s." Control Data, "Investment Community Meeting: Company Presentations, Questions and Answers," June 1984, in Minneapolis. CBI, Series 2, box 4, Folder 7.

145. There would be important degradation in performance if one had to keep switching back and forth between NOS and NOS/VE applications in a time-sharing environment. Customers were advised to batch all the NOS applications together and separately from the NOS/VE applications.

146. Interview with Slais, 7 February 2000. As the 3000 group evolved so did CDC's supercomputer aspirations. The vector-based approach that had produced the STAR 100 and that had pushed the 3000 group into the PL-50 did not lose its support within CDC. With Cray gone, CDC had to hang all its supercomputer

ambitions on the STAR approach. By the time the Cyber 180 appeared, the STAR tradition had produced the Cyber 200 series. Engineers within CDC were polarized over the merits of the STAR architecture. At times it could be a very emotional debate. One side supported it with a kind of religious fervor and the other saw it as inherently flawed. The detractors had all come from the Cyber 180 group. In 1984, the CDC spun off the STAR, vector-based, tradition into a new company, ETA Systems. The hope was that ETA Systems would take on Cray Research in the race for the ultimate supercomputer. In the end, ETA Systems ran into serious financial problems and along with Cray Research became the victims of Japanese supercomputer manufacturers.

147. The parent firm developed the higher-end models: Cyber 174, 175 and 176. It also developed multiprocessor version of the Cyber 173, Cyber 174, 175, and 176.

148. Interview with Terry Kirsch, 8 November 1999

149. Ibid.

150. There is no documentation on this figure. But several of the people interviewed for this research seem to recall a figure in the neighborhood of $30 million.

151. CDC-Canada, Annual Audited Financial Statements, November 1982. Howard Ross Management Library, McGill University, Montreal, Quebec, Canada. This library has a set of financial statements for the years 1981–1989, 1990, and 1991. Although these financial statements give the total annual R&D expenditures, one cannot pull out the total R&D costs for the S0 because Computing Devices of Canada (ComDev) and the CDC's Canadian computer operations were merged into one corporate entity.

152. Fifty percent or less of total development costs was the usual government contribution under the PAIT program.

153. Interview with Jodsaas, 2 March 2000.

154. The only man who remembers many of the details is Barry Robinson, the manager of the Canadian Development Division when the S0 project was put together. Though Robinson was instrumental in launching the S0 initiative, he refuses to speak to anyone about his time with CDC-Canada. It is regrettable that Robinson's perspective could not be a part of this story.

155. Control Data, "Investment Community Meeting: Company Presentations, Questions and Answers," June 1984.

156. Interview with Jodsaas, 2 March 2000.

157. CDC-Canada, "President's Message," winter 1986. CBI, box 4C1G, Folder "Acquired Companies—Computing Devices of Canada."

158. Paul William, "Minis and Mainframes," *IEEE Spectrum*, January 1985: 42–44.

159. Glenn Zorpette, "Minis and Mainframes," *IEEE Spectrum*, January 1988: 29–31.

160. In 1987, the companies Alliant Computers, Convex Computers, and Scientific Computer Systems introduced mini-supercomputers. Although the performance of some of these mini-supercomputers were only 25% of a CRAY X-MP, their prices were prices were only in the $500,000 to $1 million range (ibid.).

161. See entry on Control Data Corporation, *Moody's Industrial Manual*, 1986, pp. 3008–3011.

162. Ford S. Worthy, "Does Control Data Have a Future," *Fortune*, December 23, 1985: 24–26; "Control Data's Struggle to Come Back from the Brink," *Business Week*, 14 October 1985: 62–63.

163. Computing Devices of Canada, however, continued to play an important role in Ceridian's defense business.

164. Interview with Terry Kirsch, 8 November 1999.

165. Interview with Gary Glover, 9 June 1999.

166. CDC had a long history of cooperating with National Cash Register (NCR) first in peripherals, with a jointly owned company Computer Peripherals, and then to develop the "swing computer." CDC also collaborated with ICL and Nixdorf. In 1973, CDC started high-level discussions with Nixdorf in Germany to collaborate on the development of a minicomputer. That same year, CDC and Honeywell discussed setting up a jointly owned company to develop and sell magnetic disk products. An assessment of the success of CDC's pursuit of inter-firm strategic alliances is fascinating and important study that has yet to be carried out.

167. See Norris, address to shareholders, Control Data Corporation, 16th Annual Meeting: Report to Stockholders, 1 May 1974, CBI Series 2, box 1, Folder 18. It is interesting to note that, while the US government used the term "Communist Bloc," Norris consistently used the less politically charged "Socialist Bloc."

168. Ibid.

169. Under the terms of the PL-50 project, CDC-Canada was to manufacture a little over half of the computers sold globally by CDC. Whether or not Canada obtained the same level of exports under the Cyber 170 program is not known. But fragmentary data does suggest that CDC-Canada's initial sales figures for the 170 were comparable to those predicted for the PL-50. CDC had estimated that Canada would ship 65 PL-50s in the first 3 years. In the case of the Cyber 170 series, Canada shipped 50 systems in the first 3 years. Without knowledge of the total number of 170 computers CDC shipped around the world, one cannot know if the 50 systems shipped from Canada represented half of the total for these years. However, at least for the first 3 years, the value of Canada's Cyber 170 shipments far exceeded the value forecast for the PL-50. For PL-50 estimate see Ron Manning, letter to Koski, 24 June 1970. For Cyber 170 figure see "From a dud, Control Data wrought a winner," *Financial Post*, 5 November 1977.

170. With some help form CDC on tooling issues, Elco Connectors qualified to supply the 62-pin connectors. Canron was to develop and manufacture the 125-kVA motor generators for the computer. And Premier Metal had not problem supplying the chassis frames for the computer. CDC-Canada, "Third Quarterly Project Report."

171. CDC-Canada, "Fifth Quarterly Project Report, Canadian Development Division," 3 January 1973. DITC, accession 83-84/229, box 186, file P8803-EA-44, part 6.

172. Fourth Program Review Group Meeting: PL-50 Computer Systems, 12 January 1972. DITC, accession 83-84/229, box 186, file P8803-EA-44, part 6.

173. MIL's withdrawal also no doubt, determined by the problems it was having perfecting the "silicon bipolar, tantalum thin film and beam lead technologies ... to the point of commercial production." Far behind schedule in putting these technologies into practice, MIL was forced to purchase the expertise from Intel and Monolithic Memories rather than develop it in-house. Electronics Advisory Group, DITC, Minutes of Meeting, 15 June 1973. DITC, accession 83-84/297, box 2, file P. 1106-230/E1, part 2.

174. In addition to the $37 million in subsidies put out by federal government, MIL managed to lose an additional $50 million. MIL was shut down in 1975. Since an in-depth historical account of MIL has yet to be written, the reasons for MIL's collapse remain in dispute. Northern Telecomm executives claim that DITC pushed them into the chip-making business against their instincts. Evidence suggests that poor relations between Northern Electric and MIL management may have also undermined MIL's future. Electronics Advisory Group, DITC, Minutes of Meeting, 15 June 1973. Others argue that the people who were put in charge of MIL came from the slow moving telephone monopoly business and were unable to deal with the fast moving, and ruthlessly competitive nature of the chip industry. From the MIL debacle did spawn a long lineage of innovative and successful high-tech companies like MITEL, Newbridge, and Corel.

175. Electrical and Electronics Branch, DITC, The Canadian Computer Industry, July 1980.

176. Until 1994, IBM Canada's Toronto manufacturing division, called Celestica, produced circuit boards, memory devices and power supplies for hundreds of IBM's worldwide products. Then in 1994, through the initiative of its Canadian manager, Eugene Polistuk, the Toronto manufacturing facility gained its independence from IBM Canada. A wholly owned subsidiary of IBM Canada, with Polistuk as its CEO, Celestica had to compete in the global electronic component business. But until 1996, all of Celestica's output went to supply the needs of other IBM divisions. Celestica's fortunes took a dramatic turn for the better when, in 1996, the Canadian conglomerate Onex bought Celestica from IBM. Celestica embarked on an aggressive program of growth. Celestica has grown to $5.3 billion company, employing 20,000 people, in 31 facilities in 11 countries. Half of Celestica's revenues are generated from the Canadian parent operation

in Toronto. Celestica, Annual Report, 1999. As more and more of the leaders in the computer and telecommunications sector outsource their manufacturing, Celestica's business is growing. The world's third largest EMS company, after Selectron and Flextronics, Celestica produces component parts for IBM, Hewlett-Packard, Lucent, Nortel, Dell, Cisco, Sun Microsystems, and others.

177. Ibid., p. 36.

178. Ibid., p. 53.

179. Interview with Glover, 9 June 1999.

Conclusion

1. Peter Hall and Paschal Preston, *The Carrier Wave* (Unwin Hyman, 1988), p. 166.

2. After military enterprise, the US civilian space program also played an important role in sustaining the growth of that country's digital electronic industry.

3. Merritt Roe Smith, "Introduction," in Smith, *Military Enterprise and Technological Change*, p. 1.

4. The national strategic importance of automated computation was bolstered by Canada's ambitions to be a postwar pioneer in the development of nuclear energy.

5. Even if the design and construction of a full-scale version of UTEC had proceeded, it is unlikely that an operational computer would have been ready within 2 years.

6. Thomas, *Knights of the New Technology*.

7. The most intense work in real-time computer simulators was for combat aircraft. In fact, the term man/machine interface first arose in the context of the US Navy's research into flight simulators. While Computer Devices failed to go further with computer simulation, the Canadian Aviation Electronics Co. (CAE) turned its early work in this area, which was also supported by the Canadian military enterprise, into a thriving international business in flight simulators for both civilian and military applications.

8. John Chapman, who had been with DRTE since its inception and later became the first Assistant Deputy Minister of Space Programs, was quite clear on the military dimension of the Alouette I project. "The Alouette program was begun," Chapman later recounted in 1966, "in an attempt to generate a Canadian program which would give some experience in space technology to DRB [Defence Research Board], in anticipation of the ultimate military importance of satellites. At the time," Chapman went on to explain, "there was little immediate hope of developing a direct Canadian requirement for a satellite system which would justify a military project. It was therefore necessary to seek a scientific objective that

was consistent with DRTE skills and experience. A study of the high ionosphere had obvious applications both to communications and to the ICBM [Intercontinental Ballistic Missiles]." John Chapman, "The Alouette-ISIS Program," Chapman Papers, volume 12, file 13.

9. Long-distance radio communications, beyond line of sight, used the ionosphere as a reflective medium. Solar storms, which consist of highly energetic charged particles, wreak havoc with the ionosphere. The disturbances are most pronounced in the far North because the charged particles funneled in by the Earth's magnetic field. Another concern was the effect of nuclear detonations on the radio communication properties of the ionosphere.

10. When the Digital Equipment Corporation (the US firm that had pioneered the minicomputer concept) set up operations in Canada, it named Denzil Doyle as the first president. This study was carried out after he had retired from the company. (Denzil Doyle, The Communications Research Centre, Doyletech Corp., 1987). Doyle has been an important voice in high-tech policy issues. This report was based on surveys he carried out across much of the electronics industry. As soon as the Department of Communications was created in 1969, it set about trying to transfer as much of DRTE's military electronics and communications expertise to the civilian sector. To carry this out, the Department of Communications created the Communications Research Centre (CRC). CRC staff were all drawn from DRTE. See also Doyle's most recent schematic of the genealogy of Ottawa's high-tech sector, sometimes referred to as "Silicon Valley North," which occupied a full page of the *Ottawa Citizen* on 1 May 1996.

11. See Trevor Pearcy, "Australia Enters the Computer Age," in *Computing in Australia*, ed. J. M. Bennett et al. (Australian Computer Society, 1994).

12. Pearcy obtained his B. Sc. from Imperial College, London, in 1940. He stopped his graduate studies to go off and work on the radar project where he concerned himself with the program's large computational demands. After the war's end, he immigrated to Australia to join the CSIR. In 1980 he was elected as a Fellow of the Australian Academy of Technological Science and, in 1985, he retired as Dean of Engineering at Caufield Institute of Technology.

13. "CSIR's management," wrote Pearcy in 1994, "discounted the valuable expertise so early in the development from the Mark I project. Clearly, the executive's belief that Australia's main interests lay with her primary industries was a deciding factor in the eventual termination of computing at [CSIR in 1954]." Pearcy, "Australia Enters the Computer Age," 30.

14. The one area where Australian military tried to establish a technological expertise was in the development of instrumentation for gathering data in long-range missile testing. Because of its vast, relatively flat, and unpopulated desert areas, Australia became an important testing ground for British long-range missile technology after World War II. Because of the need to process the large amounts data generated during a missile test, the Australian military proposed, in 1950, the development of its own "Long Range Weapons Digital Automatic

Computer." The proposal never got to the development stage because the British members on the Joint Project Board, the group that oversaw the British-Australian testing programming, argued that there were British firms that could provide "off-the-shelf" equipment to do the job. In 1957, the British made Elliot Bros. computer was installed at Australia's Weapons Research Establishment. "The Weapons Research Establishment Digital Automatic Computer (WREDAC), as the [British made machine} was known, never reached the reliability hoped for, despite all the engineers could do for it. . . . It was replaced in 1960 by an IBM 7090 transistor machine." George Barlow, "Electronic Digital Computers in Australian Defence," in Bennet et. al., Computing in Australian, p. 123. Like the Weapons Research Establishment (WRE), DRTE also decided to buy IBM to handle the growing amount of data it was processing. But unlike WRE, DRTE had not only aggressively sought to advance Canada's ability to design digital electronic circuitry around silicon transistors, it also designed the sophisticated satellites whose top-sounding data the IBM machine processed. WRE's purpose remained a place to test British technology.

15. E. T. Robinson, who has had 40 years of experience in the Australian computer scene and is currently advisor on information technologies to the Australia's Minister of Industry, Technology, and Commerce, is emphatic about the absence of any government support, during the 1950s, to move Australia up the digital electronic leaning curve. E. T. Robinson, "Government's Participation in the Australian Computer Industry," in Bennet et. al., *Computing in Australia.*

16. Any lasting benefits from the pool of skilled scientists and engineers spawned by military enterprise during the 1950s, would take at several decades to take root. Throughout the 1960s, the 1970s, and the 1980s, and into the 1990s, the economic activity surrounding digital electronic activity was far greater in Canada than in Australia, even if one factors in differences in population. For example, during the period 1989–1993 Canada's imports of Automatic Data Processing Equipment were nearly 250% of Australia's. This attests to the much higher diffusion of digital electronic technology into the Canadian economy. At the same time, Canada's exports of Automatic Data Processing Equipment were 750% of Australia's. Of course, to calculate the total economic activity one needs statistics on the domestic production of each country. Without any easy access to Australian data, I cannot make any comparisons. Nevertheless, one can confidently conjecture that Canada's domestic production will also be much higher. For import/export data, see Annual International Trade Statistics (United Nations) under Commodity SITC Number 752 for the years in question. Similar disparities in economic activity associated with Telecommunications Equipment (SITC 764) reinforces the conclusion of Canada's greater appropriation of digital electronic technology.

17. For Krauss, the signing of the Defence Production Sharing Agreement with the US was a defining moment in Canada's retreat from self-reliance. This view is reinforced by Ernie Regehr, *Arms Canada* (James Lorimer, 1987). From a different perspective, Granatstein (*Canada 1957–1967*) also underscored the accelerated continentalization of Canadian defense during the post 1958 era, which he labeled the "Defence Debacle."

18. Michael Bliss, "Canadianizing American Business," in *Close to the 49th Parallel*, ed. I. Lumsden (University of Toronto Press, 1974).

19. Herbert Marshall, Frank A. Southard, and Kenneth W. Taylor, *Canadian-American Industry* (Yale University Press, 1936).

20. A. E. Safarian, *Foreign Ownership of Canadian Study* (McGraw-Hill Company of Canada, 1966), p. 17.

21. A. E. Safarian, *The Performance of Foreign-Owned Firms in Canada* (Canadian-American Committee, 1969), p. 103

22. Ibid.

23. Pierre Bourgault, Innovation and the Structure of Canadian Industry, Background Study for the Science Council of Canada, October 1972, Special Study No. 23.

24. Science Council of Canada, Strategies of Development for the Canadian Computer Industry. Report No. 21, Information Canada, August 1973.

25. Multinationals and Industrial Strategy (Science Council of Canada, 1980).

26. D. Rutenberg, "Global Product Mandating," in *International Business*, ed. K. Dhawan et al. (Addison-Wesley, 1981).

27. Thomas A. Poynter and Alan M. Rugman, "World Product Mandates," *Business Quarterly*, autumn 1982: 54–61.

28. Ibid., p. 59

29. For more on the network perspective, see Sumantra Ghoshal, The Innovative Multinational (doctoral dissertation, Harvard University, 1986); S. Ghoshal and C. Bartlett, "The Multinational Corporation as an Interorganizational Network," *Academy of Management Review*, October 1990: 603–625.

30. For a recent elaboration of this argument, see Julian Birkinshaw, Entrepreneurship in Multinational Corporations (doctoral dissertation, University of Western Ontario, 1995).

31. Norman McGuinness and H. Allan Conway, "World Product Mandates," in *Managing the Multinational Subsidiary*, ed. H. Etemad and L. Séguin (Croom Helm, 1986).

32. Brown, *Ideas in Exile*, p. 340.

33. Graham Taylor's study of Canadian Industries Limited (CIL) reveals similar findings. Graham Taylor, "Management Relations in a Multinational Enterprise," *Business History Review* 55, no. 3 (1981): 337–358.

34. For a nationalist critique of the role of banks in Canadian industrial development see Tom Naylor, *The History of Canadian Business, 1867–1914* (Lorimer, 1975).

35. Although the mechanism of centrally dispensed R&D funds did not exist within Sperry Rand, the parent corporation indirectly supported Sperry Canada through the vehicle of intra-firm sales. Sperry Canada financed its entire R&D program in machine-tool automation from defense contracts. As Sperry Canada's domestic military sales fell, the parent corporation channeled US contracts to its subsidiary through the vehicle of the US-Canada Defence Production Sharing agreement.

36. Ibid., p. 36.

37. Arthur Cordell, The Multinational Firm, Foreign Direct Investment, and Canadian Science Policy, Background Study for Science Council of Canada No. 22, Information Canada, 1971.

38. Ibid., p. 42, 44.

39. Ibid., p. 43.

40. Graham Taylor, "Negotiating Technology Transfers within Multinational Enterprises; Perspectives form Canadian History," *Business History* 36 (1994), January, p. 153.

41. I am now completing a history of RCA-Canada. This subsidiary developed remarkable autonomy and an ambitious R&D program.

42. John Baldwin and Petr Hanel, *Multinationals and the Canadian Innovation Process* (Statistics Canada, 2000).

43. The horizon is a consensus that represents the problem solving behavior of the individual disciplines and schools represented in the engineering community. Discussed in Felix Rauner and Klaus Ruth, "Industrial Cultural Determinants of Technological Determinants," *AI and Society* 3 (1989), April-June: 88–102. See also H. D. Hillige, "Die gesellschaftlichen und historischen Grundlagen der Technikgestaltung als Gegenstand der Ingenieurausbildung," in *Lernen aus der Technikgeschichte*, ed. U. Troitzsch and W. König (VDI, 1984).

44. Since the social, economic, cultural, and political factors, as well as basic technical knowledge change over time, the "extent" and "shape" of the "horizon" are also historical properties.

45. Throughout the war, Canada's dependence on its Allies for the sophisticated electronic technology had seriously compromised RCN's military effectiveness. "The consequences of that technological gap for this junior partner [that is Canada] were grave" writes Marc Milner. He goes on to argue that as a result, "Canadian escort services suffered perhaps the highest loss rates of the war." Marc Milner, "The Implications of Technological Backwardness," *Canadian Defence Quarterly* 19 (winter 1989): 46–52.

46. David Noble, *Forces of Production* (Oxford University Press, 1986). J. Francis Reintjes, who was director of the Servomechanisms Laboratory at MIT during the development of NC, does not share Noble's argument about the existence of

viable alternatives. See J. Francis Reintjes, *Numerical Control* (Oxford University Press, 1991).

47. On this distinction Noble (*Forces of Production*, p. 149) writes: "Both N/C and R/P were aimed at reducing the amount of skilled labor required to produce finished metal parts, but they viewed that skill differently. With N/C, machinists skills were devalued and viewed as little more than a series of straight forward operations. . . . N/C viewed machining is a process fully amenable to an abstract, formal mode of programming which eliminated the need for machinist skills altogether. The purpose of N/C was to move directly, without any manual or shop floor intervention, from the mathematical description of a part to the automatic machining of it."

48. Winner, "Do Artifacts Have Politics?" *Daedalus* 109 (winter 1980): 121–136. Noble's historical study of NC offers one of the most comprehensive replies to Winner's question.

49. Jeffrey Keefe, "Numerically Controlled Machine Tools and Worker Skills," *Industrial and Labour Relations Review* 44 (1991), April: 503–519. This paper provides the most up-to-date review of the de-skilling debate as it relates to numerical control.

50. The technological trajectory for NC in Japan was very different than the one followed in the US. Robert Mazzoleni, "Learning and path dependence in the diffusion of innovations," *Research Policy* 26, no. 4, 5 (1997): 405–428. The US Air Force's emphasis on ultra-precision led to the development and diffusion of a closed-loop approach to NC. Japan, however, the emphasis was on the development of a very low-cost system. While the US approach privileged the sophisticated user of the aeronautics industry, the Japanese NC development went after the larger less-sophisticated market. The different evolutionary paths of NC in these two countries was also manifested in very different user-producer networks in the to countries. Because of the US Air Force, "until well into the 1970s," writes Mazzoleni (ibid., p. 417), "the machine tool builders in the US had very little influence over the characteristics of the NC units supplied by the electronics firms and passively adapted to those firms offerings. In turn, the control units' suppliers were not inclined to elicit and use the informational inputs from the machine tool builders in the definition of their product development strategies." For a discussion of the NC trajectory in Germany, see Harmut Hirsch-Kreinsen, "On the History of NC Technology—Different paths of Development," in *Technology and Work in German Industry*, ed. N. Altman et al. (Routledge, 1992).

51. Harley Shaiken, *Work Transformed* (Holt, Rinehart and Winston, 1984); Barry Wilkinson, *The Shopfloor Politics of New Technology* (Heinemann, 1983).

52. Harry Braverman, *Labor and Monopoly Capitalism* (Monthly Review Press, 1974).

53. When the latter pushed a button to program a particular operation, the computer would ask him what he wanted to do next. The questions were presented

as a series of menu options. For each decision, UMAC-5 presented the machinist with the array of possible options. Each option was represented by a lit button. The push-buttons for inappropriate programming options were unlit. They could not be activated by mistake. In this way, the machinist did not have to be a parts-programming specialist. For details the reader is referred to the UMAC-5 user's manual. Sperry Gyroscope Company of Canada, Ltd., UMAC-5 Numerical Control System: Operator's Manual (1965). A copy can be found in the National Museum of Science and Technology in Ottawa; I also have a copy. For those wanting to examine the actual interface, the only existing UMAC-5 can be found in the National Museum of Science and Technology's collection.

54. Interestingly enough, Sperry Canada had included this record/playback option in its earlier more advanced NC systems.

55. Sinclair et. al., *Let Us Be Honest and Modest*, p. 2.

56. In her introductory essay in *Technology and the Wealth of Nations*, ed. D. Foray and C. Freeman (Frances Pinter, 1993), Dominique Foray raises the distinction of result and process to illuminate a debate over how to assess the economic value of R&D. But this distinction equally serves to illuminate how one assigns historical meaning to events.

57. Patrick Cohendet, Jean-Alain Héraud, and Ehud Zuscovitch, "Technological learning, economic networks and innovation appropriability," in Foray and Freeman, *Technology and the Wealth of Nations*, p. 67.

58. See section 3 "Technical learning from external source" and the references in Chris Freeman's critical survey article,, "The Economics of Technical Change," *Cambridge Journal of Economics* 18 (1994), no. 5: 463–514.

59. Keith Pavitt, "Patterns of Technical Change," *Research Policy* 13 (1984), no. 6: 343–373; Keith Pavitt, "What do firms learn from basic research?" in Foray and Freeman, Technology and the Wealth of Nations; and Bruno Amable, "National Effects of Learning, International Specialization and Growth Paths," in Foray and Freeman, *Technology and the Wealth of Nations*.

60. Porter, *The Competitive Advantage of Nations*. High productivity here refers to more than just a higher value of the ratio of output to inputs. It also refers to the ability to bring new products into the market. Thus high productivity embraces both "product" and "process" innovations.

61. Edward W. Constant, *The Origins of the Turbojet Revolution*, Johns Hopkins Studies in the History of Technology, volume 5 (John Hopkins University Press, 1980), p. 10.

62. During the 1970s and the early 1980s, Saskatoon looked as if it would earn the title of "Silicon Flats." Manufacturers of digital electronic equipment, like SED Systems and Develcon Electronics, with important development programs of their own, sprouted up in Saskatoon during the early 1980s. The University of Saskatchewan had provided an important educational and research environment

to attract and nourish the city's community of technologists. The nucleus of this university's digital electronic know-how was forged in the late 1960s and the early 1970s when people such as Arthur Porter, the man who first headed the DATAR project for Ferranti-Canada, went there to be Dean of Engineering, and Norman Moody and Richard Cobbold, who led DRTE's experimentation with digital transistor circuits, went there to head up new research programs. Moody later returned to the University of Toronto to start up the Institute of Bio-Medical Engineering, where digital electronics research became an important part of the Institute's innovations in bio-medical instrumentation Cobbold later succeeded Moody as director of the Institute.

63. Kenneth Iverson was born and raised in Camrose, Alberta. After World War II he went to university and completed his education with a doctorate at Harvard. There he worked with Howard Aitken, one of the early pioneers of digital computers. Iverson later created the programming language called APL.

64. Murray Bell, another winner of the Governor General's Medal for academic excellence, upon graduation from Queen's University went to work for the Consolidated Computer Co. When the company went into receivership, Bell eventually formed Dynalogic. After spending $1 million to develop a new personal computer called the Hyperion and $300,000 to advertise it, Bell introduced the computer at the 1982 COMDEX show in Atlantic City. The response was enthusiastic. The US Magazine Electronics described the unveiling of the Hyperion as a "20 pound bombshell." See Thomas, *Knights of the New Technology*, p. 175. David Thomas's book offers a journalistic staccato-like who's-who account of Canada's computer industry in the late 1970s and the early 1980s. Though weak in its analysis, the book nevertheless offers a fascinating picture of the web of business ties that bound the nation's computer industry in this period.

Index

Adalia, 128–129
Air Canada, 140–141
Allard, R., 258–259
Alouette I satellite, 102
American Airlines, 126, 135
American Bankers' Association, 146–147
Anti-submarine warfare, 46, 48
Aronsen, L., 9–10
ASDIC, 63–64
Australia, 280–282
Auto Pact, 178–179, 284
Automatic Surface Plot system, 62–65
Automotive Products Trade Agreement, 178–179, 284
AVRO Arrow (CF-105), 10, 12, 67, 77, 87

Baldwin, J., 294
Ballard, B. G., 21
Bell Laboratories, 24
Bell relay computer, 24–26
Belyea, J., 55–58
Bendix, 237
Benson, E., 249
Bernard, C., 165
Binary arithmetic tubes, 29–33
Blackett's Circus, 61
Bleackley, B., 165
Bliss, M., 159
Bowden, V., 37, 60–61
Braun, E., 79–80, 91
Brebner, J. B., 12
Brodeur, V., 47

Brown, J., 6, 118, 166
Bullard, B., 31
Burroughs, 224–225
Burroughs of Canada, 136
Butler, D., 165

Cabinet Committee on Research and Defence, 17
Caden, P., 200–201
Campbell, C., 134
Campbell-Kelly, M., 155
Campney, R., 71
Canadian Air Defence Automatic Reporting, 68, 77
Canadian Armaments Research and Development, 21
Canadian Armaments Research and Development Establishment, 95–96
Canadian Aviation Electronics, 92
Canadian General Electric, 85, 92
Canadian Marconi, 31, 52, 85, 92, 248, 271
Canadian military, 15–22, 50–51, 80–87, 90, 98–102, 183–184
Canadian National Railways, 120
Canadian Pacific Air Lines, 124
Canadian Pacific Railways, 124
Canadian Patents and Development Ltd., 30–31,
Canadian technological competence, 276–294, 299–302
Canadian Westinghouse, 31, 92
CARDE, 21
Casciato, L., 40–41, 129, 278

CDC-Canada, 14
Ceridian, 269
CF-105 (AVRO Arrow), 10, 12, 67, 77, 87
Chesley, L., 85
Clark, John, 4
Clark-Jones, M., 5
Claxton, B., 18, 65
Cobbold, R., 101, 103
Codd, E., 154
Compagnie Internatinal de l'Informatique, 169
Complexity and reliability, 81–83
Comprehensive Display System, 68
Computer Devices of Canada, 237, 242
Computer numerical control, 217–218
Computing Devices of Canada, 58, 60–61, 92, 128–129
Control Data Canada, 236, 241–246, 250–269, 273–274, 291–294
Control Data Corp., 14, 226–242, 250–259, 262–264, 268–270
Conway, H., 285
Cooper, A., 54
Cordell, A., 292
Coutts, R., 165
Cox, B., 197, 200
Cray, S., 226–229, 252
Crozier, J., 217–218
Cullwick, E., 23, 52, 55

Dallas, D., 217–218
Daly, B., 165
DATAR, 11, 49, 53, 57–58, 63–72, 111–112, 278–279
Davies, E., 61, 86
DeBresson, Christian, 6
Defence Research Board, 20–21, 26, 32, 38, 51–52
Defence Research Telecommunications Establishment, 80, 86, 87, 92–102, 279–280
De Gaulle, C., 170
Department of Communications, 102

Department of Defence Production, 7, 84, 183–184
Department of Forestry and Rural Development, 243
Department of Industry, Trade and Commerce, 14, 174–179, 182, 207–208, 214–220, 223–224, 236–241, 246–248, 255, 266, 270–274, 284
Department of National Defence, 7, 23, 84
Department of Regional Economic Expansion, 243–244
DEW Line, 12
Diefenbaker, J., 101
Dixon, P., 152
Drache, D., 5
DRTE, 80, 86, 87, 92–102, 279–280
Drury, C., 174, 215

Eayrs, J., 9, 18
Economic Council of Canada, 208–209
Economic nationalism, 5–7, 282–284
Edmondson, T., 144, 164
Electro-Technical Laboratory, 92, 172
EMI Electronics, 136
Engineering Research Associates, 34, 225
English Electric Computers, 170
ENIAC, 1

Federal Electric Manufacturing Co., 108–110
Federal Reserve Banking System, US, 145–149
Ferranti Electric Ltd. *See* Ferranti-Canada
Ferranti Ltd. *See* Ferranti UK
Ferranti UK, 53–55
Ferranti-Canada, 2, 13, 53–56, 63, 68–71, 77–78, 113–116, 128–129, 136–137, 143, 148–166, 279, 287–294
Ferranti-Packard Ltd. *See* Ferranti-Canada
FERUT, 40–42, 94–97, 129, 130, 133, 277–278

Florida, D., 94–97, 100, 102
Ford Instruments, 187
Foulkes, C., 50
Franklin, C., 102
Freeman, C., 4
Fujitsu, 172

Gemini computers, 138
General Electric, 224–225
Goodspeed, D., 21
Gordon, D., 124
Gotlieb, C. C., 28, 41–42
Granatstein, J., 10, 15

Hall, P., 275
Hamilton. W., 117–118
Hanel, P., 294
Hauck, H., 196–197
Herzel, P., 189–190, 193
Hitachi, 172
Honeywell, 150
Hounshell, David, 8
Howe, C., 120, 183

IBM, 29, 34, 126, 136, 167–171, 224–226
Innovation waves, 3–4, 275
International Computers & Tabulators, 163–166, 169
International Computers Ltd., 169–170
I. P. Sharp Associates, 165

Jockel, Joseph, 12
Jodsass, L., 255, 266–267
Johnson, H., 169
Johnston, B., 165
Johnston, R., 40, 129, 278

Kamp, T., 262
Kates, J., 27, 29–33, 35, 40–41, 129, 278
KCS Ltd., 278
Kierans, E., 240
King, B., 183, 214
Kirsch, T., 265
Knights, S.
Kollsmann, 183

Krause, K., 11, 277
Kusters, N. L., 27

Lacey, J., 262
LaFortune, D., 256
Lake, G., 98, 165
Lapointe, H., 117
LaPrairie, J., 165
Lawrence Livermore Laboratory, 232
Lawrie, C., 165
Laxer, G., 8, 282
Lay, H., 64, 66, 76
Learning by doing, 299–302
Levin, R., 79
Levitt, K., 6
Levy, M., 109–114
Lewis, O., 107–108
Lewis, W., 36–39, 88
Likeness, K., 83
Longstaff, F., 160
Lund, W., 26

MacCallum, D., 164
MacDonald, J., 31
Macdonald, S., 79–80, 91
Mackenzie, D., 24, 37–39
Mackenzie King, W. L., 8–9, 18–19
Mackintosh, W. A., 5
Magnetronic Reservisor, 126
Manning, R., 237, 256
Manufacturing dilemma, 195–197
McDorman, T., 165
McGregor, G., 122–127, 138
McGuiness, N., 285
McNeill, William, 8
McSherry, J., 165
MICR, 146
Microsystems International Ltd., 247–248, 270–271
Mid-Canada Line, 11
Miles, M., 65–66
Military enterprise, 2, 7–11, 15–16, 20–22, 81–82, 277, 279
Milner, M., 74–75
Minisupercomputers, 267–268
Minneapolis Honeywell, 136
MITI, 172–173

Mitsubishi Electric, 172
Moody, N., 86–94,
Moore, R., 165
Multinational corporations, 5–7, 174–175, 178, 272–273, 282–294
Multiprogramming, 154
Multiprogramming, 154

National Cash Register, 224–225, 262
National Policy, 282
National Research Council, 23
Naval Research Establishment, 96
Naval Tactical Data System, 72–74
Naylor, R., 5
New Holland Machinery, 187
Nienberg, R., 245
Nippon Electric Co., 172
Nippon Telephone & Telegraph, 172
Nixdorf, 264
Norris, W., 225–228, 235–236, 240–241, 258
North Atlantic Triangle, 2, 12, 18
Northern Electric, 52, 85, 92, 247–248
Numerical control, 188–189, 197, 219–220
Nuova San Giorgio S.P.A., 193

Oki, 172
Olivetti, 169
Olson, B., 245, 257
Ontario Hughes-Owens Co., 183

Pacific Research Laboratory, 96
Paschal, P., 275
Pepin, J.-L., 240–241, 249–250
Perroux, F., 243
Petiot, L., 85
Philco, 70–71, 85, 98, 136
Philips, 169
Picken, J., 164
Piore, E., 65–66
Plan Calcul, 170
Plessy, 170
PNPN circuit, 93–94
Porter, A., 61, 103, 111–112
Post Office Department, Canadian, 13, 106, 113–118

Poynter, T., 285
Price, R., 258, 262
Prince Albert Radio Laboratory, 101–102
Pulse-coded modulation, 56, 60

Quarterman, C., 237

Rajchman, J., 33
Rathborne, W., 134
Ratz, A., 27–28, 36
RCA, 33, 35, 49
RCA Victor, Montreal, 52, 85, 294
Redhead, P. G., 30
Reeves, H., 55–56
Reid, J., 134
Reisman, S., 209
Relay computers, 25
Remington-Rand, 186
Richardson, J., 36
Richardson, L., 128, 131, 133
Ridenour, L., 112–114
Ritchie, D., 165
Robinson, B., 264
Rocketdyne, 193–195
Rogers Majestic, 31, 85, 92
Roger, W., 58–62
Ross, F., 123
Royal Canadian Navy, 8–9, 32, 47–50, 55–63, 66, 67, 70, 74–76, 205–206
Royal Commission on Canada's Economic Prospects, 125, 178, 195–197, 208
Royal Commission on Government Organization, 157–158
Royal Navy (UK), 64–65
Rugman, A., 285
Rutenberg, D., 285

Saab, 169
SABRE 135, 139–141
Safarian, A., 283, 294
SAGE, 67, 139
Schindler, W., 65
Schmookler, J., 4, 6
Schumacher, G., 245, 257, 260
Schumpeter, J., 1–2,

Science Council of Canada, 7, 283–285
Scott, J., 97
Selectron memory, 35
Sharp, A., 165
Sharp, I., 165
Sherck, M., 245
Siemens, 169
Slais, D., 232, 245, 253, 257
Smith, M. R., 7–8, 276
Societé Genvoise d'Instrument de Physique, 193–194
Soete, L., 4
Solandt, O., 17, 50–51, 54
Sovereignty, 2, 7
Sperry Canada, 2, 13–14, 176, 179–214, 287–294
Sperry Corp., 186
Sperry Gyroscope Co. of Canada Ltd. *See* Sperry Canada
Sperry Rand, 13, 187, 190–191, 214–216, 224–225
Stanford Research Institute, 146–148
Stibitz, G., 24
Strain, T., 165
Streaming, *See* Vector processing
Supercomputers, 226
Super High Performance Electronic Computer program, 172
Superminicomputers, 267–268

Taylor, K., 53, 56
Teklogix, 165
Thomas, D., 165
Thomson, J., 54
Thornton, J., 233, 237, 252
Titsworth, J., 259
Toshiba, 172,
Transactor, 131–133
Trans-Canada Air Lines, 13, 120–141
Transistors, 82–85, 91
Tupper, K.F., 37
Turnbull, W., 107–108, 113–118
Turner, J., 249–250

UMAC. *See* Sperry Canada
Union Bank of Switzerland, 259

University of British Columbia, 278
University of Toronto, 22–24, 27, 38–43. *See also* UTEC
US-Canada Defence Production Sharing Agreement, 192
US Navy, 65–66, 72–74
UTEC, 27–29, 34–36, 39, 277–278

Vandeberg, A., 165
Vector processing, 232
Velvet Glove missile, 11, 87
Vickers, 187, 191
Vickers UMAC, 217
von Neumann, J., 20

Watkins, M., 5
Watson, T., Jr., 171
Watson-Watt, R., 129
Weinberg, G., 47
Wilde, D., 165
Williams electrostatic memory, 35
World Product Mandate, 284–285

Ziani de Ferranti, V., 53–56, 61
Zimmerman, D., 8–9, 45, 75
Zohar, A., 261